EXCURSIONS INTO MATHEMATICS

EXCURSIONS INTO MATHEMATICS
THE MILLENNIUM EDITION

ANATOLE BECK

MICHAEL N. BLEICHER

DONALD W. CROWE

UNIVERSITY OF WISCONSIN

A K Peters
Natick, Massachusetts

Editorial, Sales, and Customer Service Office

A K Peters, Ltd.
63 South Avenue
Natick, MA 01760

Copyright © 2000 by A K Peters, Ltd.

All rights reserved. No part of the material protected by this copyright notice may be reproduced or utilized in any form, electronic or mechanical, including photocopying, recording, or by any information storage and retrieval system, without written permission from the copyright owner.

Library of Congress Cataloging-in-Publication Data

Beck, Anatole.
    Excursions into mathematics / Anatole Beck, Michael N. Bleicher, Donald W. Crowe.--The millennium ed.
       p. cm.
    Includes bibliographical references and index.
      ISBN 1-56881-115-2 (pbk. : alk. paper)
        1. Mathematics. I. Bleicher, Michael N. II. Crowe, Donald W. (Donald Warren), 1927- III. Title.

QA39.2 .B42 2000
510--dc21
                                                                         99-058794

Printed in the United States of America
03 02 01 00   10 9 8 7 6 5 4 3 2 1

DEDICATED

to that fragile Hope of Peace
unto which we, a desperate Mankind,
have entrusted our Ancient Learning,
our Dreams for the Future,
and our very Lives.

PREFACE TO THE MILLENNIUM EDITION

Since the publication of the first edition of this book, much of note has taken place in Mathematics. The two most famous problems which were outstanding then, Fermat's Last Theorem and the Four Color Theorem, have been proven to one degree or another. There have been some other changes which require modification of our text. These changes have been noted in Appendix 2000, at the end of the book. The places to which they refer are marked by the icon you see in the margin here.

The Authors.

PREFACE

Ours is a mathematical age. Not only are we ever more dependent on the fruits of the physical sciences, not only is the art of data processing being developed to free man of much of his routine work, but also the ideas of mathematics are beginning to permeate our sociological, philosophical, linguistic, and artistic worlds. The Chinese compare an illiterate to a blind man. As our age develops, we may soon feel the same way of the man ignorant of mathematical thinking.

Most of high school mathematics is concerned with technical manipulation of mathematical quantities, the exception being Euclidean geometry. Thus the student does not get a real introduction to the mathematical world. The same limitation applies to introductory calculus, which at the first-year level is almost entirely a tool for further study. The average college student, seeking an insight into the mathematical world, is unlikely to find it in introductory calculus.

Excursions into Mathematics is designed to acquaint the general reader with some of the flavor of mathematics. By the general reader we mean someone who has had two or three years of high school mathematics and is either a high school graduate or senior.

Each excursion deals with a single area of mathematical interest and builds around it a body of theory. The initial problems are elementary, but as one penetrates deeper, new questions arise, new approaches suggest themselves. Each excursion takes some 40 classroom hours, but this can be shortened by making selections as indicated in the Notes to the Instructor. In this way, it may be possible to cover the essential parts of all six chapters in an academic year. However, the instructor who is aware of his students' unfamiliarity with these new ideas will take a more leisurely pace. An average class may cover most of four chapters in a year.

The conceptual level of the work is intended to be about the same as the calculus, and the student should be prepared for some hard thinking. However, he can make a fresh start with each new chapter, since the six chapters are almost completely independent. Any one of the first three can be taken first. Each of these three chapters will introduce the student to the principle of mathematical induction. Each of the last three chapters presupposes that the student has mastered this principle.

In general, the last section of each chapter, which may be skipped, is more difficult than the preceding material and is a challenge to the superior student. It is not expected that the average college student will be able to master it except by extraordinary effort. Exercises of considerable difficulty are marked by asterisks.

Each chapter has its own sequence of theorems, lemmas, and exercises, numbered consecutively throughout the chapter. With the beginning of a new chapter, the numbering starts again. Some of the results in the book have not been published before. Examples of this are found in the section on Hex in Chapter 5, and in the sections on distribution of digits and Egyptian fractions in Chapter 6. These are marked with a cross, (+), in the margin.

The oldest material in the book antedates written history in Europe; the newest is being published here for the first time.

Madison, Wisconsin ANATOLE BECK
February, 1969 MICHAEL BLEICHER
  DONALD CROWE

CONTENTS

FOREWORD   xix

NOTE TO THE INSTRUCTOR   xxi

CHAPTER 1   EULER'S FORMULA FOR POLYHEDRA, AND RELATED TOPICS

Section 1   Introduction   3

1A Prelude and Fantasy. Being an account of Kepler's first law of astrology.
1B Euler's Formula. Proof of Euler's theorem, $v - e + f = 2$, for convex polyhedra.

Section 2   Regular Polyhedra   12

Proof of the existence of at most five types of regular polyhedra. Regular tessellations. Occurrence in nature.

Section 3   Deltahedra   21

Derivation of all convex polyhedra with equilateral triangles as faces. Construction patterns.

Section 4   Polyhedra without Diagonals   31

A generalization of Euler's formula. Solids with tunnels through them. The only known such solid "without diagonals." Do-it-yourself kit.

Section 5   The n-dimensional Cube and the Tower of Hanoi   40

5A The n-cube. The n-dimensional cube and n-dimensional tetrahedron. How to visualize them. A round-trip tour in n dimensions (Hamilton circuits on the n-cube).
5B The Tower of Hanoi. A solution to the Tower finds a Hamilton circuit on the n-cube.
5C Chinese Rings.
5D An Application. How to use a Hamilton circuit to make your radar work better.

Section 6   The Four-color Problem and the Five-color Theorem   55

Every map has some region with fewer than six neighbors. No map can have five regions each bordering the other four. Every map can be colored with five colors. Some special maps which can be colored with four colors. Will four always work?

Section 7   The Conquest of Saturn, the Scramble for Africa, and Other Problems   64

7A The Conquest of Saturn. Euler's formula for a ring of Saturn (torus). Every map on the torus can be colored with seven colors. A coloring theorem for generalized pretzels (Heawood's theorem).

7B The Scramble for Africa. Twelve colors are required if every country has a colony. What about colonies on the moon?

7C Some Problems Equivalent to the Four-color Problem. Also, a list of problems—some easy, some hard, some still unsolved—about cigarettes, maps, and ping-pong balls. Bibliography for further reading.

CHAPTER 2   THE SEARCH FOR PERFECT NUMBERS

Section 1   Introduction   81

Definition and history of perfect numbers.

Section 2   Prime Numbers and Factorization   84

Unique factorization into primes.

Section 3   Euclidean Perfect Numbers   93

Mersenne primes, final digits of Euclidean perfect numbers, congruence modulo m.

Section 4   Primes and Their Distribution   102

The infinitude of primes, the number of primes less than n, construction and history of tables of primes.

Section 5   Factorization Techniques   110

Fermat's method of difference of squares, last digits of squares, casting out 9s and similar tests, Fermat's theorem, Euler's theorem on factors of Mersenne primes, known Euclidean perfect numbers.

| Section 6 | Non-Euclidean Perfect Numbers | 124 |

Sums of divisors of numbers. Odd perfect numbers; the number of divisors of an odd perfect number.

| Section 7 | Extensions and Generalizations | 137 |

Multiply perfect numbers, amicable numbers, Hadrian's headache.

CHAPTER 3   WHAT IS AREA?

| Section 1 | Introduction | 147 |

The problem of area. Greek ideas. Properties which area ought to have.

| Section 2 | Rectangles and Grid Figures | 153 |

Areas for rectangles oriented in the same direction (the class $\mathcal{R}$). Areas for combinations of such rectangles (the class $\mathcal{G}$).

| Section 3 | Triangles | 161 |

Preliminary results for special triangles. Every triangle which has an area satisfying A1 to A4 must have area $\frac{1}{2}bh$.

| Section 4 | Polygons | 166 |

Decomposition of triangles into triangles. Every convex polygon can be decomposed into triangles ("triangulated"). Every reflex polygon can be triangulated. Proof by induction.

| Section 5 | Polygonal Regions | 177 |

Triangulation of polygons with holes in them (polygonal regions). The area of a triangle is the sum of the areas of subtriangles. The area of a polygonal region does not depend on the particular triangulation.

| Section 6 | Area in General | 185 |

Area of a circle (the classical "proof"). Upper numbers and lower numbers. The figures of class $\mathcal{Q}_t$. Figures of class $\mathcal{Q}$. Proofs that various combinations of figures in Q are also in Q ($E \cap F$, $E \backslash F$, $E \cup F$). Every figure of class Q has an area. Examples of figures of class Q. Proof that every bounded convex figure is of class Q.

Section 7 Pathology   201

Are there figures which are not of class Q? The psychedelic puff-ball. A curve whose interior is not of class Q.

CHAPTER 4 SOME EXOTIC GEOMETRIES

Section 1 Historical Background   213

Euclid's contribution. Objections to Euclid.

Section 2 Spherical Geometry   217

A well-known non-Euclidean geometry, where two lines always meet.

Section 3 Absolute Geometry   221

The theorems of Euclid which do not depend on the parallel postulate. The angle sum in a triangle.

Section 4 More History—Saccheri, Bolyai, and Lobachevsky   226

An accidental discovery of non-Euclidean geometry, and two intentional ones.

Section 5 Hyperbolic Geometry   230

The geometry of Bolyai and Lobachevsky, where there are two kinds of parallels to a given line. Angle of parallelism. The angle sum in a triangle is less than 180°. Relevance to the real world.

Section 6 New Beginnings   242

Hilbert's axioms for Euclidean geometry.

Section 7	Analytic Geometry—A Reminder	248

Coordinate system, straight lines, and conics briefly reviewed.

Section 8	Finite Arithmetics	255

The integers modulo p (p prime), I_p, as a special case of a finite field.

Section 9	Finite Geometries	262

The points and lines of the analytic geometry coordinatized by a finite field. The special case of I_5.

Section 10	Application	273

How to use finite geometries to grow tomato plants. (Design of experiments.)

Section 11	Circles and Quadratic Equations	279

Second-degree equations over I_5. Interior and exterior points of a "circle."

Section 12	Finite Affine Planes	289

Section 9 redone axiomatically.

Section 13	Finite Projective Planes	292

The symmetrized geometry, where two points always determine a line and two lines always determine a point.

Section 14	Ovals in a Finite Plane	296

An axiomatic treatment of "circles." Tangents, secants, interior points, and exterior points. The strange behavior of circles in the even-order case.

Section 15	A Finite Version of Poincaré's Universe	303

The classical model of non-Euclidean geometry. A finite version of this model.

Appendix	Excerpts from Euclid	308

CHAPTER 5 GAMES

Section 1 Introduction 317

Section 2 Some Tree Games 318
The game of Nim and related games. "Natural outcomes" for these games.

Section 3 The Game of Hex 327
Analysis of the game. White should always win, but no one knows how. Beck's Hex, and the theorem which wrecks Beck's Hex.

Section 4 The Game of Nim 340
Complete analysis of Nim, including a recipe for winning. The Marienbad Game.

Section 5 Games of Chance 359
Coin tossing, dice, roulette. Mr. Mark vs. Mr. House. Analysis of some simple examples.

Section 6 Matrix Games 365
The "value" of a matrix game. Proof of the minimax theorem for 2×2 games.

Section 7 Applications of Matrix Games 378
Airplanes vs. submarines (the North Atlantic Game). Similar war games.

Section 8 Positive-sum Games 381
Analysis of a simple positive-sum game.

Section 9 Sharing 383
Further remarks about positive-sum games.

Section 10 Cooperative Games 385
Formation of coalitions to increase payoff. (The game called Production.)

CHAPTER 6	WHAT'S IN A NAME?	
Section 1	Introduction	391
	General comments on what we do and don't know about our number system.	
Section 2	Historical Background	392
	How number systems developed. Examples from different cultures.	
Section 3	Place Notation for the Base b	396
	Development of the "decimal" system of writing integers for bases other than 10.	
Section 4	Some Properties of Natural Numbers Related to Notation	402
	Divisibility criterion; casting out 9s. Perfect squares. Distribution of digits and density.	
Section 5	Fractions: First Comments	414
	Egyptian fractions and Babylonian fractions and some questions about them.	
Section 6	Farey Fractions	416
	Definition and fundamental properties. Mediation.	
Section 7	Egyptian Fractions	421
	Basic and not-so-basic properties. Fibonacci and Farey series algorithms; the splitting method.	
Section 8	The Euclidean Algorithm	435
	Greatest common divisors and least common multiples. Number of steps involved.	
Section 9	Continued Fractions	440
	Connection with Euclidean algorithm and elementary properties.	

Section 10   Decimal Fractions   444

Finite, repeating, and nonrepeating decimals. What is a decimal? The decimal algorithm. Completeness. Arithmetic properties of decimal digits: theorems of Leibniz, Gauss, Midy, and others.

Section 11   Concluding Remarks   470

What's in a name? A comparison of the different ways of naming fractions.

GLOSSARY OF SYMBOLS   477

INDEX   479

APPENDIX 2000   491

FOREWORD

Excursions Into Mathematics, the combined efforts of three University of Wisconsin mathematicians, is one of the most beautiful, most exciting textbooks about modern math to be written in recent decades. I can recall how avidly I turned its pages when I was writing the Mathematical Games department in *Scientific American*. Many of the book's new revelations found their way into my column.

Several aspects of *Excursions* merit special praise. Although written primarily for college students interested in math, the authors do not hesitate to steer readers into deeper waters, and to include elegant proofs of the sort that so often are absent from college textbooks. Another aspect which distinguishes this volume from most college textbooks is its inclusion of topics that can be considered "recreational," though far from trivial. The classic Tower of Hanoi puzzle, for example, is carefully analyzed by Donald Crowe, along with its connection to the ancient Chinese Rings puzzle, and to Hamiltonian circuits.

Anatole Beck's section on mathematical games has a good discussion of Nim, and the most complete analysis of Hex to be found anywhere, including many fresh results such as a subtle variation called "Beck's Hex." This is followed by an introduction to game theory by way of matrix games and cooperative games. A clever proof of the minimax theorem for 2 x 2 matrix games is supplied. If you imagine that such games have little practical value, you couldn't be more wrong. They have significant applications to both economic and political problems. After explaining an extremely simple cooperative game—two persons are offered $100 provided they can agree on how to share it—Beck writes: "Does it sound farfetched? Not at all. the game is played all over the world, a million times a day, as part of the game called Production."

Is it possible for a polyhedron to be free of all diagonals? Put another way, are there polyhedrons such that between any pair of corners there is an edge? Only two are known. One is the tetrahedron. The other is a nonconvex solid called the Császár Polyhedron after its discoverer. Crowe shows how to make this curiosity, which seems to have no right to exist, with cardboard. Also in this section is a wonderfully clear proof of Euler's famous formula: On any convex polyhedron the number of corners, minus the number of edges, plus the number of faces, always equals 2. Crowe uses this theorem to prove there are exactly eight possible deltahedra—convex polyhedrons with every face an equilateral triangle.

Although at the time Crowe wrote the section on map coloring, the four-color conjecture had not yet been proved, yet his discussion of graph coloring is especially informative in including problems equivalent to the four-color theorem, as well as some fascinating related problems still unsolved. The new edition includes references to the solution of the four-color problem and other problems since solved.

Michael Bleicher's section on primes, perfect numbers, and amicable pairs is a splendid survey with many simply stated but infuriating unanswered questions. Beck's section on area, and Crowe's section on exotic geometries, swarm with surprises. Exotic Geometries includes finite fields, affine and projective planes, and Poincaré's famous model of the hyperbolic plane. This model shows that hyperbolic non-Euclidean geometry is consistent if Euclidean geometry is. It provided Maurits Escher with "Circle Limit," one of his most impressive tessellations. Bleicher's section on number theory introduces Farey fractions, Egyptian fractions, continued fractions, and many other delightful topics.

Not the least of this book's attractive features are its striking illustrations, including photographs and drawings of eminent mathematicians whose discoveries are featured. Only four of these portraits are of men who were living when the book was first published. Piet Hein, who invented Hex, died in 1996. H. S. M. Coxeter, The University of Toronto's world-famous geometer, is still active in mathematics at the age of 93. Derrick H. Lehmer and Waclaw Sierpiński have also since passed away.

Excursions Into Mathematics is rich in unusual exercises for readers, and in valuable lists of references for further information. There is a glossary of all the symbols used in the book, and an excellent index. The writing is uniformly clear, informal, and free of unnecessary jargon. Correction of some printer's errors have been made, and notes added to update a few topics. Alice and Klaus Peters deserve cheers for bringing back into print this marvelous volume.

Martin Gardner
Hendersonville, NC
November 1999

NOTE TO THE INSTRUCTOR

This book is a collection of six independent mathematical essays written for the intelligent layman to give him some feeling for what mathematics is and what mathematicians do. The instructor should keep in mind that this course is not preparatory for some other course, but offers a chance for the student to discover and become interested in mathematical reasoning. Thus, comprehension is more important than technique.

What matters in *Excursions* is the careful penetration of a few problems—the book is in no sense a survey—and the student should be pleased to do two chapters each semester. (However, by making selections as suggested below, smaller portions of more chapters could be covered.) For the benefit of those instructors who prefer to teach only a portion of each chapter, we provide a guide to the chapters' construction.

Chapter 1: Sections 1 to 3 should be covered, and as many of Sections 4 to 7 as the instructor finds suitable. Sections 4 to 6 are independent of each other, all of the same difficulty, and depend on Sections 1 to 3. Section 7 is more difficult (except for Theorem 47) and depends on Section 6. Most classes, except those particularly interested in coloring problems, may omit the rest of Section 7 after Theorem 47. Exercises 58 give suggestions for further reading and do not depend on Section 7.

It is important not to rush through the material. A class might well spend a half semester on Sections 1, 2, 3, and one of 4, 5, or 6, although a stronger class can do all of these and part of 7 in the same time. The exercises should not be skipped. In particular, the ideas of Exercises 7 are of fundamental importance and may take more class time than the rest of Section 2.

A few of the "exercises" are just that and can be solved by routine calculations, but most of them should really be called "problems." The student should learn to provide good partial solutions when he cannot solve the whole problem.

There are three categories of exercises which repeatedly occur, namely:

1   Routine problems, which require little or no thought but which help fix certain ideas or relations in the solver's mind. There are not many of these since our book is designed to encourage

thought rather than teach techniques. Typical problems of this type are Exercises 1, 3 (Problems 1, 2), 17, 19, 23 (Problem 1), 27 (Problem 1), 31 (Problem 1), 32 (Problem 1).

2 Problems which ask the student to repeat an argument used in the text, but in a slightly different context or in greater generality. To solve such a problem, the student should expect to go back to the original discussion and see exactly how it must be modified or extended under the new conditions. This is likely to require considerable thought since the student is not expected to have memorized any special techniques to enable him to do this automatically. Typical problems of this type are Exercises 7 (Problems 7, 8), 14 (Problem 3), 15, 16, 37 (Problems 1, 2).

3 Experimental problems. These are probably the most important problems in this chapter. The reader is expected to draw pictures or look at models he has already made in order to guess a general theorem or to disprove some natural conjecture. This experimental method is of the utmost importance, for it shows the way in which mathematics actually grows and by which many of the results of this chapter were actually discovered. The instructor must encourage this method at every opportunity, for most students are unfamiliar with its importance. Students should be encouraged to invent problems of this type for themselves and test them out by experiment. Typical problems of this type are Exercises 3 (Problem 3), 14 (Problems 4, 6), parts of 15 and 16, 24 (Problems 2, 3, 4), 25 (Problems 1, 2), 27 (Problem 2(a)), 30, 40 (Problem 3), 42 (Problem 4(a)), 58 (Problem 8).

Chapter 2: Sections 1 to 3 should be covered. The material on congruences will be helpful in the second half of the book in parts of Chapters 4 and 6. Some people may wish to assume unique factorization into primes, which is proved in Section 2.

The end of Section 5 is another reasonable stopping place. Those who continue may find it well, unless the class is unusually interested, bright, or motivated, to skip the proof (or at least part of the proof) of Theorem 89.

Section 7 is an open-ended section, throwing out ideas which can be followed up or ignored as time and interest dictate.

Chapter 3: The basic work of the chapter is to be found in Sections 1 to 5, culminating in the remarks at the end of Section 5 to the effect that area is well defined for polygonal regions. The work is of approximately uniform difficulty, except for the proof of Theorem 19, together with its lemmas. The teacher might easily skip this proof, merely sketching it to provide the flavor of the technique. The five sections are about half a semester's work for an average class. The chapter's work can easily be shortened by omitting Section 5; in this case, Theorem 23 and the material following it should be attached to Section 4.

Section 6 is largely new work for the average student, and it is here that he may find the flavor of modern measure theory. Superior students should be able to negotiate this section without difficulty, and it should prove exciting. Section 7 is difficult and demands the best efforts of the best students, yielding a sense of the intricacies and brilliance of this century's mathematics.

Most of the problems are directed down the side corridors of the text, picking up estimates or subsidiary questions. In this approach, they reflect the modus operandi of the working mathematician, who lays aside questions in the main thrust of his work only to return to them later. Some of the "exercises" are, in fact, thought questions and are answered by lemmas and theorems coming later in the text. The teacher should not distinguish between these but should encourage the students to at least try all the problems. The remarks on the problems in Chapter 1 apply here as well.

Chapter 4: This chapter is in three distinct parts. Part 1 is composed of Sections 1 to 6, Part 2 of Sections 7 to 11, and Part 3 of Sections 12 to 15. Part 1 is a self-contained simple introduction to classical non-Euclidean (hyperbolic) geometry. It provides useful historical and mathematical motivation for Parts 2 and 3. Part 2, also self-contained, describes a particular finite geometry in detail, with application to design of experiments. Part 3 repeats some of the results of Part 2 in a more general context and shows how some of the ideas of Part 1 can be incorporated into the context of finite geometries.

It is very important that the student make many drawings to illustrate the results of Sections 9 and 11. He must become as completely at home in $AG(F_5)$ as he is already at home in the real Euclidean plane. Only then can he use this geometry to give concrete meaning to the abstractions of Part 3.

For a very short course (say, to fill in the last few weeks of a semester), Part 1 alone or Part 2 alone would be suitable. For a somewhat longer short course in finite geometries, Parts 2 and 3 (omitting Section 15) are suitable.

Chapter 5: A minimum for this chapter is the first four sections. This will give an introduction to the theory of games with full information, where probability is not employed. These sections are of about uniform difficulty, except that the section on Hex is somewhat harder and the part on Beck's Hex is rather subtle. The part on Beck's Hex compensates somewhat for its subtlety by giving the student the experience (which should be exhilarating) of seeing brand-new mathematics, published here for the first time. The section on Games of Chance should be given only if Matrix Games are also to be given, and the one is a prerequisite for the other. Actually, Section 6 loses some of its flavor if Section 7 (applications) is not also included.

Sections 8, 9, and 10 are only slightly harder, but they can be left out without any serious detriment to the other sections.

As an alternative, Sections 1, 5, 6, and 7 could form the core of a short chapter, possibly with Sections 8 and/or 9 and/or 10. This shortened version would tend to be more "applied," from the linear-programming point of view, but would be less interesting as mathematics for the general student.

Chapter 6: Many of the sections of this chapter can be expanded, contracted, or skipped altogether without serious effect. The logical dependence is shown on the next page.

In order to cover the first part of Section 4, it is necessary to have an elementary knowledge of the theory of congruence, which can be obtained in Chapter 2, paragraphs 33 to 39. This background is necessary for paragraphs 12 through 23 of Section 4. The remainder of the section does not depend on these paragraphs, so that you may skip these rather than fill in the background from the above-mentioned paragraphs of Chapter 2.

Note to the Instructor

In Section 10 there are a number of classical, but relatively unknown, results on the digits of repeating decimals. Many of these results are surprising and beautiful, and they show us patterns which have been right under our noses for years but have gone unnoticed.

The section on continued fractions is the least complete, but if the class is an exceptional one that can handle the many recursions necessary to develop this topic further, there are several good sources in the references.

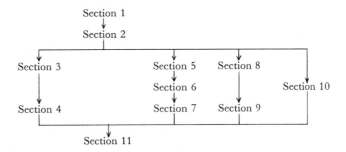

EXCURSIONS INTO MATHEMATICS

CHAPTER 1 **EULER'S FORMULA FOR POLYHEDRA, AND RELATED TOPICS**

BY DONALD W. CROWE

Illustration 1

OCTAHEDRON
Air

CUBE
Earth

TETRAHEDRON
Fire

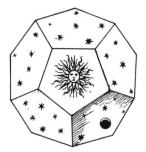

DODECAHEDRON
the Universe

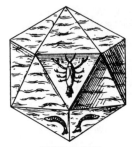

ICOSAHEDRON
Water

Section   1   Introduction

1A   Prelude and Fantasy

Patterns and regular figures have played a varied role in many cultures, from the decorations in Egyptian tombs and the beauty of the Alhambra to the present day. It is, therefore, not surprising that the *mathematical* study of regularity is of great antiquity. The regular polygons and regular solids (polyhedra) were the object of special study by the ancient Greeks. Still today the five familiar regular polyhedra (tetrahedron, cube, octahedron, dodecahedron, and icosahedron) are often called *Platonic solids*, while certain other, less regular polyhedra are called *Archimedean solids*.

It has even been suggested [5, p. 13]† that Euclid's famous treatise, the 13 "books" constituting *The Elements*, was designed as a text on regular figures. This is because the very first Proposition of Book I shows how to construct an equilateral triangle—the simplest of regular polygons—and the last Propositions of Book XIII deal with the construction of the regular dodecahedron and regular icosahedron, the most complex of the five regular polyhedra in space.

Certain of the ancient Greeks attached a mystical significance to the five regular polyhedra, associating them with the four "elements" and the universe. This mystical point of view persisted through the centuries to influence Johannes Kepler (1571–1630), whose scientific work was one of the starting points of modern physics. Kepler's explanation for the association of regular polyhedra with the four elements and the universe is shown graphically in Illustration 1, which is an exact reproduction of Kepler's own drawing. Note that each solid is decorated with the appropriate symbols for the element to which it corresponds. The tetrahedron and icosahedron have, respectively, the smallest and largest volumes for their surface. They correspond to dryness and wetness, that is, fire and water. The cube, because it stands so firmly on its base, corresponds to the stable earth, while the octahedron, which rotates freely when held by opposite vertices, corresponds to the mobile air. Unfortunately, this leaves out the dodecahedron, but Kepler's astrological skill avoided this difficulty by noting that its 12 faces can

† Numbers in square brackets refer to the references at the end of each chapter.

Illustration 2 — Size and position of the Planetary orbits given by the five regular polyhedra. (Kepler was so impressed with this (false!) discovery that he proposed that a gold decanter be made in this shape, delivering seven kinds of wine from the six spheres and the sun.)

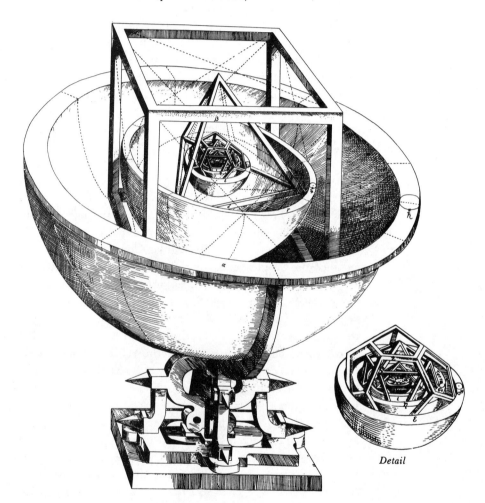

Detail

α SPHERE OF SATURN	β CUBE	γ SPHERE OF JUPITER
δ TETRAHEDRON	ε SPHERE OF MARS	ζ DODECAHEDRON
η ORBIT OF EARTH	θ ICOSAHEDRON	ι SPHERE OF VENUS
χ OCTAHEDRON	λ SPHERE OF MERCURY	μ SUN

From Kepler's *Mysterium Cosmographicum*, in Vol. I of The Complete Astronomical Works of Kepler, Ed. Ch. Frisch.

be thought of as the 12 signs of the zodiac. Hence the dodecahedron represents the universe itself.

Kepler also found less fanciful reasons for the importance of the regular polyhedra. Influenced by Copernicus' teaching that the planets move in orbits about the sun, the young schoolteacher Kepler had tried to discover numerical relations which would explain why there were exactly six planets (Mercury, Venus, Earth, Mars, Jupiter, Saturn) and the reason for their particular distances from the central sun. Every attempt failed until one memorable day in July, 1595. Then, a drawing he had made on the blackboard, for his students, of a figure suggesting an equilateral triangle together with its inscribed and circumscribed circles gave him the necessary clue. For he saw that the orbits of planets about the sun should not be explained *arithmetically* but *geometrically*. He soon realized that the equilateral triangle and the other regular polygons in the *plane* were still not quite the proper tool, for planets move in *space*, not in the plane. Furthermore, there were an infinite number of regular polygons but only six known planets. From this, it was only a small step to see that the five gaps between the six planets must correspond exactly to the five regular polyhedra. This explained two puzzles at once: why there are only five regular polyhedra and why there are only six planets.

After trying various ways of arranging the five regular polyhedra, Kepler found the right arrangement, which is shown in Illustration 2. An outer sphere (the sphere on which Saturn moves) contains a cube, in which is inscribed the sphere of Jupiter. In it is inscribed a tetrahedron, containing the sphere of Mars, and in that is a dodecahedron, then an icosahedron, and then an octahedron, whose inscribed spheres are the spheres of Earth, Venus, and Mercury, respectively.

We can imagine the excitement felt by Kepler when all the pieces of this puzzle finally fell into just the right places. The care with which his drawing (Illustration 2) is executed must reflect the pride he felt in his discovery. In fact, this pride went even further. In his eagerness to let the world know of his explanation of the mystery of the universe, Kepler proposed to his patron, the Duke of Württemberg, that a gold model be made in the form of an elaborate decanter from which seven kinds of beverage would pour from the seven spheres representing the six planets and the sun. Although the

Illustration 3

TETRAHEDRON

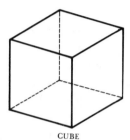
CUBE

OCTAHEDRON

DODECAHEDRON

ICOSAHEDRON

negotiations for this project dragged out for several years, it was apparently never actually carried out except as a paper model made by Kepler himself.

But the epilogue is the most remarkable part of this remarkable story. For, as we know now, this whole brilliant insight, conceived by one of the great minds of his time, *is absolute nonsense*. For one thing, the correspondence between the spheres and the planetary orbits is not really exact. There are slight discrepancies, which Kepler himself tried to account for by making the spherical shells of varying thicknesses but for which there is no real explanation. But more important, although there are only five regular polyhedra (as we shall prove in Section 2), there are at least three more planets: Uranus, Neptune, and Pluto. Hence there is no possible simple correspondence of the kind Kepler described. Finally, of course, modern science places little faith in such "accidental" relationships, and even if no more planets had been found, Kepler's remarkable discovery would no longer be accepted.†

Having seen the failure of Kepler's bold attempt to relate the regular polyhedra to the real, physical world, we shall turn in Section 2 to the *mathematical* reason why there can be no more than five regular polyhedra. Our main tool will be a deceptively simple formula, named after the Swiss mathematician Leonhard Euler (1707–1783), although it was already known to Descartes (who was born in 1596, the year Kepler published his great discovery).

1B Euler's Formula

Before we state, and prove, Euler's formula, the reader should look more carefully at the five regular polyhedra shown in Illustration 3. Although the ancient Greeks had proved that these were the only regular polyhedra, they apparently never looked closely at the numbers of vertices, edges, and faces of these solids. For, if they had, it is difficult to imagine that they could have failed to note the curious relation apparent from the last column of Table 1 below.

† For a highly readable account of Kepler's life and of the role played by his mystical and astrological notions in his scientific discoveries, the student should read the chapter on Kepler in Arthur Koestler's *The Sleepwalkers*.

Table 1

	v	e	f	$v - e + f$
Tetrahedron	4	6	4	2
Cube	8	12	6	2
Octahedron	6	12	8	2
Dodecahedron	20	30	12	2
Icosahedron	12	30	20	2

We are tempted to guess that the 2's appearing in this last column are due to the regularity of the solids in question. But this is not the case. In fact, we shall prove a general theorem to the effect that a 2 is *always* obtained. This theorem is named for Euler, rather than Descartes, because Euler published a proof in 1758.

1   EXERCISES

1   Verify all the entries in Table 1.
2   Calculate $v - e + f$ for a triangular prism, for a pyramid with a square base, and for a cube with its corners cut off one-third of the way along the edges. (See Illustration 4.)

Euler's theorem, as we prove it, deals with arbitrary *convex polyhedra*, of which the regular polyhedra are just special cases. By a *polyhedron*, we mean a solid (or sometimes just the surface of such a solid) with plane faces and straight edges, arranged so that every edge joins two vertices and is also the common edge of two faces. We also make the

Illustration 4   *Each of these convex polyhedra satisfies the condition* $v - e + f = 2$.

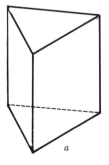
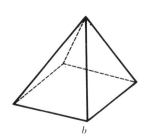
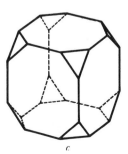

a   b   c

Illustration 5 *This nonconvex polyhedron also satisfies* $v - e + f = 2$.

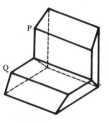

special requirement that the boundary of each face have a single component, so that, for example, the ring-shaped region between two concentric squares is not permitted as the face of a polyhedron. A polyhedron is called *convex* if it is always possible to join any two points in it by a line segment lying completely in the polyhedron. Thus, an ordinary cube is convex, but the L-shaped solid in Illustration 5 is not convex because, for example, the segment between points P and Q will lie entirely *outside* the L (except for the end points P and Q themselves).

2 Theorem (Euler's Theorem)

If v, e, and f are the numbers of vertices, edges, and faces, respectively, of any convex polyhedron whatsoever, then $v - e + f = 2$.

A simple proof of this theorem goes as follows. We first hold our convex polyhedron so that one face is horizontal and all the rest of the polyhedron hangs below this face. (For a cube this is its natural position. However, a regular tetrahedron held this way would seem to be upside down, as in Illustration 6.) Now, remove that horizontal face on top, and project the resulting figure downward from a point slightly above this face onto a horizontal plane below the polyhedron. The result will be a network in the horizontal plane, as shown at the bottom of Illustration 7. For the regular polyhedra the results of this operation are shown in Illustration 8. (Note that the point from which the projection is made should be taken close enough to the removed face so that no overlapping of edges or regions occurs in the plane network. If the polyhedron is convex, this is always possible, whereas for the polyhedron in Illustration 5, it would not be possible to find a suitable projection point.)

Illustration 6   *Tetrahedron, in position ready for projection.*

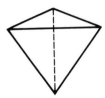

Illustration 7   *Imagine the tetrahedron and cube to be made of glass, with black edges. The shadows they cast on a plane are called the* projections *from the light source.*

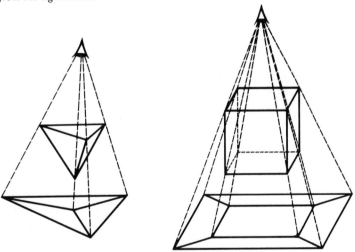

Illustration 8   *Projections of the five regular polyhedra onto a plane.*

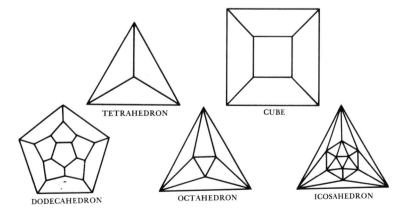

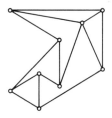

Illustration 9

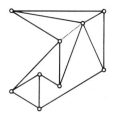

Illustration 10

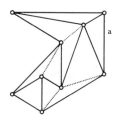

Illustration 11

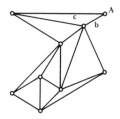

Illustration 12

Each region in such a network represents a face, each line segment an edge, and each intersection of the line segments a vertex. The only omission is the top (horizontal) face, which we took away before we made the projection and which we now think of as the infinite region surrounding the network. So, considering the finite regions in the plane as faces, we need only show that $v - e + f = 1$ for this network in the plane. In fact, it is true that $v - e + f = 1$ *for any connected network in the plane at all.* However, to simplify the proof, let us assume that the edges of the network are straight-line segments and that no fewer than two edges meet at each vertex. We also assume that the deletion of a single vertex and all the edges meeting at it does not separate the network into two distinct parts. The network does not have to be the projection of a convex polyhedron.

Consider an arbitrary network of this type, as in Illustration 9. Note first that if we add any line segment in the interior of one of these regions which joins two vertices of the region we do not change the number $v - e + f$. (See Illustration 10.) This is because adding a line segment does not change v at all, but it adds 1 to each of e and f. Since e and f have opposite signs in the expression $v - e + f$, these extra 1's just cancel out. Now, this process can be repeated—always without changing the number $v - e + f$—until all the regions have been converted into triangles. The network of Illustration 9 has now become that of Illustration 11. Thus, we need only show that $v - e + f = 1$ for networks composed entirely of triangles. For such a triangulated network we now turn around and reverse the process, this time beginning from the outside boundary of the network and deleting edges one at a time. The first step in this particular example does not cause any trouble; by deleting edge a, say, we get Illustration 12. And, of course, $v - e + f$ has not changed. Now, however, if we delete edge b we have c sticking out into the air. So let us agree that whenever a vertex (such as A) on the outside boundary of the network has exactly two edges meeting at it, we shall delete it along with both the edges meeting at it. Moreover, we agree that whenever there is a vertex like A, we shall delete it, and its two edges, *before* we delete any single edges like a. What does this do to the number $v - e + f$? Clearly, v is decreased by 1, e is decreased by 2, and f is decreased by 1. Taking account of the signs, this means that the number $v - e + f$ still remains unchanged. But we can repeat this process of deleting edges from our network until we end up with a single triangle, as in Illustration 13. For a triangle, however, we easily find that $v - e + f = 1$. Since $v - e + f$ has not changed at all, this proves the theorem.†

Note that although our proof used the convexity of the original polyhedron, the theorem is also true for certain nonconvex polyhedra—in particular, for the L-shaped solid in Illustration 5. In fact, the preceding proof of Euler's theorem can be modified to prove that $v - e + f = 1$ for an arbitrary connected network in the plane. The network may even have curved edges, so that some regions may have only one or two edges (for example, see Illustration 14). This fact will be used when we discuss map-coloring problems in a later section.

Illustration 13

Illustration 14

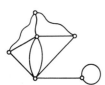

3 EXERCISES

1. Verify that $v - e + f = 2$ for the solid in Illustration 5, and that $v - e + f = 1$ for the networks in Illustrations 11 and 14.
2. Draw the figure corresponding to Illustration 8 for the *cuboctahedron*, that is, a cube each of whose vertices has been cut off by a plane through the midpoints of the 3 edges which meet at that vertex. Calculate $v - e + f$ for this solid.
3. What is $v - e + f$ if the network in question is not connected, but has, say, n components?

There are other instructive ways of proving that $v - e + f = 2$. One of them, due to von Staudt, is given picturesquely in [17, p. 75], and more prosaically in [5, p. 9]. The von Staudt proof, and the one just given, are "topological" proofs which do not depend on lines being straight. An interesting geometric proof, using angle sums in polygons, is given in [10, p. 28].

---

† An ingenious reader can discover that in spite of the elaborate precautions we have taken to ensure that no edge sticks out by itself, it may still happen that such an edge appears. Although awkward, this fact offers no essential difficulty, for deleting such an edge, and the single vertex at the end of it, changes both v and e by 1, leaving f unchanged. Hence $v - e + f$ is unchanged.

Although Leonhard Euler (1707–1783) was Swiss, his entire mathematical life was spent in St. Petersburg (1725–1741, 1766–1783) and Berlin (1741–1766). In the 530 books and articles he published in his lifetime, and in the 220 published after his death, he made fundamental contributions to every branch of mathematics, setting trends that were followed for 150 years. Generalizations of his elementary formula $v - e + f = 2$ are a recurring theme in modern topology. Some of his contributions to number theory appear in Chapter 2. (Courtesy of the Smith Collection, Columbia University.)

Section 2 Regular Polyhedra

A great deal is known about the five regular polyhedra of the Greeks. Are there any other regular polyhedra? The answer to this depends on what we mean by *regular*. Certainly to be called regular, a polyhedron ought to have all its faces alike and all its vertices alike. That is,

(a) all faces have the same number $p \geq 3$ of edges, and
(b) the same number $q \geq 3$ of edges meet at each vertex.

However, there are polyhedra satisfying these two conditions which we would not want to call regular, for example, an ordinary solid rectangular box. Hence we include as a further condition

(c) all faces are regular polygons.

A convex polyhedron is said to be *regular of type* $\{p, q\}$ if it satisfies these three conditions. The five regular polyhedra already discussed have types $\{3, 3\}$, $\{4, 3\}$, $\{3, 4\}$, $\{5, 3\}$, $\{3, 5\}$, respectively.

Now, suppose we have a convex polyhedron satisfying conditions (a) and (b). Let us count its edges, in two ways. Since there are f faces, each of them having exactly p edges, the product pf is twice the total number of edges of the polyhedron. (Every edge is counted twice, once for each of the two faces which meet at it.) Hence,

4 $$pf = 2e.$$

Similarly, by counting the q edges at each of the v vertices,

5 $$qv = 2e.$$

From Equations 4 and 5 we have $f = \dfrac{2e}{p}$ and $v = \dfrac{2e}{q}$. Substituting these into Euler's formula $v - e + f = 2$ we get $\dfrac{2e}{q} - e + \dfrac{2e}{p} = 2$. Dividing by $2e$, and rearranging, we have

$$\frac{1}{p} + \frac{1}{q} = \frac{1}{2} + \frac{1}{e}.$$

What integers p, q, and e satisfy this so-called "Diophantine"† equation? If $p = 6$ in this equation, then q must be less than 3, contrary to condition (b). If p is larger than 6 (written $p > 6$), then q must be still smaller, again contrary to condition (b). Hence p can only be 3, 4, or 5. For the same reason, q can only be 3, 4, or 5. Checking the nine combinations of these values shows that only five of them make $1/p + 1/q$ greater than $\frac{1}{2}$ (as required by the equation). These five solutions are listed in Table 2.

Table 2

p	q
3	3
3	4
3	5
4	3
5	3

We have thus proved:

6 Theorem

There are no other types of regular convex polyhedra than those of the five classical regular polyhedra already known to the Greeks. It is remarkable that we were able to get this information as a purely arithmetic consequence of Euler's formula with essentially no further reference to geometry at all! The attentive reader will have noticed that we have only proved that there are *at most* five types of regular polyhedra. A strict proof can be given that each of these five numerical possibilities can actually be realized geometrically [4, pp. 148–149]. However, the student will probably also be convinced of this mathematical fact if he actually constructs each of the five, as in Problem 1 of Exercises 7.

† An equation for which only integer solutions are desired is called *Diophantine*. Such equations will occur frequently in this chapter, as well as in Chapter 2.

Illustration 15 *The three regular tessellations of the plane.*

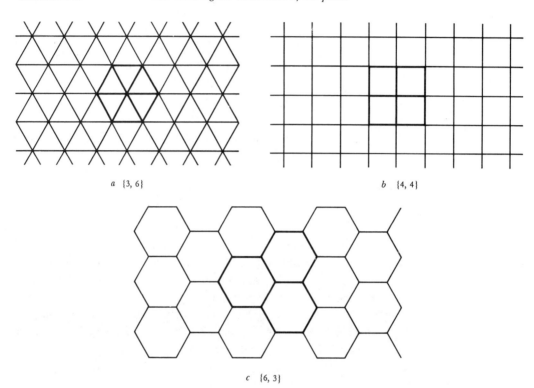

a {3, 6}

b {4, 4}

c {6, 3}

It is instructive to note what happens if we take p and q so that $\frac{1}{p} + \frac{1}{q} = \frac{1}{2}$. This can be thought of as taking e to be infinite, in the formula $\frac{1}{p} + \frac{1}{q} = \frac{1}{2} + \frac{1}{e}$, so that $\frac{1}{e} = 0$. Then there are three more solutions {3, 6}, {4, 4}, and {6, 3} to this modified equation. These can be thought of as "degenerate polyhedra," and they have a very clear geometric meaning. For {3, 6} means "3 edges to each face and 6 edges at each vertex" or "six triangles around each vertex." Trying to put 6 equilateral triangles around a point in space shows immediately what happens: the whole figure flattens out. And in fact, {3,6} is a convenient symbol for the "tessellation" of the plane by equilateral triangles shown in Illustration 15(a). This shows why e was infinite.

Section 2   Regular Polyhedra

In fact, of course, v and f are also both infinite. Likewise, $\{4, 4\}$ is the familiar chessboard pattern of squares shown in Illustration 15(b). The remaining tessellation $\{6, 3\}$ is the honeycomb pattern of Illustration 15(c) which is often found in tile floors. So even in these cases the arithmetic calculations have a real geometric significance.

It is remarkable that the regular tetrahedron, the cube, and the regular octahedron, dodecahedron, and icosahedron are approximated very closely in the skeletons of some of the minute marine animals called *radiolaria*. These are shown in Illustration 16, taken

Illustration 16   *Good approximations to the five regular polyhedra are found in the skeletons of minute marine animals called radiolaria.*

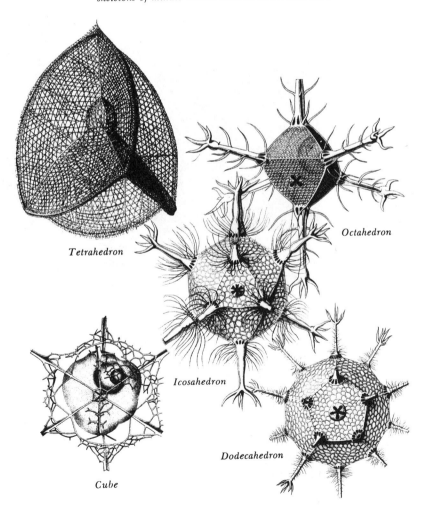

from Plates 18, 63, and 117 of E. H. Haeckel, *Report on the Scientific Results of the Voyage of H.M.S. Challenger*, Vol. 18 (London, 1887).

Among inanimate forms in nature the crystals of various substances show regularities of various types, without the face curvature shown by the radiolaria. (The crystallographer A. E. H. Tutton remarked that "the beauty of the crystals lies in the planeness of their faces.") However, the regular dodecahedron and icosahedron cannot occur as crystals. Illustration 17 shows the tetrahedron and cube as crystals of sodium chlorate and the octahedron as a crystal of chrome alum.

Illustration 17

The cube and regular tetrahedron occur as crystals of sodium chlorate, and the regular octahedron as a crystal of chrome alum.

(From *Crystals and Crystal Growing* by Alan Holden and Phylis Singer. Copyright © 1960 by Educational Services Incorporated. Reprinted by permission of Doubleday & Company, Inc.)

It is possible to derive several other properties of polyhedra by making numerical deductions from Euler's formula, like those which were used to prove Theorem 6. Several of these are included below as Problems 4-11. Note that the result of Problem 9 is somewhat surprising, for at first glance it does not appear to be very hard to construct a polyhedron with exactly seven edges. G. B. Halsted has put this picturesquely by suggesting that anyone who thinks that empty space doesn't impose any restrictions should "take a knife and a raw potato and try to cut it into a seven-edged solid."

To illustrate the methods to be used in these problems, we give two examples worked out in detail as follows:

Example 1

(a) Show that every convex polyhedron has at least two more vertices than half its number of faces. That is, $v \geq 2 + \frac{1}{2}f$. (b) Show also that $f \geq 2 + \frac{1}{2}v$.

SOLUTION TO (a): We know that $v - e + f = 2$. If there were *exactly* three edges for each face, we would have $3f = 2e$ (as in Equation 4). Since we only know that there are *at least* three edges on each face, the number $3f$ may omit some edges. So we only have $3f \leq 2e$ ("$3f$ is less than or equal to $2e$"). That is, after dividing by 2 and writing in opposite order, $e \geq 3f/2$. We substitute this relation into Euler's formula, $v + f = 2 + e$, to eliminate e (since we notice that e does not appear in the desired result, $v \geq 2 + \frac{1}{2}f$). This substitution gives $v + f \geq 2 + (3f/2)$, so that $v \geq 2 + (3f/2) - f$. That is, $v \geq 2 + \frac{1}{2}f$, which is the required result.

SOLUTION TO (b): The solution to part (b) follows by exactly the same kind of deduction. The only difference between parts (a) and (b) is that v and f are interchanged. Interchanging v and f in Euler's formula, $v - e + f = 2$, leaves it unchanged. So we need only note that the other relation we used, $3f \leq 2e$, also holds when v and f are interchanged. That is, we need only note that $3v \leq 2e$ holds—which it does because there are at least three edges at every vertex. From then on, the argument is exactly like that in part (a), and the corresponding conclusion, $f \geq 2 + \frac{1}{2}v$, is obtained.

Example 2

A certain convex polyhedron has faces that are squares and regular hexagons. Each square is surrounded by hexagons, and each hexagon is surrounded alternately by three squares and three hexagons. Determine the numbers f_4, f_6, e, and v of squares, hexagons, edges, and vertices of the polyhedron.

SOLUTION: We know that there are exactly three faces meeting at each vertex, for if there were as many as four, the total angle at that vertex would be at least $4 \cdot 90° = 360°$. But this is impossible for a convex polyhedron (since even for $360°$, the vertex flattens out completely). Hence $3v = 2e$, as in Equation 4. Solving this for v, substituting into $v - e + f = 2$, and clearing fractions, we obtain $3f - e = 6$.

Since the total number of faces is the number of squares plus the number of hexagons, we have $f = f_4 + f_6$. If we count the number, 4, of edges on each square and the number, 6, of edges on each hexagon, we shall count every edge twice. That is, $2e = 4f_4 + 6f_6$. Substituting the values for f and e from these last two equations into $3f - e = 6$ (in the form $6f - 2e = 12$) gives $6(f_4 + f_6) - (4f_4 + 6f_6) = 12$. This simplifies to $2f_4 = 12$, which says that the number of squares is 6.

Now we can determine the number of hexagons by counting their occurrences around the square faces, namely, $4f_4$. But this counts each hexagon three times, since it is counted on each square of the three which surround it. That is, $4f_4 = 3f_6$. Since $f_4 = 6$, this yields $24 = 3f_6$, which says that the number of hexagons is 8. Finally, substituting the total number of faces, $f = 14$, into $3f - e = 6$ gives $e = 36$. Substituting these values of e and f into Euler's formula gives $v - 36 + 14 = 2$, so that $v = 24$.

7 EXERCISES

1 Construct models of the five regular polyhedra. A pattern for the dodecahedron is given in Illustration 18. Patterns for the tetrahedron, octahedron, and icosahedron appear in Illustrations 19, 25(a) and 25(e). The cube can safely be left for the reader. (For hints for constructing these and other interesting models, see the excellent book *Mathematical Models* [9].)

2 Prove that it is not possible to color the faces of a regular dodecahedron with three colors in such a way that no two

Illustration 18

Two caps made from this pattern fit together to form a regular dodecahedron.

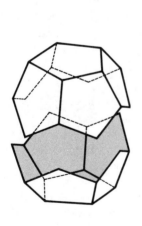

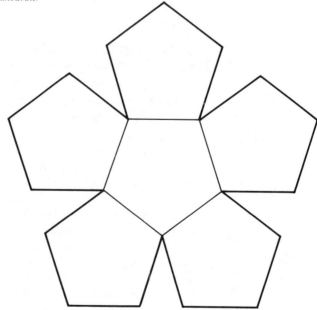

adjacent faces have the same color. Determine the minimum number of colors needed to color the faces of each of the regular polyhedra. Answer: 4, 3, 2, 4, 3.

3 A *Hamilton circuit*† in a network is a closed path (consisting of edges of the network) which passes through each vertex of the network exactly once. Find such a Hamilton circuit for the network of vertices and edges of each of the regular polyhedra.

4 (a) Since there are at least three edges meeting at each vertex of a convex polyhedron, we know that $3v \leq 2e$. Use this fact, and Euler's formula, to show that $3f \geq 6 + e$.

(b) Show also that $3v \geq 6 + e$.

5 (a) Deduce from Problem 4 that $6v \geq 12 + 2e \geq 12 + 3v$.

(b) Show that $6f \geq 12 + 3f$.

(c) From (a) and (b) show that no convex polyhedron has fewer than 4 vertices and 4 faces. Prove also that none has fewer than 6 edges.

6 If the average number of edges per face of a convex polyhedron is at least 6, then $6f \leq 2e$. Combine this fact with the result of Problem 4(a) to prove that every convex polyhedron must have at least one face with 5 or fewer edges. [Hint: If you show that $6f \leq 2e$ is impossible, then you will know that the average number of edges per face is *less than* six. But if the *average* is less than six, then certainly at least one face must have fewer than six edges.]

7 A European "football" is made of pieces of leather in the shapes of regular pentagons and regular hexagons. These are sewed together so that each pentagon is surrounded by hexagons and each hexagon is surrounded (alternately) by three pentagons and three hexagons. Determine the numbers of pentagons and hexagons of such a football. [Hint: Follow worked Example 2.] Answer: 12 pentagons and 20 hexagons.

8 A certain convex polyhedron has faces which are equilateral triangles and regular pentagons, and four faces meet at each vertex. Each triangle is surrounded by pentagons, and each pentagon is surrounded by triangles. Determine the number of triangles, pentagons, edges, and vertices. Answer: $f_3 = 20$, $f_5 = 12$, $e = 60$, $v = 30$.

† Hamilton circuits are discussed in more detail in Section 5 of this chapter.

9 (a) Prove that there is no convex polyhedron with exactly seven edges. [Hint: This can be done entirely by formula. For, if $e = 7$ then $3f \leq 2e = 14$ and $3v \leq 14$. But also $f = 9 - v$, from Euler's formula. Hence, $3f = 27 - 3v \geq 13$. Why is this impossible?]

*(b) For which n is there a convex polyhedron having exactly n edges?

*10 Let e be the number of edges and f the number of faces of an ordinary convex polyhedron.

(a) If all the faces of the polyhedron are triangles, determine the ratio f/e. If f/e has this value for some convex polyhedron, are its faces necessarily all triangles? If not, find a convex polyhedron with this value of f/e for which some faces are not triangles. [Hint: Remember that no face can have fewer than 3 edges.]

(b) If all the faces of a polyhedron are quadrilaterals, what is the ratio f/e? If f/e has this value for some convex polyhedron, are the faces necessarily all quadrilaterals? If not, find a convex polyhedron with this value of f/e for which some faces are not quadrilaterals. [Hint: Can you construct a polyhedron whose 7 faces are a triangle, 5 quadrangles, and a pentagon?]

11 Prove that every convex polyhedron has at least two faces with the same number of edges. [Hint: The "smallest" face must have at least 3 edges. If no two faces have the same number of edges, how many edges must the largest of the f faces have? Is it possible to construct a polyhedron with f faces if one face has so many edges?]

12 We have excluded the cases $p = 2$ or $q = 2$ from the eligible solutions to $\dfrac{1}{p} + \dfrac{1}{q} = \dfrac{1}{2} + \dfrac{1}{e}$. Why?

13 If we permitted $p = 2$, what figure would result from the solution $p = 2$, $q = e$? From $q = 2$, $p = e$? [Hint: Try to draw the vertices and edges on a sphere, using arcs of great circles as edges.]

Section 3 Deltahedra

A regular polyhedron is one having both congruent regular faces and the same number of faces at each vertex. A common procedure in mathematics is to *generalize*. That is, we drop one restriction and try to see how much of what we proved *with* the restriction can still be proved *without* it. Thus, we might ask what sort of polyhedra there are having congruent regular faces but not necessarily the same number at each vertex. In particular, polyhedra whose faces are all equilateral triangles have been called *deltahedra*, since the faces are the shape of the Greek letter delta, Δ. We propose to enumerate all possible convex deltahedra.

Since Euler's formula holds for any convex polyhedron, we have

8 $$v - e + f = 2.$$

Now, there are 3 edges for each face, so we can count the number of edges by multiplying the number of faces by 3—except that this counts each edge twice. Thus,

9 $$3f = 2e.$$

Furthermore, at each vertex there are 3, 4, or 5 edges, since if there were as many as 6 at some vertex the total angle would be $6 \cdot 60° = 360°$, and the polyhedron would flatten out—making, at that vertex, a part of the tessellation $\{3, 6\}$ of equilateral triangles discussed in the preceding section (Illustration 15(a)). So we can count the total number of edges by counting the edges at each of these three types of vertex. In fact, if we let

v_3 = number of vertices at which 3 edges meet,
v_4 = number of vertices at which 4 edges meet,
and v_5 = number of vertices at which 5 edges meet,

we have

10 $$3v_3 + 4v_4 + 5v_5 = 2e.$$

We note also that

11 $$v_3 + v_4 + v_5 = v.$$

Table 3

v_3	v_4	v_5
0	0	12*
0	1	10
0	2	8*
0	3	6*
0	4	4*
0	5	2*
0	6	0*
1	0	9
1	1	7
1	2	5
1	3	3
1	4	1
2	0	6
2	1	4
2	2	2
2	3	0*
3	0	3
3	1	1
4	0	0*

Now, putting Equation 9 into Equation 8 yields
$$v - e + \tfrac{2}{3}e = 2.$$
Replacing e and v by their values from Equations 10 and 11 gives an equation which simplifies to

12 $\qquad 3v_3 + 2v_4 + v_5 = 12.$

The 19 solutions of this Diophantine equation are readily verified to be only those in Table 3.

We now show by more geometric arguments that the 11 cases not marked by an asterisk in the table are, in fact, impossible. First, we consider those cases for which $v_3 \neq 0$, that is, those cases for which there is at least one vertex—call it P—at which exactly three faces meet. Let these three faces be named PAB, PBC, PCA. Then there are at least two things we can do:

(a) We can close off the trihedron $PABC$ with a single face ABC to obtain a tetrahedron, as in Illustration 19.

(b) We can attach equilateral triangles to each of the edges AB, BC, CA and bring them together at a point D to form two tetrahedra, base to base, as in Illustration 20.

In fact, these are the *only* two things we can do. For, suppose we try to add more equilateral triangles and *not* close up the figure at stage (b). Then there would be triangles ABD and ACE, say, attached at edges AB and AC.

But there is not enough room for another equilateral triangle between these faces. For, if E and D are pushed out so far that they lie in the same planes as PAC and PAB, respectively, then (and only then) ADE is an equilateral triangle (see Illustration 21). The reason for this is that BD and EC are equal and parallel (both being equal and parallel to PA), and hence $BDEC$ is a parallelogram. Consequently, the sides BC and DE of this parallelogram are also equal and parallel, which proves that triangle ADE is congruent to the equilateral triangle PBC. Since, of course, E and D *cannot* be pushed out this far, ADE can never be an equilateral triangle. (A good way to see this is to make a three-dimensional model using the pattern in Illustration 21.) This means that the 10 remaining cases where $v_3 \neq 0$ are all impossible.

Only one case remains to be eliminated, namely, $v_3 = 0$, $v_4 = 1$, $v_5 = 10$. This again can be done easily by trying to make a model. Begin with the one vertex having 4 edges (Illustration 22(a)).

Section 3   Deltahedra   23

Illustration 19

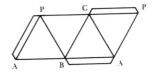

Illustration 20

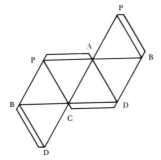 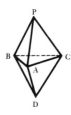

Illustration 21

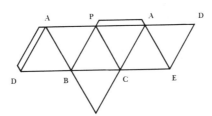 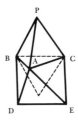

Illustration 22

a   *b*

Each of the vertices adjacent to it must have 5 edges, so in the next step we must attach triangles as shown in Illustration 22(b). We now have 9 vertices altogether, and it can be easily seen that it is impossible to have exactly 2 more vertices. In fact, the natural way to finish this model is to add *one* vertex at the bottom, connected to each of the last 4 vertices. This is the case $v_3 = 0$, $v_4 = 2$, $v_5 = 8$. A pattern for it is given in Illustration 23. (A full-size pattern appears as Illustration 24.)

Having eliminated 11 of the 19 possibilities for deltahedra, the easiest way to see that the remaining 8 can actually be constructed is to construct them. Patterns for 3 of them have already been given in Illustrations 19, 20, 23 (and 24). Illustration 25 gives patterns for the remaining 5. It will be seen that among them are the regular octahedron (Illustration 25(a)) and regular icosahedron (Illustration 25(e)).

We can state part of our results as:

13   Theorem

There are exactly eight convex polyhedra all of whose faces are equilateral triangles.

The student may be interested to note that the relatively simple Theorem 13 was not discovered until 1947 and only then by two very prominent mathematicians, namely, B. L. van der Waerden and H. Freudenthal. ("Over een bewering van Euclides," *Simon Stevin*, **25** (1946–1947), pp. 115–121. See also [2], Chapter 6.)

Illustration 23   (See Illustration 24 for full-size pattern.)

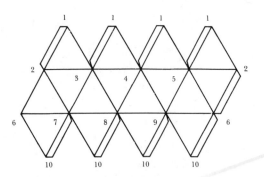 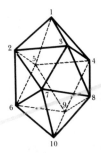

Section 3   Deltahedra

Illustration 24

An exact copy of this pattern can be cut out and glued along the flaps to make the deltahedron shown in Illustration 23.

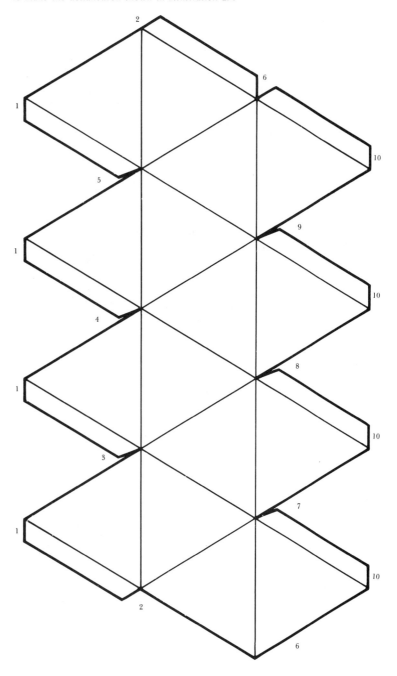

Illustration 25   *These are five of the only eight convex deltahedra. (The others are shown in Illustrations 19, 20, 23.)*

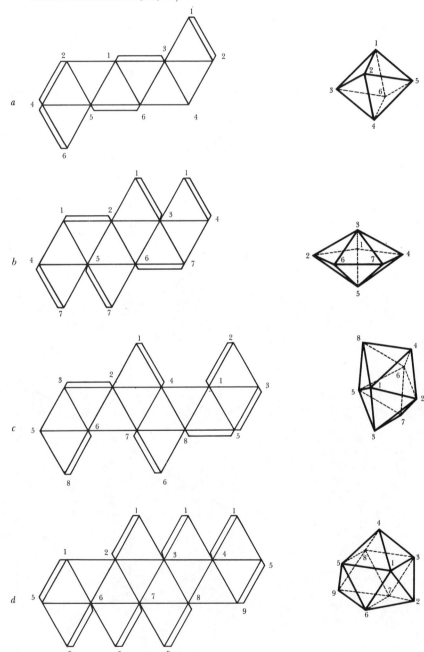

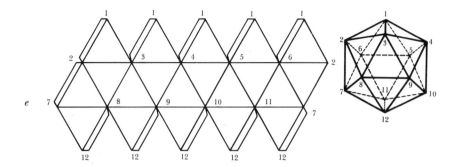

14 EXERCISES

1. Construct the eight convex deltahedra. (Patterns are given in Illustrations 19, 20, 23–25.)

2. If the restriction of convexity is removed, how many deltahedra are there?

3. (a) Enumerate all the convex polyhedra whose faces are congruent squares. [Hint: Follow the argument of Equations 8 to 12, making the appropriate changes along the way so that Equation 12 becomes simply $v_3 = 8$.]

   (b) Enumerate all the convex polyhedra whose faces are congruent regular pentagons.

4. Show that a polyhedron can be constructed having faces which are all triangles (not necessarily equilateral) and having $v_3 = v_4 = v_5 = 2$. (This shows that numerical arguments alone would not be enough to rule out all the potential deltahedra which our geometric arguments ruled out.)

5. (a) Show that if all the faces of a convex polyhedron are quadrangles it must have at least 6 faces. [Hint: Use $3v \leq 2e = 4f$ to eliminate e and v from Euler's formula.]

   (b) Show that if all the faces of a convex polyhedron are pentagons it must have at least 12 faces.

6. Show that no convex polyhedron can have exactly 7 faces, all of them quadrangles. [Hint: Determine the numbers v, e, f for such a polyhedron. Show that $v = v_3 + v_4$, since if, say, $v_5 \neq 0$ there would have to be at least 11 vertices altogether. Determine v_3 and v_4. Then try to construct the polyhedron, beginning at a vertex where 4 faces meet.]

7. Prove that every convex polyhedron has either a triangular face or at least eight vertices where only three edges meet.

[Hint: Put equations $3v_3 + 4v_4 + 5v_5 + \cdots = 2e$ and $v_3 + v_4 + v_5 + \cdots = v$ and the corresponding equations for faces into Euler's formula $4v - 2e - 2e + 4f = 8$ and solve for $v_3 + f_3$.]

*8 For which of the 11 cases in Table 3 ruled out by geometric arguments is it nevertheless possible to construct a polyhedron with (not necessarily equilateral) triangular faces satisfying the given numerical conditions? [Hint: This is really 11 separate problems, several of which require considerable patience. Of the 11, only 3 can be constructed. To show that the other eight cannot be constructed, you may assume the following facts: (a) The network consisting of five points, each joined by an edge to the other four, cannot be drawn in the plane without extra crossings. (b) The network consisting of two sets of three vertices, each vertex joined by an edge to the three vertices in the other set, cannot be drawn in the plane without extra crossings. Hence, of course, any network in which one of these two special networks appears *cannot* be the network of vertices and edges of a convex polyhedron. (Facts (a) and (b) are proved later, as Problems 4(c) and 6 of Exercises 24.) On the other hand, some cases, such as $v_3 = 3$, $v_4 = 1$, $v_5 = 1$, can be disposed of much more simply. For, if you try to draw this network, beginning with the vertex at which five edges meet, you will need five more vertices at the ends of these five edges. But there are only four more available!]

We could generalize still further and put no explicit restriction at all on the types of faces and vertices in our polyhedron. This might lead to the following sort of question, raised in the *Scientific American* of July, 1961, p. 161. How many polyhedra are there with exactly 4 faces, exactly 5 faces, exactly 6 faces, etc.? The student is invited to answer these questions himself, using the techniques of the preceding sections, as outlined in Exercises 15 and 16. We use the notation $f_3, f_4, f_5, \ldots$ to denote the numbers of faces having 3, 4, 5, ... edges, respectively. For the purpose of these Exercises we say that two polyhedra are of the same *kind* if they have the same number of triangles, quadrangles, etc., as faces and if, furthermore, these polygons are arranged in the same way around corresponding vertices of the two polyhedra. For example, an ordinary solid rectangular box is of the same kind as a cube, and the solid obtained by cutting off the corners of this box one-third of the way along the edges (thus replacing the corners by triangular faces) is of the same kind as the polyhedron shown in Illustration 4(c).

15 EXERCISE

A *pentahedron* is defined to be a convex polyhedron with exactly 5 faces, not necessarily regular. Prove there are exactly two kinds of pentahedra.

OUTLINE OF SOLUTION: Show that for a pentahedron $f_3 + f_4 = 5$ and $3f_3 + 4f_4 = 2e$. Hence, tabulate the possible values of f_3 and f_4, listing beside each pair of values the corresponding value of e. By an argument similar to that for the deltahedra, obtain the relation $v_3 + 2v_4 = 6$. Hence, tabulate the possible pairs v_3, v_4, listing beside each pair the corresponding value of e. By considering the columns of e's show that there are only two possible combinations of pairs v_3, v_4 with pairs f_3, f_4, and make drawings or models of the two corresponding pentahedra. Note that in fact there is a model of each of these having only regular polygons for faces. (See Illustration 26.)

Illustration 26

These are the only two kinds of pentahedra.

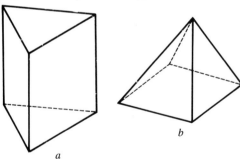

a   b

*16 EXERCISE

Prove that there are exactly 7 kinds of convex hexahedra. (obvious definition).

OUTLINE OF SOLUTION:

(a) Show that $f_6 = f_7 = \cdots = 0$ for any hexahedron (also $v_6 = v_7 = \cdots = 0$).

(b) By counting the edges, using the numbers at each of the various types of faces, derive an expression for $2e$ in terms of f_3, f_4, f_5. Similarly, derive an expression for $2e$ in terms of v_3, v_4, v_5.

(c) Show that a hexahedron has exactly 4 more edges than vertices.

(d) From (b) and the formula $f_3 + f_4 + f_5 = 6$ construct a table of all 16 "possible" values of the f's ($i = 3, 4, 5$), listing beside each set the corresponding value of e. (Note that e must be a whole number.)

(e) Show that $v_3 + 2v_4 + 3v_5 = 8$. Use this result to construct a table of all 10 "possible" values of the v's ($i = 3, 4, 5$), listing beside each set the corresponding value of e. (Note that e must be a whole number.)

(f) By considering the e's in (d) and (e), reduce the number of possibilities in (d) and (e) to 10 and 6, respectively.

(g) By considering which hexahedra are possible if $v_5 \neq 0$ eliminate one more of the possibilities in (e). Using all the possible arguments you can think of (such as drawing networks) eliminate the cases:

f_3	f_4	f_5
1	4	1
4	0	2
3	0	3

(h) Among the 5 remaining possibilities for the v_i's certain ones have the *same number of edges* as certain ones of the 7 remaining possibilities for the f_i's. These combine to give 9 possibilities altogether. Eliminate two of them by geometric arguments showing that if $v_5 = 1$ then $f_5 = 1$, and that if $f_5 = 1$ and $f_3 = 5$ then $v_5 = 1$. Show that each of the 7 remaining possibilities does yield a convex hexahedron by actually constructing it (by a drawing or by making a model).

According to calculations made by O. Hermes about 1900, there are exactly 34 kinds of septahedra and 257 kinds of octahedra. For obvious reasons we omit the proofs of these facts!

Section 4 Polyhedra without Diagonals

We have already noticed that Euler's formula holds for certain polyhedra which are not convex. In fact, it holds for all polyhedra which are topologically equivalent to a sphere. Roughly speaking, this means that if a balloon were made in the shape of such a polyhedron, and then blown up, it would eventually become spherical. We say that a polyhedron has *one tunnel in it* (or is a *sphere with one handle*) if it would look like an automobile inner tube when blown up in the same way. An example of such a polyhedron is a cube in which a square tunnel has been drilled all the way through from one face to the directly opposite face. (See Illustration 27(a).) Note that this "cube with a tunnel in it" could be obtained by pasting two of the L-shaped figures of Illustration 28 together along faces a and b.

It must be noticed that by polyhedron we mean a surface whose faces are ordinary polygons. If we permit faces to be ring-shaped regions, such as the front face in Illustration 27(b), then the modified Euler Formula 18 will no longer hold.

Illustration 27

A "cube with a tunnel in it."

$$v - e + f = 0$$

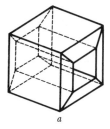

a

This is not called a polyhedron because one face is ring-shaped.

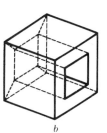

b

Illustration 28

Two of these fit together to make a "cube with a tunnel in it."

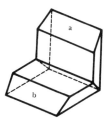

17 EXERCISES

1. Verify that $v - e + f = 0$ for the cube with a tunnel in it (Illustration 27(a)). (Verify that for the "nonpolyhedron" of Illustration 27(b), $v - e + f = 1$.)
2. Draw an "interior equator" and an "exterior equator" on one of Saturn's rings. Then draw two circles of longitude, each crossing both of the "equators." Calculate $v - e + f$ for the resulting network on this ring.
3. Verify that $v - e + f = -2$ for a cube with two tunnels side by side.

The results of the preceding exercises suggest that for a polyhedron with h tunnels the formula corresponding to Euler's formula is

18 $$v - e + f = 2 - 2h.$$

The number $k = 2 - 2h$ is called the Euler *characteristic* of a surface. The following argument suggests how Formula 18 could be proved: Consider a network on a given surface, with $v - e + f = k$. Take a "handle" in the form of a bent prism whose cross section is an n-gon. (See Illustration 29.) Attach it to the original surface at two regions which are n-gons. (In particular, we could first introduce diagonals in the regions to reduce them all to triangles, as in the proof of Euler's theorem. Then we could take n to be 3.) For the new map, on the new surface, v is unchanged. But e has increased by n, and f has increased by $n - 2$. (Since, although n new faces were added, 2 of the original ones were covered up.) So for this new network we have $v - (e + n) + (f + (n - 2)) = v - e + f - 2 = k - 2$. That is, adding one handle decreases the characteristic by 2. Hence, adding h handles decreases the characteristic by $2h$, so that $v - e + f = 2 - 2h$ for a surface obtained by adding h handles to a sphere. (Of course, this "proof" just shows that for the particular network the number $v - e + f$ has decreased by 2. For an outline of a simple proof that a given surface really does have a *characteristic* k, that is, that $v - e + f$ has the same value for all suitably chosen networks on the surface, see [1, pp. 232–233].)

The branch of mathematics known as *topology* originated from the study of surfaces of the type we have just discussed. In particular, the topologist considers two surfaces to be equivalent if, when they are imagined to be made of completely elastic material, one can be deformed to coincide with

Section 4 Polyhedra without Diagonals

Illustration 29   *A "sphere with a handle," for which* $v - e + f = 0$.

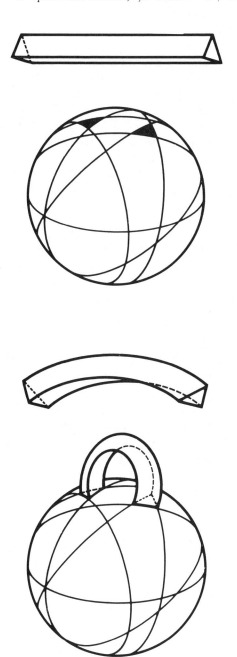

the other without puncturing. This is the origin of the popular definition of a topologist as "a man who doesn't know the difference between a coffee cup and a doughnut," since either of these can be deformed into the other. The Euler characteristic, $v - e + f$, of a surface is a *topological property* because it does not change under such deformations.

The converse is not necessarily true. That is, there are surfaces with the same characteristic which cannot be deformed into each other. For example, if a bicycle inner tube is cut open (the short way), tied into a knot, and then sealed back together, the resulting knotted inner tube has the same characteristic—namely, zero—as an ordinary inner tube. But it is impossible to deform one of them into the other without cutting one of them apart again. However, the surfaces in the following exercises are not so complicated. The pyramid with a tunnel through it can be deformed to a cube with a tunnel in it, and the cube with a split tunnel can be deformed to a cube with two tunnels side by side (or to a sphere with two handles).

19 EXERCISES

1 Compute $v - e + f$ for the pyramid with a tunnel through it in Illustration 30.

2 Compute $v - e + f$ for the cube with a split tunnel in Illustration 31. Hence show that $v - e + f$ has the same value as for a "sphere with two handles."

We propose to apply Equation 18 to the problem of finding polyhedra which "have no diagonals." This means that we look for a polyhedron (such as the tetrahedron, but not the octahedron or any other regular polyhedron), possibly with h tunnels in it, which has the property that *each pair of its vertices is joined by an edge*, that is, it has *no diagonals*. Since each of the v vertices is joined by an edge to each of the other $v - 1$ vertices, we have

20 $$v(v - 1) = 2e,$$

because each edge is counted *twice* by $v(v - 1)$. Furthermore, since each face must be a triangle, we will count the edges twice by counting the faces and multiplying them by 3; that is,

21 $$3f = 2e.$$

If we substitute for f, from Equation 21, in Equation 18 we get $v - \frac{1}{3}e = 2 - 2h$. Then substituting for e, from Equation 20, we get an equation which reduces to

22 $$12h = v^2 - 7v + 12 = (v - 3)(v - 4).$$

Section 4   Polyhedra without Diagonals   35

Illustration 30   *A pyramid with a tunnel through it.*

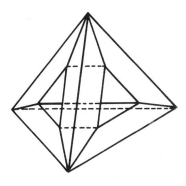

Illustration 31   *A cube with a split tunnel through it.*

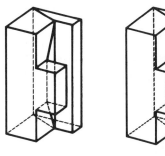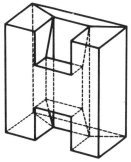

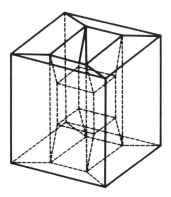

We can only use *integers* as solutions to Equation 22, since our polyhedron cannot very well have a fractional number of holes. So $h = (v - 3)(v - 4)/12$ must be an integer. The values of $v > 3$ which make h an integer ≤ 50 are given in Table 4.

The first row in this table of purely numerical results tells us that the only possible ordinary polyhedron ($h = 0$) without diagonals has 4 vertices, and we know that such a polyhedron (the tetrahedron) actually exists. What about the second entry? Is it really possible to construct a polyhedron with one tunnel through it, having 7 vertices, each joined to all the others? The Hungarian mathematician, A. Császár [8], showed in 1949 that it *is* possible, by the simple method of constructing one. We give the (x, y, z) coordinates of its 7 vertices in Table 5.

Table 4

v	h
4	0
7	1
12	6
15	11
16	13
19	20
24	35
27	46
28	50

Table 5

Vertex	x	y	z
1	3	3	0
2	3	−3	1
3	1	−2	3
4	−1	2	3
5	−3	3	1
6	−3	−3	0
7	0	0	15

The 14 faces are determined by the following sets of 3 vertices:

```
126   235   356   346   467   237   267
156   245   124   134   137   457   157
```

Illustration 32 is a drawing of this *Császár polyhedron*.

From Table 4, the next value of h is 6, so if there are any other polyhedra without diagonals, they must have at least six holes. It is an unsolved problem whether such a polyhedron (or any other polyhedron without diagonals, except the two already discussed) can actually be constructed. After the first five entries in Table 4, there are more tunnels than vertices. It seems very likely that such a polyhedron *cannot* be constructed.

23 EXERCISES

1 What are the next two entries in Table 4?

Section 4 Polyhedra without Diagonals

2 Construct the Császár polyhedron, using the patterns for the faces given in Illustrations 33(a) and (b). The following step-by-step directions may be useful. (It is essential to use some quick-drying glue, such as model airplane cement.)

(a) Cut out two copies each of the figures shown in Illustrations 33(a) and (b) from light cardboard about the weight of a manila folder. Fold *back* on all solid lines and forward on all dotted lines. Label one copy as in Illustration 33 and the other with A', B', etc. Cut off the flaps on A', C' and the dotted flap on D'.
(b) Turn all pieces over.
(c) Glue A to A' by the flap (1) on A.
(d) Glue the flap (2) on D' to the base of F.
(e) Glue the long flap (3) of E to G'. (You are now at Illustration 33(c).)
(f) Pull triangles C and C' together and glue by the flap on C as in Illustration 33(d). (This is not impossible, just tricky!)
(g) Fold triangles D and D' over C and C' and glue by the flap on D as in Illustration 33(e).
(h) Glue the flap of E to C, and the flap of G' to A'.
(i) Glue the flap of D to F', and the flap of E' to C'.
(j) Finally, glue the flap of G to A, and the flap of E' to G.

Illustration 32

Császár's polyhedron with a tunnel, in which every pair of vertices is joined by an edge.

Illustration 33

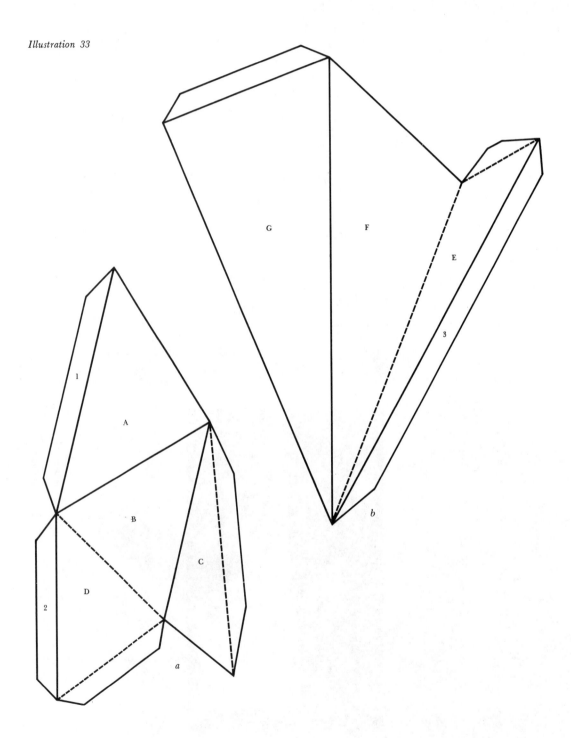

Section 4 Polyhedra without Diagonals

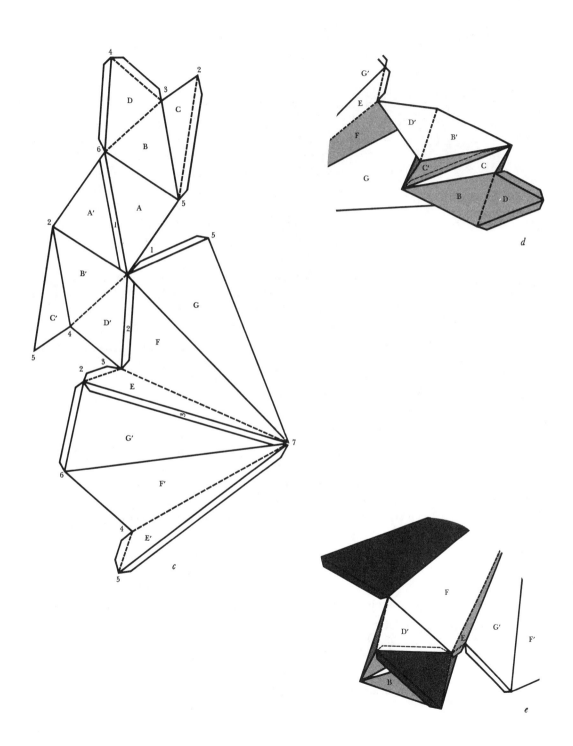

Section 5   The n-Dimensional Cube and the Tower of Hanoi

5A   The n-Cube

The discussion of the polyhedron with a tunnel in it showed how Euler's theorem could be extended to include such tunneled polyhedra. Euler's theorem can be extended in another way, namely, to apply to polyhedra in other dimensions. For example, in the plane (that is, in two-dimensional space) a convex "polyhedron" is just an ordinary polygon. For every polygon, certainly $v - e = 0$. Comparing this with the formula $v - e + f = 2$ for the polyhedra in three dimensions suggests that some such formula as $v - e + f - s = \text{constant}$ might hold in four dimensions, where s is the number of three-dimensional faces, or "solids," in the figure. This is in fact true, and the reader who has drawn pictures of, or built a model of, a four-dimensional cube—sometimes called a *tesseract*—can verify that $v = 16, e = 32, f = 24, s = 8$, so that $v - e + f - s = 0$. This suggests that the corresponding formula in five dimensions is $v - e + f - s + t = 2$ (where t is the number of "four-dimensional faces"). This is correct. The reader can now guess for himself the appropriate formula in n dimensions, namely, $s_0 - s_1 + s_2 - s_3 + \cdots + (-1)^{n-1} s_{n-1} = 1 + (-1)^{n-1}$, where s_i denotes the number of i-dimensional faces, $i = 0, 1, \ldots, n - 1$. A proof of this is given in [5, Chapter 9].

Are there regular figures in higher dimensions? Since there are infinitely many regular polygons in two dimensions, and only five regular polyhedra in three dimensions it might be guessed that there are very few regular figures in higher dimensions. In fact, there are exactly six in four dimensions, but in five or higher dimensions there are always exactly three. These three in each dimension are analogues of the regular tetrahedron, cube and regular octahedron. (See Illustration 34.) We shall look more closely at one of these, the n-dimensional cube. However, we shall simplify the problem by examining only its *skeleton*, that is the network of its vertices and edges.

To help visualize the skeleton of an n-dimensional cube it is convenient to begin at the very lowest dimension, $n = 0$. By a 0-dimensional cube is meant a *single point*. By a one-dimensional cube is meant *two points joined by an edge*. The skeleton of a *two-dimensional*

Illustration 34   Only three regular polyhedra occur in space of more than four dimensions. This shows what they look like in four dimensions.

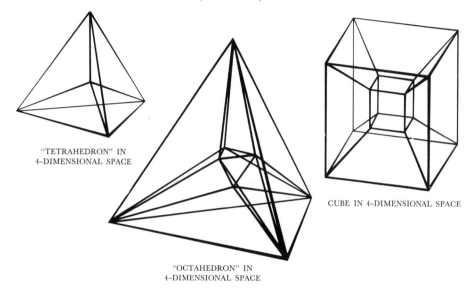

"TETRAHEDRON" IN 4-DIMENSIONAL SPACE

"OCTAHEDRON" IN 4-DIMENSIONAL SPACE

CUBE IN 4-DIMENSIONAL SPACE

cube (ordinarily called a *square!*) is obtained by taking two copies of a one-dimensional cube and joining corresponding vertices by edges. The skeleton of an ordinary *three-dimensional cube* is obtained by taking two copies of the skeleton of the two-dimensional cube and joining the corresponding vertices by edges. In general, the skeleton of a cube of any dimension is obtained by making two parallel copies of the one of the next lower dimension and joining corresponding vertices in the two copies. (See Illustration 35.)

In particular, from this point of view, it is no harder to draw a picture of a four-dimensional cube than of a three-dimensional cube, providing we are satisfied to draw only the vertices and edges. And a five-dimensional cube is still no harder, though the picture will be rather messy because so many edges must be drawn.

Of course, strictly speaking, a "cube" should have all its edges meeting at right angles. Our drawings will not be able to show this, just as the usual drawings of an ordinary cube do not show all angles as right angles. Since this right-angle property is of no importance for our purposes, we are willing to sacrifice it in order to get *some* picture to focus our attention on the properties which do interest us.

Illustration 35

0-DIMENSIONAL CUBE

1-CUBE

2-CUBE

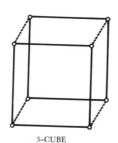

3-CUBE

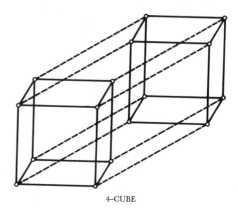

4-CUBE

24 EXERCISES

1. How many edges are there in a five-dimensional cube?
2. How many vertices are there in an n-dimensional cube?
3. How many edges are there in an n-dimensional cube?
4. The network consisting of $n + 1$ vertices and all edges joining them in pairs is the skeleton of an *n-simplex* (or *n-dimensional tetrahedron*).
   (a) How many edges does the five-simplex have?
   (b) How many edges does the n-simplex have?
   (c) Prove numerically that the skeleton of the four-simplex cannot be drawn on a sphere without extra crossings. [Hint: Recall that $3f \leq 2e$ for any network on a sphere. Show that this contradicts the value of f obtained from Euler's formula.]
5. By an argument like that in Problem 4(c), but using $h = 1$ in Formula 18, prove that the skeleton of the seven-simplex cannot be drawn on a "sphere with one handle" without extra crossings. (But note that the skeleton of the six-simplex is exactly the same as the skeleton of Császár's polyhedron. Hence, it *can* be drawn on a sphere with one handle.)
6. Prove, numerically, that a network consisting of two sets of 3 vertices, each joined by an edge to the 3 vertices in the other set, cannot be drawn on a sphere without extra crossings. [Hint: The network cannot have any triangles. (Why?) Hence, $4f \leq 2e$. Show that this contradicts the value of f obtained from Euler's formula.]

Section 5   The n-Dimensional Cube

An interesting game was devised about 1850 by the mathematician William Rowan Hamilton. He asked that the vertices and edges of a regular dodecahedron be thought of as the towns and roads of a small planet. The problem is then whether or not it is possible to begin at one of the towns and travel (along roads) to every other town, visiting each town exactly once, and ending up at the original starting point. It is not hard to find such a *Hamilton circuit* (see Problem 3 of Exercises 7, Section 2) on the dodecahedron, or on any of the five regular polyhedra, but, as usually happens in mathematics, the problem was immediately generalized. The general problem reads as follows: Given an arbitrary network (not necessarily on a sphere or in the plane), is there a systematic way to determine whether or not there is a Hamilton circuit through the network? Surprisingly enough, no satisfactory general answer to this question is known.

Illustration 36   *The seven bridges of Königsberg, as originally drawn by Euler in the Proceedings of the St. Petersburg Academy of Sciences for 1736.*

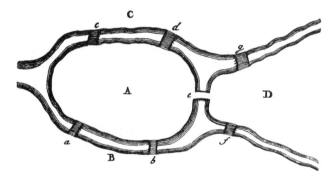

The reader should be careful not to confuse this problem with a similar problem invented by Euler concerning the bridges of Königsberg. Euler asked whether it was possible to begin a walk at some point in Königsberg, pass over each of the seven bridges exactly once, and end up at the starting point. Illustration 36 shows Euler's original map. The bridges are represented schematically by the network of Illustration 37. More generally, Euler's question is the following: Given an arbitrary network, is there a systematic way to determine whether or not there is a path through the network, beginning at some vertex and ending at that same vertex, which traverses each edge exactly once? Such a closed path through every edge of a network is called an *Euler circuit*.

Theorem. There is an Euler circuit in a network if and only if the network is connected and every vertex has an even number of edges coming from it.

Illustration 37   *A schematic drawing of the seven bridges of Königsberg.*

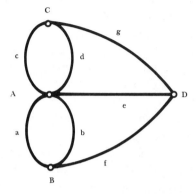

Illustration 38   *The theorem about Euler circuits tells which of these can be drawn without lifting pencil from paper and without retracing.*

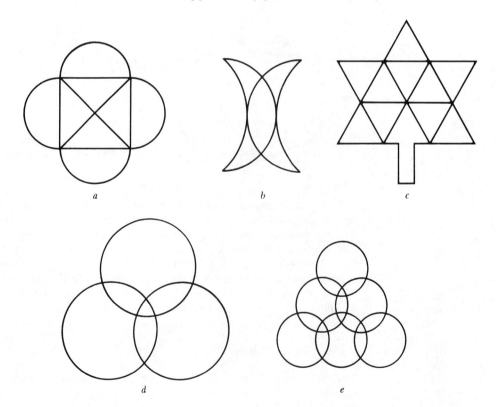

Section 5   The n-Dimensional Cube   45

25   EXERCISES

1. Which of the networks in Illustration 38 can be drawn without retracing any line or lifting the pencil from the paper? What if it is also required to begin at a vertex and end at that same vertex?

2. Is there a way of traversing the 15 bridges of the imaginary map (due to Euler) of Illustration 39 in a single trip without passing over any bridge twice? What if it is not required to return to the starting point?

*3. Prove the theorem preceding these exercises. [Hint: When you arrive at a vertex, you must also be able to leave it.] (Also see [1, pp. 242–247].)

Illustration 39

Euler also drew this imaginary map in his 1736 article.

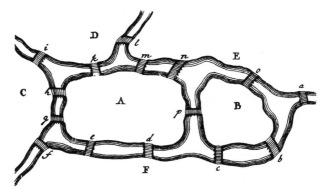

Although we cannot answer the question about the existence of a Hamilton circuit in general, we can answer it in a special case, namely, for the network of vertices and edges of an n-dimensional cube (abbreviated n-cube from now on). Certainly there is such a circuit for a square (2-cube). Using this fact we can find a Hamilton circuit on the 3-cube as follows:

(1) Start at any vertex, say 0, of one of the squares which were copied to make the cube and traverse all but the last edge of a Hamilton circuit on it. (See Illustration 40(a), edges a, b, c.)

(2) Go along an edge, say d, to the other square and traverse all but the last edge of the corresponding Hamilton circuit on it. See Illustration 40(b), edges c', b', a'.

(3) Complete the Hamilton circuit by returning to 0 along the edge e.

Illustration 40   *A Hamilton circuit on the edges of the cube.*

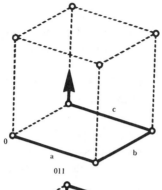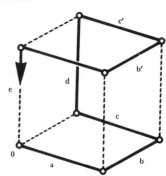

Illustration 41

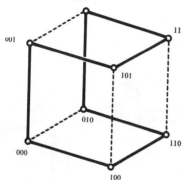

Illustration 42   *A Hamilton circuit on the edges of the 4-cube.*

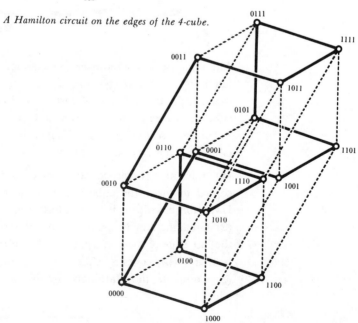

It should be evident that this same process can now be used to find a Hamilton circuit on the 4-cube, by first following all but the last step of the circuit just found on the 3-cube, then going to the other copy of the 3-cube and retracing the circuit in reverse, and finally returning to the starting point. From this result, a Hamilton circuit can be found on the 5-cube, and so on. We summarize this as:

26 Theorem

There is a Hamilton circuit on the skeleton of the n-cube, if $n \geq 2$.

If the vertices of the 3-cube are given ordinary rectangular coordinates, a sequence of vertices defining a Hamilton circuit is given in Table 6, and the corresponding circuit is shown in Illustration 41.

Note that the last column changes from 0 to 1 when we go from one square to the other and that the first two columns then repeat, in reverse order, the entries in the first two columns for the first square. For the 4-cube, we would write down the above eight coordinates, each with a 0 attached on the end and then write them down again, in reverse order, with a 1 attached on the end, as in Table 7.

Table 6. A Hamilton circuit for the 3-cube

0	0	0
1	0	0
1	1	0
0	1	0
0	1	1
1	1	1
1	0	1
0	0	1

Table 7. A Hamilton circuit for the 4-cube

0	0	0	0
1	0	0	0
1	1	0	0
0	1	0	0
0	1	1	0
1	1	1	0
1	0	1	0
0	0	1	0
0	0	1	1
1	0	1	1
1	1	1	1
0	1	1	1
0	1	0	1
1	1	0	1
1	0	0	1
0	0	0	1

27 EXERCISES

1. Write down a sequence of vertices defining a Hamilton circuit on the 5-cube, following the pattern of Tables 6 and 7.

2. (a) Is there a Hamilton circuit on the network whose vertices and edges are those of a 7 × 7 chessboard (that is, 49 squares in all, instead of the usual 64)?

   (b) What about an ordinary chessboard? Prove your answer. [Hint: Show that an even number of horizontal edges must be included in such a circuit, and an even number of vertical edges. Compare this with the known total number of edges in the circuit.]

Is there a Hamilton circuit on this 7 x 7 chessboard?

5B The Tower of Hanoi

The Tower of Hanoi puzzle consists of a number, say n, of disks of decreasing size placed on one of three pegs. The problem is to transfer all the disks to one of the other pegs by moving them one at a time from one peg to another, never placing a larger one on top of a smaller one. (See Illustration 43.)

This puzzle was brought out in 1883 by E. Lucas, a professor at the Lycée Saint-Louis in France. At about the same time De Parville published the following description of the Tower, attributed to "the mandarin Claus of the College of Li-Sou-Stian" (anagrams for Lucas, and Saint-Louis!), quoted from [1, p. 308].

Section 5 The Tower of Hanoi

In the great temple at Benares, beneath the dome which marks the centre of the world, rests a brass plate in which are fixed three diamond needles, each a cubit high and as thick as the body of a bee. On one of these needles, at the creation, God placed sixty-four discs of pure gold, the largest disc resting on the brass plate, and the others getting smaller and smaller up to the top one. This is the Tower of Bramah. Day and night unceasingly the priests transfer the discs from one diamond needle to another according to the fixed and immutable laws of Bramah, which require that the priest on duty must not move more than one disc at a time and that he must place this disc on a needle so that there is no smaller disc below it. When the sixty-four discs shall have been thus transferred from the needle on which at creation God placed them to one of the other needles, tower, temple, and Brahmins alike will crumble into dust, and with a thunderclap the world will vanish.

Now, it is not hard to show that no fewer than $2^n - 1$ moves are needed to transfer n disks from one diamond needle to another. Hence, $2^{64} - 1$ transfers are needed before the world ends. Now, $2^{64} - 1 = 18,446,744,073,709,551,615$, and Kasner and Newman have taken the trouble to calculate that if the Brahmins made one transfer per second, 24 hours a day for 365 days a year it would take them 58,454,204,609 centuries plus slightly more than six years to finish the job. Hence, it would appear that we are in more danger from other possible catastrophes in Hanoi than from the danger that the priests will complete their transfer!

Illustration 43

From "The Scientific American Book of Mathematical Puzzles and Diversions," by M. Gardner. Copyright © May 1957 by Scientific American, Inc. All rights reserved.

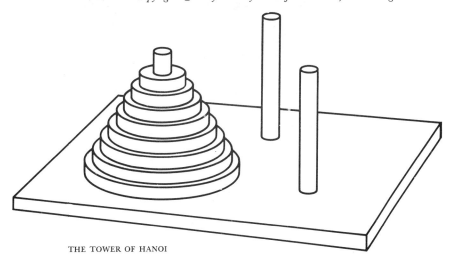

THE TOWER OF HANOI

28   EXERCISE

Prove that n disks can be transferred in $2^n - 1$ moves. [Hint: Note that a solution of the puzzle for n disks can be accomplished by (a) solving for $n - 1$ disks, then (b) moving the nth disk, and finally (c) re-solving for $n - 1$ disks.]

(A formal proof would use *mathematical induction*, which is discussed in detail in Chapter 2, and later in the present chapter, for those readers who are not familiar with that method of proof.)

Let us look at a simple tower, with only 3 disks. (The reader should try this with a dime, a nickel, and a quarter.) The sequence of moves is shown in Illustration 44.

At the left we have assigned "coordinates" to each position. The left-hand coordinate represents the smallest disk and is 0 or 1 according to whether this disk has moved an even or an odd number of times. Similarly, the second and third coordinates (y and z) are 0 or 1 according to whether the second and third disks (from the top) have moved an even or odd number of times. *This sequence is exactly the sequence of vertices of a Hamilton circuit for the 3-cube*. Moreover, this relation continues to hold if we take 4 disks, for in order to solve the puzzle for 4 disks we must first remove the top 3 disks, that is, we solve the puzzle for 3 disks. This corresponds to our list of coordinates but with a 0 inserted at the end since the bottom disk has not moved yet. Then we move the bottom disk, that is, put a 1 in the last coordinate position, and re-solve the puzzle for 3 disks, which simply means rewriting the 8 coordinates already appearing, but in reverse order, this time with a 1 in the fourth place since the bottom disk never moves again.

This discussion clearly applies for any n disks, once we have written the solution for $n - 1$ disks. Comparison with the analogous discussion for a Hamilton circuit of the n-cube shows that we have proved the following:

29   Theorem

A simplest solution of the Tower of Hanoi with n disks describes a Hamilton circuit on the n-cube.

The reader will note, however, that the converse is not true. It is easy to find a Hamilton circuit on the n-cube which does *not* represent

Section 5   The Tower of Hanoi   51

Illustration 44   *The solution, in 7 moves, of a Tower of Hanoi puzzle having only three disks.*

x   TOP DISK
y   SECOND DISK
z   THIRD DISK

x y z
0 0 0
1 0 0
1 1 0
0 1 0
0 1 1
1 1 1
1 0 1
0 0 1

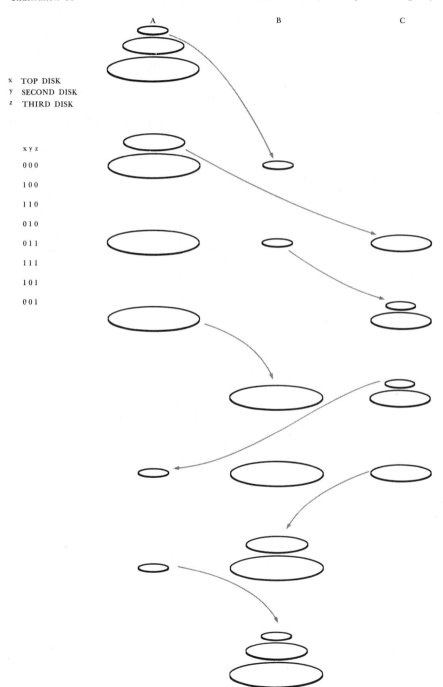

a solution to the Tower of Hanoi. Nor is it always possible just to change the coordinate system so as to turn a non-Hanoi Hamilton circuit into one which *is* a solution to the Tower problem. In fact, if $n > 3$ there are at least two essentially different Hamilton circuits on the n-cube, that is, two circuits which cannot be made the same just by changing the coordinate system. It appears to be an unsolved problem, of considerable practical interest to the telephone industry, just exactly how many essentially different Hamilton circuits there are on an n-cube.

30   EXERCISES

1 Find a Hamilton circuit on the 3-cube which does not represent a solution to the Tower of Hanoi.
2 Find two essentially different Hamilton circuits on the 4-cube. [Hint: If the edges of a Hamilton circuit are removed from a 4-cube, the remaining edges will form either (a) another Hamilton circuit, or (b) a number of smaller circuits. The cases (a) and (b) clearly correspond to essentially different Hamilton circuits.]

5c   Chinese Rings

It should be mentioned in passing that the old Chinese rings puzzle, described, for example, in [1, p. 305], is essentially the same as the Tower of Hanoi. That is, the solution to the Chinese rings—suitably interpreted—gives the same Hamilton circuit on the n-cube as the solution to the Tower of Hanoi.

5d   An Application

The Hamilton circuits on the n-cube have the following interesting application to a very specific engineering problem. We imagine that a radar sweeps around a complete circle, divided into, say, eight sectors, as in Illustration 45. For example, the radar can be thought of as being located on a hill above the mouth of a harbor and the radar is on the lookout for attack from the sea. Let us take sector 7 to coincide with the harbor mouth. Then when a foreign object is sighted in sector 7 the radar triggers an alarm.

Section 5   An Application

Illustration 45

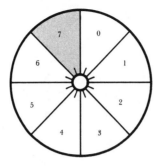

Illustration 46

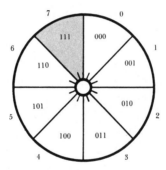

Illustration 47

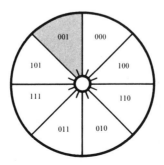

However, in this age of computers which can only comprehend 0's and 1's, the sectors are not labeled 0, 1, 2, . . . , 7 but with the corresponding binary digits

| 0, | 1, | 2, | 3, | 4, | 5, | 6, | 7 |
| 000, | 001, | 010, | 011, | 100, | 101, | 110, | 111 |

represented by 3 switches and the various ways in which they can be off ($= 0$) or on ($= 1$). Thus, for example, if the three switches are on, off, on, this represents sector 101, that is, sector 5. In Illustration 46 the sectors are named with the proper binary digits.

As the radar sweeps around, the various sectors are represented by the various positions of the switches. However, the switches may not be perfectly synchronized, and as the scanner moves from sector 3 to sector 4, for example, the 011 may become 111 en route to 100. That is, the first switch changes to "on" just before the other two change to "off." If, at the same time, the scanner is passing by a house on a hill or any other "foreign object" it will send off the alarm it reserves for sector 7, since 111 represents 7.

The problem is: How to avoid this? One answer would be to arrange the symbols so that each two adjacent sectors differ by only a single throw of one switch. Thus, 011 and 100 must never correspond to adjacent sectors, whereas 011 and 010 could be adjacent. Can this be accomplished? The answer is *yes*, for this is exactly the property possessed by a Hamilton circuit on the n-cube, namely, the coordinates of adjacent vertices differ by a single entry. Referring to the Hamilton circuit for the 3-cube in Table 6 or Illustration 41, we can label the sectors as in Illustration 47. Clearly, for more sectors we just go to a higher-dimensional cube.

What if the number of sectors is not a power of 2 (but still even)? This makes no difference, for we can always take a shortcut and omit any even number of vertices out of the exact middle of a Hamilton circuit. For example, if there are only six sectors to be labeled, we eliminate the two middle entries 010 and 011 and get a suitable sequence.

31 EXERCISES

1. Write down a sequence of 16 binary digits suitable for use with 16 sectors.
2. Write down a sequence of 12 binary digits suitable for use with 12 sectors, and a sequence of 10 suitable for use with 10 sectors.

Section 6 The Four-Color Problem and the Five-Color Theorem

One of the most famous unsolved problems of mathematics is: Can every map (such as a map of the countries of Europe) be colored with four colors? This question was first asked in 1852 by Francis Guthrie, a graduate student at University College, London. He passed the question on to his brother who asked his teacher, the mathematician Augustus de Morgan. Eventually, the problem was publicized by the most famous English mathematician of his time, Arthur Cayley, and in 1879 a "proof" was published by A. B. Kempe, showing that the answer was *yes*. However, in 1890 it was pointed out by P. J. Heawood [14] that this "proof" was *false!* To this day the problem remains unsolved. In fact, Heawood himself was still publishing on the subject in 1950 at the age of 88.† Although almost all efforts have been directed toward proving that the answer is *yes*, it has recently been suggested by some prominent mathematicians that there may in fact be a map, having a large number of regions, which cannot be colored with four colors. If this is so, it might partially explain the long years of failure to solve the problem! (The most thorough examination of the problem is contained in the recent book by Ore [16]. The best reference for the related coloring problems discussed in Section 7 is [18].)

Following his education at Oxford, Percy J. Heawood (1861–1955) was appointed Lecturer in Mathematics at Durham in 1887. He retired there in 1939, at the age of 78, after a distinguished career. (Retirement rules were different then!) His chief mathematical work was the four-color problem, and his seven papers on this subject span a period of 60 years. Although Heawood's early insights and contributions have never been surpassed, the problem remains unsolved to this day, in spite of heavy attack in recent years. (The photograph was taken in 1910 and is courtesy of T. J. Willmore, University of Durham.)

† However, Heawood actually presented this paper to the London Mathematical Society in 1947, when he was a mere 85.

Some preliminary agreements about what is meant by a "map" are necessary before we proceed. In particular we must agree that no country can be composed of disjoint pieces. Thus, we have to consider Alaska and Hawaii as countries separate from the rest of the U.S.A. This agreement also implies, of course, that no color can be reserved for any special use the way blue is reserved for water on geographical maps. An ocean or lake is to be thought of as just another country.

It is fairly clear that if all the disjoint regions of a country (like the U.S.A. or Pakistan) must be colored the same color then more than four colors will be required. Already by 1890 Heawood had shown that if every country has exactly two components then no more than 12 colors are needed to color a map of the world. We give a proof of this fact under the heading "The Scramble for Africa" in Section 7. Conversely, Heawood invented the map in Illustration 64, having 12 such countries, each touching all the others, which requires 12 colors. So this apparently much more complicated problem (where each country has exactly two components instead of exactly one) is completely solved, while the original four-color question is still unanswered.

Illustration 48   *This map cannot be colored with fewer than four colors.*

32 EXERCISES

1. Color the (slightly simplified) map of Europe in Illustration 48 with four colors, so that no two regions of the same color touch along a common boundary. Note that you can also color the ocean, but some inland country will have to be colored the same color as the ocean.
2. Prove that the map in Illustration 48 cannot be colored with *fewer* than four colors, even if it is not required to color the ocean.
3. Prove that if France annexes Luxembourg and Russia annexes Czechoslovakia, the new map of Europe, excluding the ocean, can be colored with three colors.
4. Construct a map with some countries having two disjoint regions which requires more than four colors. Is it possible to construct such a map having only one such country?

We should also agree that the map is *connected*, that is, that we can get from any point on any boundary to any point on any other boundary by walking along boundaries. In geographical maps this condition is not always satisfied. For example, from a point on the boundary of Spain one cannot get to points on the boundary of Florida by traveling along shorelines. (This was especially relevant in 1492!) This is, of course, no essential restriction, for if we find a way of coloring all connected maps (for example, those of Europe or North America) then we can color all maps. Hence, we shall only consider maps of a single island (such as North and South America together).

Furthermore, we do not have to consider maps having vertices where fewer than three regions meet, since no such vertex need appear on any actual geographical map. And finally, we note that on any geographical map, crossing a border brings you to a new country. Hence, we can assume that no edge separates a region from itself.

We can summarize these agreements in the following:

33 *Definition*

An *ordinary map on the sphere* is a connected network on the sphere which has the properties: (a) no vertex has fewer than three edges meeting at it, (b) no region meets itself at an edge, and (c) no edge fails to have a vertex in it.

Illustration 49

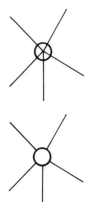

Condition (c) just means, for example, that the "map" consisting of the Northern and Southern Hemispheres, separated by the equator, is not ordinary. We emphasize that when an ordinary map is drawn in the plane, the outside (infinite) region (the "ocean") must also be thought of as a region to be colored.

We shall repeatedly use the fact already mentioned in Section 1 that Euler's formula, $v - e + f = 2$, holds for any ordinary map.

Proposition 34 will be of crucial importance in our proof that every ordinary map can be colored with at most five colors.

34 Proposition

Every ordinary map on the sphere has at least one region with no more than five edges (and hence no more than five different neighbors).

PROOF: As before, we let v, e, f be the number of vertices, edges, and faces (that is, regions) of the map. Let p be the average number of edges of such a region; that is,

Illustration 50

$$p = \frac{f_1 + 2f_2 + 3f_3 + 4f_4 + 5f_5 + \cdots}{f}$$

where f_1 denotes the number of regions with 1 edge, f_2 the number of regions with 2 edges, etc. Then since pf counts each edge twice, once for each of the 2 faces it lies on, we have $pf = 2e$. (Note that this follows from Definition 33(b).) Similarly, if q is the average number of edges at a vertex, $qv = 2e$. (Note that if $f_1 > 0$, so that some edge has the same vertex at each end of it, then that edge itself will be counted twice at its vertex.) Substituting these values for f and v into Euler's formula $v - e + f = 2$ yields (exactly as in our proof that there are at most five regular polyhedra)

$$\frac{2e}{q} - e + \frac{2e}{p} = 2 \quad \text{or} \quad \frac{1}{p} + \frac{1}{q} = \frac{1}{2} + \frac{1}{e} \quad \text{or} \quad \frac{1}{p} + \frac{1}{q} > \frac{1}{2}.$$

But since $q \geq 3$ (by Definition 33(a)) we have

$$\frac{1}{q} \leq \frac{1}{3}, \quad \text{so} \quad \frac{1}{p} + \frac{1}{3} > \frac{1}{2}, \quad \text{hence} \quad \frac{1}{p} > \frac{1}{6}$$

and, finally, $p < 6$.

This shows that the *average* number, p, of edges per region is less than 6. Hence, at least one region must have fewer than 6 edges, which is what we wanted to prove.

Section 6   The Five-Color Theorem

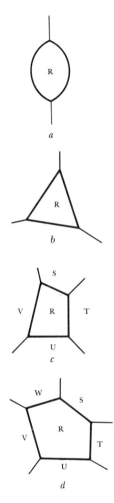

Illustration 51

We can simplify our subsequent proofs if we agree to consider only those ordinary maps for which condition 33(a) is strengthened to read: 33(a*) Each vertex has exactly three edges meeting at it. Such maps have been called *standard*. Proposition 35 shows that we do not lose anything by restricting our attention to standard maps.

35 Proposition

If every standard map on the sphere can be colored with k colors, then every ordinary map can be colored with k colors.

PROOF: Any ordinary map can be made standard by the simple device of drawing a small circle around each vertex where more than three edges meet and then erasing everything inside the circle, as in Illustration 49. We now color the resulting map with k colors, then shrink the small circles to points to get the original map. Regions which did not touch before this shrinking do not touch in the restored original map (except possibly at a single point, which doesn't matter). Hence the k coloring of the standard map is still valid for the restored ordinary map. This completes the proof.

We note that in a standard map no country has exactly one boundary, for if a country C had exactly one boundary, and hence exactly one vertex on that boundary, it would look like Illustration 50. But this is impossible, since country D is precluded by condition 33(b) from touching itself along an edge.

We now state and prove the "five-color theorem."

36 Theorem

Every standard map on the sphere can be colored with at most five colors.

PROOF: We shall use the method of induction. That is, we shall show that the problem of coloring any given standard map M (with five colors) reduces to coloring a certain standard map having fewer regions. Hence, by repeated reductions, the problem reduces to coloring a standard map having five or fewer regions. Since this can certainly be done (by giving each region a different color!), this will prove that M can be colored.

Suppose that the standard map M has exactly f regions. From Proposition 34 we know that M has some region R with exactly (a) two, (b) three, (c) four, or (d) five edges. (See Illustration 51.)

In case (a) or (b), we simply delete one of the edges of R to get a new standard map M^*, having $f - 1$ regions. Obviously, if we can color M^*

Illustration 52

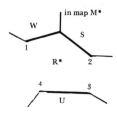

Illustration 53

then we can color M, just by using one of the colors which is not used in the regions neighboring R. (Of course, we must also delete the vertices at the ends of the deleted edge, so that M^* will still be regular.)

In case (c) we must be somewhat more careful. For conceivably S and U might be one and the same region, so that deleting the edge between S and R would form a region S' which meets itself along an edge, as in Illustration 52. That is, the reduced map is not standard. We can avoid this difficulty by noting that if S and U are the same region, then certainly V and T are *not* the same region, since they are separated from each other by the big region $S\ (=U)$.† Hence, we can delete, say, the edge between T and R to obtain a standard map M^* having $f-1$ faces. If M^* can be colored with five colors then so can M, just by using the fifth color (not used for S, T, U, or V) for R.

Case (d) is somewhat similar. Some pair of opposite regions, say T and V, are really different regions, and also do not touch each other along any edge. (If they did they would separate S and U from each other, so that we could choose S and U instead.) This time we delete *both* the edges separating T and V from R to form a new large region R^* in a standard map M^* having only $f-2$ regions, as in Illustration 53. (Again, the superfluous vertices 1, 2, 3, 4, must be deleted before M^* is actually standard.) If M^* can be colored with five colors then so can M. For whatever color is used for R^* in M^* can be used for *both* T and V in M. Thus, R is surrounded by at most four colors, and the fifth color can be used for R itself.

This completes the proof.

37 EXERCISES

1. Draw the standard map M of seven regions obtained by cutting off three corners of a tetrahedron, and show how to color it with five colors by reducing it to a standard map of five or fewer regions, as described in Theorem 36. (Note that one method of reduction, beginning with a triangular region, takes two steps, while another, beginning with a pentagonal region, takes only one step.)
2. Reduce the standard map defined by the regular dodecahedron (Illustration 8) to a standard map of six or fewer regions, and show how a coloring of that map with five colors gives a coloring of the faces of the dodecahedron with five colors.

† The reader will probably agree that this step in the proof is justified. However, a complete proof of this step would require the celebrated "Jordan-curve theorem" of topology.

*3 Prove that every standard map on the sphere which has no countries with more than four boundaries can be colored with four (or fewer) colors. [Hint: By an argument like that leading to Formula 12 in Section 3 show that $4f_2 + 3f_3 + 2f_4 = 12$. Show that only five of the seven solutions to this equation are realizable, and that each of the five resulting maps can be colored with four colors.]

Attempts to prove that *four* colors are enough to color every standard map on a sphere often proceed by showing that maps with fewer than a certain number of regions have faces of a certain type, and then show that maps with faces of these types are necessarily colorable with four colors. In this way it has been shown by C. E. Winn in 1938 [16, Ch. 12] that if a map is not colorable with four colors it must have at least 36 countries. According to Ore (in the preface to [16]) this is the best result so far.† Proofs of such facts are often very complicated, and we shall be satisfied with proving, in Theorem 41, that if an ordinary map on the sphere is not colorable with four colors then it must have at least 12 countries. We need two preliminary results, which we now prove.

38 Proposition
Let q be the average number, $2e/v$, of edges at each vertex of an ordinary map on the sphere. Then $(2 + q)f_1 + 4f_2 + (6 - q)f_3 + (8 - 2q)f_4 + (10 - 3q)f_5 + (12 - 4q)f_6 + (14 - 5q)f_7 + \cdots = 4q$.

PROOF: Since $q = 2e/v$ we have $v = 2e/q$. Put this into Euler's formula, and also put $f = f_1 + f_2 + f_3 + f_4 + f_5 + \cdots$. This gives $(2e/q) - e + f_1 + f_2 + f_3 + f_4 + \cdots = 2$. But $2e = f_1 + 2f_2 + 3f_3 + 4f_4 + \cdots$ by counting the edges at each type of face and remembering that each edge is counted twice in the process. Substituting this value of e into the preceding equation, and collecting terms in f_1, f_2, f_3, f_4, etc., gives the required result.

39 Theorem
If an ordinary map on the sphere has no regions with fewer than 5 edges then it has at least 12 regions. In fact, it has at least 12 regions with exactly 5 edges.

† In the *Notices* of the American Mathematical Society of January, 1968 (p. 196), Ore and Stemple announce that they have been able to raise this number to 40.

PROOF: Since $f_1 = f_2 = f_3 = f_4 = 0$ the equation of Proposition 38 reduces to

$$(10 - 3q)f_5 + (12 - 4q)f_6 + (14 - 5q)f_7 + \cdots = 4q.$$

But since $q \geq 3$ all terms after the first term on the left are negative or zero. That is $(10 - 3q)f_5 \geq 4q \geq 12$. This means that $10 - 3q$ is certainly positive, and $f_5 \geq 12/(10 - 3q)$. But we know $10 - 3q \leq 1$ (since $q \geq 3$), so that $f_5 \geq 12$, as required.

40 EXERCISES

1 Prove that if the average number of edges at each vertex of an ordinary map is as large as $3\frac{1}{3}$, then there must be regions with fewer than 5 edges. [Hint: Recall that under the conditions of Theorem 39, $10 - 3q > 0$.]
2 Under the condition of Problem 1 above, prove that if $f_1 = f_2 = 0$, then $2f_3 + f_4 \geq 10$.
3 Prove (or disprove) that there is an ordinary map for which $2f_3 + f_4 = 10$.

As a corollary to Theorem 39 we can prove:

41 Theorem

If a standard map on the sphere is not colorable with four colors, then it must have at least 12 regions.

PROOF: We proceed by induction again. Certainly all standard maps with no more than 5 regions can be colored with (at most) four colors. We have only to show how to reduce the problem of coloring a standard map M (having fewer than 12 regions) to the problem of coloring a map M^* with fewer regions.

From Theorem 39 we know that M has a region with 4 edges, as in Illustration 54. (Or 2 or 3 edges. We leave these simpler cases to the reader.) As before, at least one pair of opposite regions, say T and V, are really different regions and do not touch along an edge. By deleting the 2 edges separating T and V from R we form a new large region R^* in a map M^* having only $f - 2$ regions, as in Illustration 54(b). (We delete the vertices 1, 2, 3, 4 as before.) If M^* can be colored with four colors, so can M. For if we restore the original edges, R is surrounded by only three colors and hence can be colored the fourth color. This completes the proof.

Illustration 54

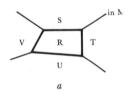

a

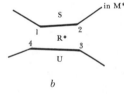

b

Illustration 55

SKELETON S
OF THE 4-SIMPLEX

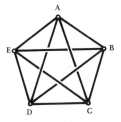

Illustration 56

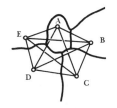

Furthermore, if a standard map has exactly 12 faces then either (a) it has some region with 3 or 4 edges, in which case the proof of Theorem 41 applies to show it can be colored with four colors, or (b) it consists of 12 pentagonal regions and is essentially the same as the regular dodecahedron, which by experiment can be colored with four colors. This proves the slightly stronger result that if there is a standard map on the sphere which cannot be colored with four colors then it must have at least 13 regions.

42 EXERCISES

1 Complete the proof of Theorem 41 in case R has 2 or 3 edges.

*2 Prove that if a map M on the sphere can be colored with only two colors then there is an Euler circuit along its vertices and edges. [Hint: Show that if any vertex of M has an odd number of regions meeting at it, then more than two colors are required. Then apply the Theorem immediately preceding Exercises 25.]

3 It was shown in Problem 4(c) of Exercises 24 that the skeleton S of the 4-simplex (Illustration 55) cannot be drawn on a sphere without extra crossings. Criticize the following "proof" of the four-color theorem: We have only to show that there can be no five countries on a map on a sphere, each one touching all the others. Suppose, if possible, there are five such countries A, B, C, D, E, as in Illustration 56. Since each touches all the others we can take a point in each and join it to a point in each of the others, without crossing. (Make a road from the capital of each country to the capital of each other country, crossing the common border in each case.) This network $ABCDE$ is just the network S, drawn on the sphere without any extra crossings, which is impossible. Hence, there cannot be five countries, each touching the other four. (An entertaining discussion of this problem is contained in [20, Chapters 4 and 11].)

4 A finite number of pennies are placed flat in the plane, so that no two touch in more than a single point. The pennies are to be painted so that no two touching pennies have the same color.

*(a) Show that there is at least one such way of placing the pennies so that four colors are needed.

**(b) Prove that *every* such arrangement of pennies can be colored with (at most) four colors.

Section 7 The Conquest of Saturn, the Scramble for Africa, and Other Problems

7A The Conquest of Saturn

It is interesting to consider the map-coloring problems on other surfaces, say on a ring of Saturn. For this purpose we can define the terms *ordinary map* and *standard map* on other surfaces in the same way they were defined for the sphere. It might be supposed that if the coloring problem for maps on a sphere has remained unsolved for so long, then the corresponding problems on surfaces of "spheres with handles" must be even more difficult. Surprisingly enough, this is often not so. In particular, the problem has been completely solved for regular maps on a ring of Saturn. (The technical name for such a ring is *torus*.)

Since, on the sphere, Euler's formula was a fundamental tool for proving the five-color theorem, we can expect the analogous Formula 18 of Section 4, in the form

43 $$v - e + f = 0,$$

to be fundamental for investigating the torus. But there is a slight complication when we want to apply this formula to *maps* instead of to the *polyhedra* for which it was established. For, even in a *standard map*, some region may very well circle the ring of Saturn completely, as in Illustration 57, which shows a standard map consisting of three countries, a, b, c which together occupy the entire ring of Saturn. Note that the boundary of region a actually consists of two distinct "triangles." In this map, $v = 4$, $e = 6$, and $f = 3$, so that $v - e + f = 1 (\neq 0)$. The reader should verify for himself that the map of 13 countries represented by Illustration 27(b) in Section 4 also fails to satisfy Formula 43.

How do we have to change Formula 43 to apply to the case of ordinary maps (and hence, in particular, to standard maps)? The reason Formula 43 fails is that we can insert a new edge, for example, an edge AB in Illustration 57, on such a map without changing the number of faces. That is, we can *decrease* the number $v - e + f$. Of course, this process can only be repeated a certain number of times, after which there will be no more ring-shaped regions. From then on, any added edge will cut some face in half, thus also increasing the number of faces by 1 and hence keeping $v - e + f$ constant. But for

Section 7   The Conquest of Saturn   65

such a map with no ring-shaped regions, we know from Section 4 that $v - e + f = 0$. This is the essence of the proof of

44 Theorem

If v, e, and f are the number of vertices, edges, and regions of an ordinary map on the torus, then $v - e + f \geq 0$. Equality holds if and only if there are no ring-shaped regions in the map.

From this, we can follow step-by-step the proof of Proposition 34 to prove:

45 Proposition

Every ordinary map on the torus has at least one region with no more than 6 edges (and hence no more than 6 neighbors).

46 EXERCISE

Prove Proposition 45. (At the essential step, of course, you must use $v - e + f \geq 0$ instead of $v - e + f = 2$.)

Illustration 57   A standard map on a torus, having one region which completely circles the torus.

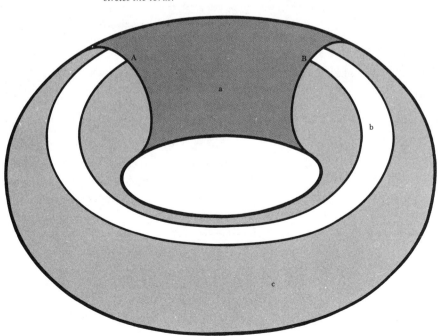

Hence in much the same way as in the proof of Theorem 36 (the five-color theorem for the sphere) we can prove:

47 Theorem

Every standard map on the torus can be colored with at most seven colors.

PROOF: We shall use the method of induction again. We consider a standard map M on the torus, having exactly f regions. From Proposition 45 we know that M has some region R with exactly 2, 3, 4, 5, or 6 edges. If we delete one of the edges of R, as in Illustration 58, to form a new map M^*, then some region S, say, together with R, forms a new region R^*. (Of course, as before, we also delete the vertices 1 and 2 where otherwise only two edges would meet.) However, it may be that M^* is not yet standard. For, perhaps, the regions T, U, or V are really the region S. In that case, the new region R^* would touch itself along an edge, which is not permitted. We can avoid this by deleting any such edges, which just separate a region from itself. This will give a standard map M^{**}. In M^{**} there may be a ring-shaped region, but this doesn't matter, since the formula $v - e + f \geq 0$ takes this into account.

Now, if we color M^{**} with the seven colors at our disposal, and put back whatever edges were removed, there will be at most six colors surrounding R (in the original map M). Hence, the seventh color can be used to color R. This completes the induction, since the coloring of any standard map M has been reduced to the coloring of a standard map M^{**} with fewer regions, and hence eventually to a standard map of seven or fewer regions, which can certainly be colored with (at most) seven colors.

We note that the separate cases which arose in the proof of Theorem 36 do not arise here, since we are willing to permit ring-shaped regions in M^{**}.

Now, to complete the solution of the coloring problem for maps on the torus it is only necessary to show that *some* map requires as many as seven colors. This can easily be done. In fact, there is a map on the torus consisting of seven hexagons, each touching the other six and, hence, obviously requiring exactly seven colors. It is evident that such a map of seven regions, each touching the other six, can be constructed from the Csàszàr polyhedron of Section 4 (with a tunnel in it) as follows: Let the vertices of the map be the midpoints of the faces of the Csàszàr polyhedron. Then join each of these 21 midpoints by an edge to the 3 midpoints of the adjacent faces. This will make a

Illustration 58

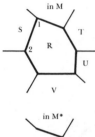

Section 7   The Conquest of Saturn   67

hexagon about each of the 7 vertices of the Császár polyhedron. Since each of these vertices was joined to each of the others, each hexagon is adjacent to each of the others. A beautiful picture of this map is found on the jacket, and Plate V, of Tietze's book *Famous Problems of Mathematics*.

A convenient way to draw this map is shown in Illustration 59. Here a rectangle (imagine it made of sheet rubber or some other very flexible material) is marked as shown in Illustration 59(a) and painted with seven colors 1, 2, 3, 4, 5, 6, 7. Then it is rolled up, and the two long edges AB and $A'B'$ are joined, to make a cylinder, as in Illustration 59(b). Finally, the ends of this cylinder are joined, to make a torus, as in Illustration 59(c). The reader should verify that each of the resulting 7 regions on the torus actually touches each of the other 6.

Illustration 59   *A map on the torus which cannot be colored with fewer than seven colors.*

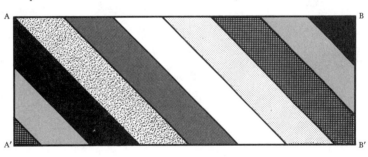

a

b

c

We conclude this section with Heawood's estimate as to the number of colors needed to color any standard map on a surface formed by adding h handles to a sphere. We let $[N]$ denote the largest integer less than or equal to N. We shall need the analog of Formula 43 for spheres with h handles, namely,

$$\text{48} \qquad v - e + f \geq 2 - 2h,$$

the proof of which is essentially the same as the proof of Formula 43, if we assume that $v - e + f = 2 - 2h$ (Formula 18, Section 4) is true for all ordinary maps having no ring-shaped regions. (The reader should note that very complicated regions may occur if h is large, such as the region of Illustration 60. In that face as many as four new edges can be drawn, as in the dotted lines, without changing the total number of faces. Thus, $v - e + f$ for a map containing such a face is at least $2 - 2h + 4$.)

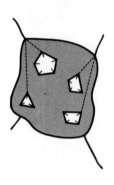

Illustration 60

49 Theorem (Heawood)

A standard map on a sphere with $h \geq 1$ handles can be colored with at most $[N]$ colors, where $N = (7 + \sqrt{1 + 48h})/2$; that is, N satisfies the equation

$$\text{50} \qquad N - 1 = 6 + \frac{12(h - 1)}{N}.$$

PROOF: We consider a given surface with $h > 1$. (We have proved the theorem for $h = 1$, the torus, in Theorem 47, and the question of its truth for $h = 0$ is exactly the four-color problem!) If any standard map on the surface has no more than $[N]$ regions, then it can certainly be colored with no more than $[N]$ colors. So we can assume that the map M in question has f regions, where $f > [N]$ and hence

$$\text{51} \qquad f > N,$$

and proceed by induction on f. That is, we assume that all standard maps on our surface having *fewer* than f regions can be colored with $[N]$ colors, and we demonstrate how to color the map M with $[N]$ colors.

First, recall that

$$\text{48} \qquad v - e + f \geq 2 - 2h.$$

Section 7   *The Conquest of Saturn*

Then, since M is standard, we know that

$$v = \frac{2e}{3}.$$

Putting $v = 2e/3$ into Formula 48 gives $(2e/3) - e + f \geq 2 - 2h$, and hence

52
$$\frac{2e}{f} \leq 6 + \frac{12(h-1)}{f}.$$

Now, the average number of edges per region is

$$p = \frac{f_1 + 2f_2 + 3f_3 + \cdots}{f} = \frac{2e}{f}.$$

Hence,

$$p = \frac{2e}{f} \leq 6 + \frac{12(h-1)}{f} < 6 + \frac{12(h-1)}{N} = N - 1.$$
$$\quad\quad\quad (52) \quad\quad\quad\quad (51) \quad\quad\quad\quad (50)$$

This shows that the average number p of edges per region is less than $N - 1$. Hence, there must be at least one region R with fewer than $N - 1$, and hence no more than $[N] - 1$, edges. (From here on the proof is just like the proof of Theorem 47, the seven-color theorem for the torus.)

Delete one of the edges of R. If this creates a region which touches itself along other edges of R, then delete those edges also, to form a new map M^*, having fewer regions than M. Now, color M^* with $[N]$ colors. Then put back the deleted edges, to get the original map M, whose region R is surrounded by at most $[N] - 1$ colors (since R has at most $[N] - 1$ edges). Then use the $[N]$th color for R itself. This completes the proof.

This estimate of Heawood's is a very good one, so good that no "sphere with h handles" is known to have all maps on it colorable with *fewer* than $[N]$ colors. In 1953 it was shown by G. Ringel that Heawood's estimate cannot be wrong by more than *two*. That is, for each h there is always some map on a "sphere with h handles" which requires as many as $[N] - 2$ colors.

Finally, in 1968, Ringel and J. W. T. Youngs completed a proof that in fact Heawood's N is exact. That is, for each $h \geq 1$, there is always some map on a sphere with h handles which requires as many as $[N]$ colors.

7B The Scramble for Africa

There are many other more or less natural ways to modify the original four-color problem to get new and almost equally interesting problems. We might ask, for example, what is the maximum number of colors required to color a map of the earth if each country has at most one colony, which must be colored the same color as the corresponding country. Or, to be more timely, what if each country on earth has one territory on the moon which is to be colored the same color as the home country? Or, what if each country on one of the rings of Saturn has a colony on the same ring—or on a different ring? Except for the earth and moon† the answers to these particular four questions are known. (They are 12, 13, 13. See the book by G. Ringel [18, pp. 25, 77, 78].) We present Ringel's proof for the first one.

53 Theorem (Heawood, 1890)

(a) Let each country on a sphere have at most one colony, which is to be colored the same color as the corresponding country. Then every such map can be colored with at most twelve colors.

(b) There is a map, under conditions (a), which requires twelve colors.

Illustration 61

A map having five countries $(1, 2, 3, 4, 5)$ in the northern hemisphere, each with a colony $(1', 2', 3', 4', 5')$ in the southern hemisphere.

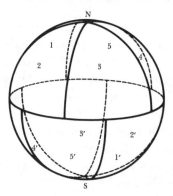

† In fact, however, very little is known about the case of the earth and the moon. At present, according to Ringel, it is only known that the answer is 8, 9, 10, 11, or 12, with the strong suspicion that 8 is correct.

Section 7   *The Scramble for Africa*   71

PROOF: Consider a map as described. Represent each country by a vertex of a network, and each colony by a vertex of the same network. (We call these networks "graphs" in the sequel to distinguish them from the original network of vertices and edges of the map.) Let two vertices be joined by an edge if the corresponding regions are adjacent. For example, suppose there are five countries in the Northern Hemisphere meeting at the North Pole, N, and five colonies similarly placed in the Southern Hemisphere, as in Illustration 61; the corresponding graph is shown in Illustration 62. We call such a graph a "colony graph." Certainly this graph can always be drawn on the sphere, without having any edges meet except at the appropriate vertices. Now, from a given colony graph we construct a new graph, with vertices which represent country-colony pairs, and with all superfluous edges deleted. (In particular, we do not join any vertex to itself by an edge even though the home country touches its colony.) This new graph, for the map of Illustration 61, is shown in Illustration 63. In this case, each vertex is joined to each other, since every country (or its colony) has a border in common with every other country or its colony. Such a graph can be called an "empire graph." Obviously, an empire graph cannot always be drawn on the sphere without some extra crossings of edges. In particular, the empire graph just drawn is essentially the skeleton of the 4-simplex, which we showed in

Illustration 62   Each region is represented by a vertex, and two vertices are joined by an edge if the two regions have a common boundary.

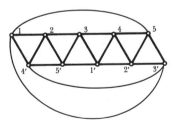

Illustration 63   The empire graph derived from the colony graph of Illustration 62.

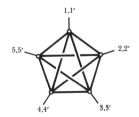

Problem 4(c) of Exercises 24 cannot be drawn on the sphere without extra crossings. An empire graph (or any *subgraph*, that is, any graph obtained by deleting some edges and possibly vertices, of an empire graph) can be thought of as having been obtained from a colony graph (or a subgraph of a colony graph) by coalescing certain vertices and certain edges.

Our problem now is to color the vertices of an empire graph G so that no two vertices joined by an edge have the same color. Suppose that G requires κ colors, and cannot be colored with fewer than κ colors. We call any such graph κ-*chromatic*, and say κ is the *chromatic number* of G. We now delete as many edges and vertices as possible from G without changing its chromatic number. This new graph G^* has at least $\kappa - 1$ edges at each vertex. (For, a vertex with fewer than $\kappa - 1$ edges could be deleted, with all its edges, the vertices of the new graph colored with $\kappa - 1$ colors, and then the deleted vertex restored and colored with one of the $\kappa - 1$ colors not used for the vertices joined to the deleted vertex. This would contradict the fact that G^* requires κ colors.) Hence, if v^* and e^* are the numbers of vertices and edges respectively, of G^*, we have

54 $$(\kappa - 1)v^* \leq 2e^*.$$

Now, recall that G^*, being a subgraph of the empire graph G, was obtained by coalescing certain vertices and edges of a subgraph $\bar{G}^*$ of the colony graph $\bar{G}$ which coalesced to form G^*.

Let $\bar{v}^*$ and $\bar{e}^*$ denote the numbers of vertices and edges of $\bar{G}^*$. Then certainly

55 $$e^* \leq \bar{e}^* \quad \text{and} \quad \bar{v}^* \leq 2v^*,$$

since at most two vertices are coalesced to a single one. It is also true that

56 $$\bar{e}^* \leq 3\bar{v}^* - 6.$$

($\bar{G}^*$ is a graph on a sphere. Its regions correspond to vertices of a map, and at least 3 edges meet at each vertex of a map. That is, $3\bar{f}^* \leq 2\bar{e}^*$. Putting this into Euler's formula $\bar{v}^* - \bar{e}^* + \bar{f}^* = 2$ yields Formula 56.) It follows from Formulas 55 and 56 that

57 $$e^* \leq 6v^* - 6.$$

Combining Formulas 54 and 57 we have $(\kappa - 1)v^* \leq 2e^* \leq 12v^* - 12$. That is, $(\kappa - 1) \leq 12 - (12/v^*)$, and hence $\kappa - 1 < 12$ and $\kappa \leq 12$, since κ is a whole number. This completes the proof that at most 12 colors are needed to color the empire graph G.

We still have to prove part (b) of the theorem, namely, that there is a map of countries and colonies which requires twelve colors. Such a map was discovered by Heawood. Heawood's map is shown in Illustration 64.

Section 7   The Scramble for Africa

Illustration 64   Heawood's map of twelve countries, and their twelve colonies, each country (or its colony) touching every other country (or its colony).

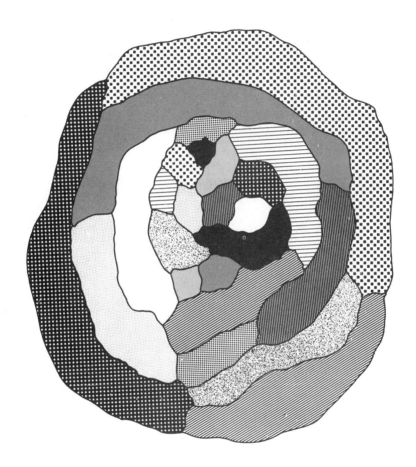

7c Some Problems Equivalent to the Four-Color Problem

There are certain apparently simple problems which turn out to be *equivalent* to the four-color problem. That is, an affirmative answer to any one of them would show that every ordinary map on the sphere can be colored with four colors. And conversely, if every ordinary map on the sphere can be colored with four colors then these problems all have the answer *yes*. Four such problems are:

(1) Let P be a convex polyhedron. Is it always possible to repeatedly cut off corners from P so that a polyhedron eventually is obtained, all of whose faces have a number of edges which is divisible by 3? (H. Hadwiger, in *Elemente der Mathematik*, 1957, p. 61. Heawood posed essentially the same problem in 1898.) See Illustration 65, where this is done for the cube and dodecahedron.

Illustration 65

Some corners of the cube and dodecahedron can be cut off so that the number of edges of each face of the new polyhedron is divisible by 3.

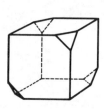 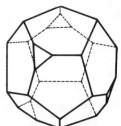

(2) Consider an indicated sum, $a_1 + a_2 + \cdots + a_n$. Suppose parentheses are inserted to explain completely the order in which the addition is to be carried out. (For example, if $n = 5$ this could be done as $(((a_1 + a_2) + a_3) + a_4) + a_5$ or as $(a_1 + a_2) + ((a_3 + a_4) + a_5)$, say.) For each two such arranged n-fold sums is it always possible to choose the a's to be integers in such a way that no partial sum of either arranged sum is divisible by 4? (H. Whitney, *Monatsheft für Mathematik und Physik*, 1937, pp. 207–214.)

(3) Is it always possible to color the *edges* of a regular map on the sphere with three colors so that no two edges of the same color meet at a vertex? [10, p. 25.]

(4) Is it always possible to color the *vertices* of a regular map on the sphere with two colors $+1$ and -1 so that the sum of the colors of the vertices of each face is a number divisible by three? [13, pp. 77–89.]

58   FURTHER PROBLEMS

Some of the following problems could be considered topics for short essays or further reading. Many of them are quite difficult. The student can expect to spend several hours on some of them. Others can be done more quickly. No student should expect to do those marked**, but he may be interested in reading the references given. The first four are taken from the excellent booklet by Dynkin and Uspenskii, *Multicolor Problems* [10].

*1 Let n circles be drawn in the plane. Prove that the map formed by these circles can be colored with two colors.

*2 Show that every map all of whose vertices have an even number of edges meeting at them can be colored with two colors.

*3 Suppose n circles are drawn in the plane. In each circle draw a chord so that the chords of two different circles have at most one point in common. Prove that the resulting map can be colored with three colors.

*4 Prove that a standard map can be colored with three colors if and only if every region has an even number of edges.

*5 If a square is partitioned into n convex polygons show that the total number of edges present is at most $3n + 1$. Show furthermore that it is possible to have exactly $3n + 1$ edges. [Hint: Get a lower bound for the total number of edges by adding the number at each vertex. This gives $2e \geq 3(v - 4) + 2 \cdot 4$. Then apply Euler's formula to get the final result.] [*American Mathematical Monthly*, 1962, p. 1010.]

A region is called *odd* if it has an odd number of edges.

*6 Prove that a standard map on the whole sphere can be colored with no more than four colors if and only if the number of odd regions colored any two colors (say, red and green) is even. [*American Mathematical Monthly*, 1966, p. 204.]

7 What is the maximum number of solids which can be arranged in space so that each touches all the others? Notice that it is

not required that the solids be convex. Illustration 29 may help you guess the answer. [20, pp. 77–78.]

8 This is a restriction of Problem 7 above to solids which are convex and (in (a) to (d)) congruent.

(a) Arrange 4 billiard balls so that each touches the other 3.
(b) Arrange 5 half-dollars so that each touches the other 4.
(c) Arrange 6 cigarettes so that each touches the other 5.
(d) Arrange 7 cigarettes so that each touches the other 6.
(e) Arrange 8 (not necessarily congruent) tetrahedra so that some face of each touches a face of each of the other 7. [Hint: First arrange 4 triangles in the plane so that each pair touch along an edge.]

(For solutions to (a)–(d) see [12, pp. 110, 114–115]. For (e) see V. J. D. Baston, *Some Properties of Polyhedra in Euclidean Space*, Pergamon Press, New York, 1965, or F. Bagemihl, in *American Mathematical Monthly*, 1956, pp. 328–329. It is an unsolved problem whether 9 tetrahedra can be arranged so that a face of each touches a face of each of the other 8.)

9 What is the maximum number of *convex* solids that can be arranged in space so that each touches all the others? (For the surprising answer, see Besicovitch, *Journal of the London Mathematical Society*, **22 (1947), pp. 285–287.)

*10 If one end of the rectangular strip of paper in Illustration 66(a) is given a half turn before gluing (so that AA and BB are glued together), the *Möbius strip* of Illustration 66(b) results. (Without the half-turn, we would get a stubby cylinder.) Draw a map on the Möbius strip containing 6 regions each touching the other 5. (For one solution, see [20, Plate VI].)

Illustration 66

When you give this strip a half turn before gluing the ends together, the result is a Möbius strip.

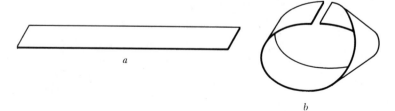

a

b

*11 A finite number of ping-pong balls are placed in space. The problem is to paint them so that no two which touch have the same color. Prove that four colors is not enough.

***12 In the preceding problem, find the least number of colors which will color *every* such arrangement of ping-pong balls. (This is an unsolved problem! However, it is known that the answer is either 5, 6, 7, 8, or 9. The author offers a $10 prize for the first correct proof that any one of these is impossible. Compare this with Problem 4 of Exercises 42 for pennies.)

***13 (This is an unsolved problem, from [18, p. 26].) A finite number of circles, not necessarily congruent, are drawn in the plane, so that at no point do more than two circles touch. What is the least number of colors which will color all such systems so that no two touching circles have the same color?

Professor H. S. M. Coxeter, F.R.S., (1907–), of the University of Toronto, is one of the world's most distinguished geometers. As a schoolboy in England, he became fascinated by regular and nearly regular polyhedra and their analogs in higher dimensions. His book REGULAR POLYTOPES *is now the standard treatise on that subject. He has made contributions to the study of regular figures in non-Euclidean geometry and in the application of geometric techniques to modern algebraic problems. Much of the recent revival of interest in geometry in North America can be credited to Coxeter's many textbooks, films, and new discoveries about the subject. (Courtesy of the Canadian Journal of Mathematics, from an issue dedicated to Professor Coxeter on his sixtieth birthday, in 1967.)*

References

1. W. W. Rouse Ball, *Mathematical Recreations and Essays*, 11th ed., Macmillan & Co., Ltd., London, 1959, Chaps. 5, 8.
2. H. Behnke, H. Bachmann, K. Fladt, and W. Süss, *Grundzüge der Mathematik*, Vol. 2, *Geometrie*, Vandenhoeck and Ruprecht, Göttingen, 1960, Chap. 6.
3. R. Courant and H. E. Robbins, *What Is Mathematics?* Oxford University Press, Fair Lawn, N.J., 1953, Chap. 5.
4. H. S. M. Coxeter, *Introduction to Geometry*, John Wiley & Sons, Inc., New York, 1961, Chaps. 10, 21.
5. ———, *Regular Polytopes*, The Macmillan Company, New York, 1963.
6. ———, "Map-coloring problems," *Scripta Mathematica*, **23** (1957), pp. 11–25.
7. ———, "The four-color map problem," *The Mathematics Teacher*, **52** (1959), pp. 283–289.
8. A. Császár, "A polyhedron without diagonals," *Acta Scientiarum Mathematicarum*, **13** (1949–50), pp. 140–142.
9. H. M. Cundy and A. P. Rollett, *Mathematical Models*, Oxford University Press, London, 1961.
10. E. B. Dynkin and V. A. Uspenskii, *Multicolor Problems*, D. C. Heath and Company, Boston, 1963.
11. Philip Franklin, "The four-color problem," *Galois Lectures, Scripta Mathematica Library*, **5** (1951), New York, pp. 49–85.
12. M. Gardner, *The Scientific American Book of Mathematical Puzzles and Diversions*, Simon and Schuster, Inc., New York, 1959, Chap. 6.
13. H. Hasse, *Proben mathematischer Forschung in allgemeinverständlicher Behandlung*, Otto Salle Verlag, Frankfurt am Main-Pinneberg, 1955, pp. 55–101.
14. P. J. Heawood, "Map-colour theorem" and "On the four-colour map theorem," *Quarterly Journal of Mathematics*, **24** (1890), pp. 332–338, and **29** (1898), pp. 270–285.
15. D. Hilbert and S. Cohn-Vossen, *Geometry and the Imagination*, Chelsea Publishing Company, New York, 1952.
16. O. Ore, *The Four-color Problem*, Academic Press Inc., New York, 1967.
17. H. Rademacher and O. Toeplitz, *The Enjoyment of Mathematics* (trans. H. Zuckerman), Princeton University Press, Princeton, 1957, Chaps. 12, 13.
18. G. Ringel, *Färbungsprobleme auf Flächen und Graphen*, VEB Deutscher Verlag der Wissenschaften, Berlin, 1959.
19. H. Steinhaus, *Mathematical Snapshots*, Oxford University Press, Fair Lawn, N.J., 1960, Chaps. 7, 8, 12.
20. H. Tietze, *Famous Problems of Mathematics*, Graylock, New York, 1965, Chaps. 4, 11.

EXCURSIONS INTO MATHEMATICS

CHAPTER   2   THE SEARCH FOR PERFECT NUMBERS

BY MICHAEL N. BLEICHER

Pythagoras (580?–500? B.C.) founded his own academy, a secret numerological society, at Croton in southern Italy, and began to bridge the gap between mysticism and number theory. We probably should credit this group with the definition of prime, *many of the theorems later found in Euclid, and some of the rules for solving the equation* $x^2 + y^2 = z^2$ *in whole numbers. The Pythagorian belief that all natural phenomena could be* explained *in terms of whole numbers and their ratios led to the development of the musical scale as pictured below in this 15th century Italian woodcut. The discovery by later Pythagorians that the* $\sqrt{2}$ *could not be so expressed was a shattering experience. (Courtesy of the New York Public Library.)*

Section 1 Introduction

Who has not been fascinated with numbers? Perhaps you have added them: $1 + 1 = 2$, $2 + 2 = 4$, $4 + 4 = 8$, etc., and noticed how quickly the sums grow. Or perhaps you were mystified the first time you were told to pick a number with two or three distinct digits, change the order of the digits, subtract these two numbers, and add the digits of this difference, leaving you with nine. Or perhaps you pondered over the mystic qualities of numbers: Why is lucky seven lucky? Why is thirteen unlucky? Why are there the three fates in Greek mythology or the three weird sisters in Shakespeare?

On the other hand, you may have wondered as you read the Bible why an omnipotent God should choose for creation to take six days, or why 28 days should be chosen as the full cycle of the moon, or why there should have been eight people in Noah's ark. All these things have been explained in terms of the properties of the numbers involved by some of the greatest scholars in history, from prehistoric times to the present. Indeed, in his recent book, Shanks [14, p. 130] makes an amazing case for Pythagorean numerology. We shall begin with the study of perfect numbers.

A number is called *perfect* if it is the sum of all the numbers (except itself) which divide it; that is, if it is the sum of its aliquot parts. It is very important in mathematics that everyone means the same thing when he uses mathematical terms and therefore rather than define things as above, it is more common and accurate to say explicitly when something is being defined so that the reader will know he is not expected to be familiar with these terms from a previous discussion and so that future reference can be made more easily. Thus:

1 Definition

A whole number d is said to *divide* a whole number n or be a *factor* of n if and only if $\frac{n}{d}$ is a whole number.

It should be clear that this is the same thing as saying d divides n if and only if there is a whole number $q \left(= \frac{n}{d} \right)$ such that $n = q \cdot d$. We note that 1 divides every number since $\frac{n}{1} = n$ is always a whole number, and also that every number divides 0 since $\frac{0}{d} = 0$ is always a whole number.

2 Examples

The chart below gives some numbers and their positive divisors.

n	Positive divisors of n
1	1
2	1, 2
4	1, 2, 4
5	1, 5
6	1, 2, 3, 6
7	1, 7
8	1, 2, 4, 8
9	1, 3, 9
12	1, 2, 3, 4, 6, 12
20	1, 2, 4, 5, 10, 20
28	1, 2, 4, 7, 14, 28

3 Definition

A divisor d of a whole number n is called an *aliquot part* or *proper divisor* if and only if d is between 0 and n, $0 < d < n$.

4 Definition

A whole number is *perfect* if and only if it is equal to the sum of its aliquot parts.

From Example 2 we see that 6 and 28 are perfect numbers.

5 EXERCISES

1 Verify that 6 and 28 are the first two perfect numbers.
2 Find the positive divisors of 72 and 1,001.
3 Find the negative divisors of 72 and 1,001.
4 Find all the divisors, positive and negative, of 1,144.
5 Show that d divides n if and only if $-d$ divides n.
6 Show that if a divides b and b divides c, then a divides c.
7 Show that if n is a perfect number then the sum of the reciprocals of the positive divisors of n is two; for example, $2 = \frac{1}{1} + \frac{1}{2} + \frac{1}{3} + \frac{1}{6}$; $2 = \frac{1}{1} + \frac{1}{2} + \frac{1}{4} + \frac{1}{7} + \frac{1}{14} + \frac{1}{28}$.
8 Show that the converse of Problem 7 is also true; namely: If the sum of the reciprocals of the positive divisors of n is two, then n is perfect.

*9 Carry out the addition of the reciprocals of the positive divisors of 6 and 28 in binary expansion. Notice that it is never necessary to carry in performing this addition. Perfection! (Shanks [14, p. 25]. See also Section 3 of Chapter 5.)

A consequence of Problem 5 is that the divisors of n come in pairs $(d, -d)$, so that if we know the positive divisors of a number, we can immediately get all the divisors. For this reason, we shall concentrate our attention on the positive divisors; in listing divisors or counting how many there are, we shall only list or count the positive ones. However, if at some point all we know about a number d is that it divides n, we shall be unable to determine if d is positive or negative without some additional information.

The first four perfect numbers 6, 28, 496, 8,128, were certainly known to the Greeks, and were probably known earlier. (Can you show that they are perfect?) By looking at these numbers the ancients were led to make several conjectures. First, they conjectured that there were infinitely many perfect numbers. Second, that these numbers ended in 6 and 8 alternately. Third, that there was exactly one perfect number with a given number of digits (see Section 3). We quote from Nicomachus's *Arithmetica* (c. 100 A.D.):

But it happens that, just as the beautiful and the excellent are rare and easily counted but the ugly and the bad are prolific, so also excessive and defective numbers are found to be many and in disorder, their discovery being unsystematic. But the perfect are both easily counted and drawn up in a fitting order: For only one is found in the units, 6; and only one in the tens, 28; and a third in the depth of the hundreds, 496; as a fourth the one, on the border of the thousands, that is short of the ten thousand, 8128. It is their uniform attribute to end in 6 or 8, and they are invariably even.

We shall soon see what light more recent investigations can throw on these conjectures.

Before entering our investigation let us look at a few of the supernatural aspects associated with these notions. In classic Greek numerology (the study of the mystic power of numbers), 6 represented marriage, health, and beauty, being the sum of its own parts. It was also said to represent Venus, for it is the product of 2, which is female, and 3, which is male. Judeo-Christian philosophy also includes comments on perfect numbers. For example, Aurelius

Augustinus (354–430 A.D.) made the first recorded statement that creation took six days because both God's creation and 6 are perfect. According to St. Augustine,

Six is a number perfect in itself, and not because God created all things in six days; rather the converse is true; God created all things in six days because this number is perfect, and it would have been perfect even if the work of the six days did not exist.

Alcuin (735–804), Charlemagne's tutor, elaborated that the second creation from Noah was not perfect inasmuch as there were 8 souls in Noah's ark and 8 is a deficient number (the sum of its aliquot parts is less than itself).

We stop here, but those readers interested in pursuing these avenues should consult Cajori, Dantzig, Dickson, or Ore [3, 4, 5, 7, 12].

Section 2 Prime Numbers and Factorization

It is evident that if we are to study properties of sums of divisors we should first make a study of divisors themselves as a basis for more advanced work.

6 *Definition*

A positive number greater than one is called *prime* if and only if its only positive divisors are itself and one. Other numbers greater than one are called *composite*.

Thus 28 is not prime since $28 = 14 \cdot 2$ yielding divisors other than itself and one. The reader should verify that 31, for example, is prime. Notice further that the factor 14 of 28 can itself be factored into $14 = 7 \cdot 2$ and both 7 and 2 are prime. Thus $28 = 7 \cdot 2 \cdot 2$ can be decomposed into a product of prime factors. Can this be done with every number? Try it with a few numbers. How do you go about it?

(Courtesy of Johnny Hart Enterprises.)

7 *Definition*

If $n = p_1 \cdot p_2 \cdot p_3 \cdots p_k$, where $p_1, p_2, p_3, \ldots, p_k$ are all primes, then we call $p_1 \cdot p_2 \cdot p_3 \cdots p_k$ a *prime factorization* of n.

8 Theorem

Every whole number greater than 1 has a prime factorization.

PROOF: The theorem is certainly true when applied to all numbers less than 6. We proceed by mathematical induction. Suppose the theorem is true for all numbers between 1 and m (the induction hypothesis). We shall show it is also true for all numbers between 1 and $m + 1$. Let n denote a number, and suppose $1 < n < m + 1$. If $n < m$, then by the induction hypothesis it has a prime factorization. If $n = m$, then either it is prime and $n = m$ is its prime factorization or it is composite. If it is composite, it has at least one proper divisor other than 1, say p. Thus, $m = p \cdot q$. Since $1 < p < m$, we see that $1 < q < m$. By the induction hypothesis, p is either a prime, say $p = p_1$, or has a prime factorization, say $p = p_1 \cdot p_2 \cdots p_k$. Similarly, either q is prime, $q = q_1$, or $q = q_1 \cdots q_l$, where $q_1, \ldots, q_l$ are primes. Replacing p and q either by themselves if they are prime or by their factorizations if they are not, we obtain $m = p_1 \cdots p_k \cdot q_1 \cdots q_l$ (where possibly $k = 1$ or $l = 1$; in which case, $p_1 \cdots p_1$ is to be read p_1), a prime factorization of m. The truth of the theorem now follows from the principle of mathematical induction. For the reader unfamiliar with this principle, it is explained, differently, in each of the next two paragraphs.

Since by arithmetic verification the induction hypothesis is true for $m \leq 6$, it follows from the preceding paragraph that the theorem is true for all numbers between 1 and 7. Thus the induction hypothesis holds true with $m = 7$, whence the theorem is true for all numbers between 1 and 8, and the induction hypothesis holds for $m = 8$. We can proceed one step at a time by means of the last paragraph until we prove that the theorem is true for any particular integer we choose. But if the statement is true about every particular integer, then the theorem is true. The explanation of mathematical induction could have proceeded differently by arguing as outlined below.

If there is a number greater than 1 without a prime factorization, then there is a smallest such number without a prime factorization. We call this number m. Then, since every n, where $1 < n < m$, has a prime factorization, the induction hypothesis is satisfied and the argument of the second paragraph back yields a prime factorization of m. This is a contradiction since we assumed m had no factorization. Thus, our assumption that there is a number greater than 1 without a factorization into prime factors is contradictory and therefore false. The theorem is true.

Usually these last two paragraphs will be replaced by one sentence, for example: "By induction, the theorem is true."

To obtain a prime factorization of 28, we first break it up as $28 = 14 \cdot 2 = 7 \cdot 2 \cdot 2$. Suppose we break it differently, $28 = 4 \cdot 7$, do we still get the same prime factors? We will call two factorizations the same if they disagree only in the order of writing the factors. Is this only because 28 is perfect, or would this work for every number, even large ones, for example, 510,510? Before going on, try factoring a few composites by different methods. Try to guess whether or not the factorization is independent of how you go about it. Can you prove your conjecture?

9 EXERCISES

1 Find all the primes less than 50.
2 How many divisors do 6, 10, 14, 15, 21, 22, and 35 have? Do you see a connection with their prime factorization?
3 A number is called *abundant* if the sum of its aliquot parts (proper divisors) is greater than itself. Find some abundant numbers. Can you find an odd one?
4 How many factors does a square of a prime have? Can you prove your conjecture without knowing whether such a number has more than one prime factorization? Can you prove it assuming a number has only one prime factorization?

From the preceding exercises we can get a glimpse of the importance of the fact that every number has only one factorization into prime factors. The theorem stating this fact is so basic that it is normally called the Fundamental Theorem of Arithmetic.

10 The Fundamental Theorem of Arithmetic

Every number greater than one can be written as a product of prime factors. Except for the order in which the factors are written there is only one way to write a number as the product of prime factors. More simply, every number has a unique prime factorization.

The simplicity and intuitive correctness of this theorem tend to obscure the subtlety and difficulty involved in its proof. There is no simple proof known.

Consider for the moment the set of all positive numbers which have remainder 1 upon division by 3, that is, the numbers obtained by starting with 1 and counting by three's. We shall call this the Three's System. Namely, the numbers 1, 4, 7, 10, 13, 16, 19, 22, 25, 28, 31, 34, 37, 40, . . . , 97, 100, 103, ... The product of any two of these numbers yields another number in the system. Thus this system is multiplicatively similar to the ordinary system of positive integers. Analogous to our definition of primes, we define atomic numbers in this system. Atomic numbers are numbers in this system which are greater than one and have no proper divisors *in this system* except themselves and one. Thus the first few atomic numbers are 1, 4, 7, 10, 13, 19, 22, 25, 31, 34, 37, 43, 46, 55, etc., since none of these numbers has proper divisors *in the Three's System*. Analogously, 3 is a prime in the ordinary system of integers even though it can be factored $3 = 2(1\frac{1}{2})$ since $1\frac{1}{2}$ is not in the system of positive integers. The number 100 has two atomic decompositions, namely,

$$100 = 4 \cdot 25 \quad \text{and} \quad 100 = 10 \cdot 10.$$

Thus our proof of the fundamental theorem, if correct, must at some point use the fact that the usual number system has some properties that the three's system doesn't have.

Before proving the fundamental theorem we must marshall our forces and prepare for the attack. In mathematics this is usually done by proving a number of easier theorems and/or lemmas. A lemma is a statement the truth of which will be needed in the proof of some theorem, but which the author feels is not worth proving or remembering in and of itself. It may be noted, historically, that the author's judgments on these matters are often bad. In some cases a lemma, which an author has proved very casually as an unimportant step toward proving some theorem, is remembered long after the theorem is forgotten.

We prove now several lemmas and theorems which will be needed to prove the fundamental theorem.

11 Theorem

If n divides both the numbers a and b then n also divides both $a + b$ and $a - b$.

PROOF: Suppose that n divides a and b: then $\frac{a}{n}$ and $\frac{b}{n}$ are integers. But

$$\frac{a+b}{n} = \frac{a}{n} + \frac{b}{n} \text{ and } \frac{a-b}{n} = \frac{a}{n} - \frac{b}{n}$$

hence both $\frac{a+b}{n}$ and $\frac{a-b}{n}$ are integers. The theorem is proved.

12 Theorem

If n divides c but not d, then n divides neither $c + d$ nor $c - d$.

PROOF: If n divides c, and n does not divide d, then $\frac{c}{n}$ is a whole number while $\frac{d}{n}$ is not. It follows that neither $\frac{c}{n} + \frac{d}{n}$ nor $\frac{c}{n} - \frac{d}{n}$ can be whole numbers, hence n divides neither $c + d$ nor $c - d$. The theorem is established.

Because of the frequency with which the phrase "n divides b" occurs, it pays to have a concise symbol for it.

13 *Notation*

The phrase n *divides* b is denoted by $n|b$, and the phrase n *does not divide* b is denoted by $n \nmid b$.

14 Lemma

If n is a number with unique prime factorization and p is a prime which divides n, then p is one of the factors in the prime factorization of n.

PROOF: Suppose $n = q_1 \cdot q_2 \cdot \cdots \cdot q_k$ is the prime factorization of n. Since $p|n$, we have $n = p \cdot r$ for some number r. By Theorem 8, r has a prime factorization, say $r = r_1 \cdot r_2 \cdot \cdots \cdot r_l$, where $r_1, \ldots, r_l$ are primes. Thus $n = p \cdot r_1 \cdot r_2 \cdot \cdots \cdot r_l$ is a prime factorization of n. Since the hypothesis is that n has a unique factorization, we know that $p, r_1, \ldots, r_l$ is a rearrangement of $q_1, q_2, \ldots, q_k$ and hence $p = q_i$ for some choice of i. The lemma is proved.

15 Lemma

If p is a prime and each number greater than one and less than p^2 has a unique prime factorization, then $p|a \cdot b$ implies that either $p|a$ or $p|b$.

Before proving this we note that the hypothesis that p is a prime is essential, since $6|9 \cdot 2$, but $6 \nmid 9$ and $6 \nmid 2$.

PROOF: We divide a and b by p and get quotients and remainders:

$$a = kp + r \quad \text{and} \quad b = lp + s$$

where $0 \leq r < p$ and $0 \leq s < p$. If $r = 0$ or $s = 0$, then $p|a$ or $p|b$, respectively, and the lemma follows. We shall show that the supposition that neither is zero leads to a contradiction. First, we note that $ab = (kp + r)(lp + s) = p(klp + ks + lr) + rs$. Since $p|p(klp + ks + lr)$ and $p|ab$, it follows from Theorem 11 that $p|rs$. Since $rs < p^2$, it follows from the hypothesis that rs has unique factorization, and by Lemma 14 we conclude that p is one of its prime factors. Since the unique prime factors of rs are just the prime factors of r and the prime factors of s combined, we conclude that either p is a prime factor of r or p is a prime factor of s. But this is impossible since $1 \leq r < p$ and $1 \leq s < p$. Thus either $r = 0$ or $s = 0$, and $p|a$ or $p|b$, respectively.

We are now in a position to prove the fundamental theorem of arithmetic (Theorem 10).

PROOF OF THE FUNDAMENTAL THEOREM: We note that the number 2 has unique prime factorization, namely, $2 = 2$. We make the induction hypothesis that each number between 1 and N exclusive has a unique prime factorization and show that N also has a unique factorization. If N is a prime, $N = N$ is the only factorization and we are done. If N is not prime suppose $N = p_1 \cdots p_k$ and $N = q_1 \cdots q_l$ are two factorizations of N. We shall show that the q_i's are simply a rearrangement of the p_i's. We renumber the p_i's in increasing order so that p_1 is the smallest. Since N is not prime, $k \geq 2$. Hence, $N = p_1 p_2 \cdots p_k \geq p_1 p_2 \geq p_1^2$. By the induction hypothesis all numbers exceeding one and less than p_1^2 have unique factorization. Since $p_1|N$ we obtain from Lemma 15 applied to $a \cdot b = q_1(q_2 \cdots q_l) = N$ the conclusion that either $p_1|q_1$ or $p_1|q_2 \cdots q_l$. If $p_1|q_1$, then $p_1 = q_1$ since q_1 is prime and $p_1 \neq 1$. If $p_1|q_2 \cdots q_l$, we apply Lemma 14 with $p = p_1$ and $n = q_2 \cdots q_l$. Since $n < N$, we know n has unique factorization and the hypotheses of Lemma 14 are satisfied. Thus, in any case, $p_1 = q_i$ for some i. Renumber the q's so that $p_1 = q_1$. Thus $N' = p_2 p_3 \cdots p_k = q_2 q_3 \cdots q_l$. Since $N' < N$ it follows from the induction hypothesis that $p_2, \ldots, p_k$ are simply $q_2, \ldots, q_l$ rearranged. Thus $p_1, \ldots, p_k$ are simply $q_1, q_2, \ldots, q_l$ rearranged. Thus any two prime factorizations of N are rearrangements of one another. It follows that N has a unique prime factorization. The theorem follows by induction.

Before continuing, the student is advised to reread the proof until he discovers exactly where we have done something which fails for our three system.

Now that we have proved the fundamental theorem, we can return to the lemmas used in its proof and strengthen them.

16 Theorem

If p is a prime and $p|N$, then p is one of the factors in the prime factorization of N.

17 Theorem

If p is a prime and $p|a \cdot b$, then $p|a$ or $p|b$.

18 EXERCISES

1 Prove Theorem 16 from Theorem 10 and Lemma 14.
2 Prove Theorem 17 from Theorem 10 and Lemma 15.
3 Does either Lemma 14 or Lemma 15 hold in the Three's System?
4 Prove that every number except 1 in the Three's System has an atomic factorization.
5 Prove that the square of a prime has exactly three divisors. Give an example to show that in the Three's System, a square of an atom can have more than three divisors in the Three's System.
6 How many divisors does a product of two distinct primes have? Prove your answer.
7 Show that if $n|a$ and $n|b$, then for any integers r and s, $n|ar + bs$.

We return for a moment to the question of why the fundamental theorem holds for the positive integers but not for the Three's System. The real crux of the matter is that to prove this theorem it is necessary to go outside the multiplicative structure and use additive properties of whole numbers. There are many different proofs of this theorem; all make use of addition.

For further discussion of this point, the reader is referred to either Davenport or Stein [6, 17].

Section 2   *Prime Numbers and Factorization*

19   *Definition*

A number n greater than one is presented in *standard form* $n = p_1^{\alpha_1} p_2^{\alpha_2} \cdots p_k^{\alpha_k}$ if and only if $p_1, p_2, \ldots, p_k$ are distinct primes, $p_1 < p_2 < p_3 < \cdots < p_k$, and $1 \leq \alpha_i$, $i = 1, 2, \ldots, k$.

The fundamental theorem tells us that each number greater than one has a unique standard form presentation. Since we shall often deal with numbers in standard form it pays to develop some notation.

20   *Notation*

We denote by $q^\alpha \| n$ the phrase $q^\alpha | n$ but $q^{\alpha+1} \nmid n$.

For example, $2 \| 6$, $3 \| 6$, $5^2 \| 100$, $3^2 \| 18$, $2^4 \| 48$. Also the symbol p will always denote a prime henceforth.

21   *Notation*

We use the symbol Π (capital Greek π, for product) to denote products of things. Its use is best illustrated by example.

$$\prod_{i=1}^{k} p_i = p_1 p_2 \cdots p_k.$$

$$\prod_{i=1}^{k} p_i^{\alpha_i} = p_1^{\alpha_1} p_2^{\alpha_2} \cdots p_k^{\alpha_k}.$$

$\prod_{d|n} d$ = product of all divisors (not necessarily prime) of n.

$\prod_{d|10} d = 1 \cdot 2 \cdot 5 \cdot 10 = 100.$

$\prod_{d|60} d = 1 \cdot 2 \cdot 3 \cdot 4 \cdot 5 \cdot 6 \cdot 10 \cdot 12 \cdot 15 \cdot 20 \cdot 30 \cdot 60$
$= 46{,}656{,}000{,}000.$

$\prod_{p|n} p$ = product of all primes which divide n.

$\prod_{p|10} p = 2 \cdot 5 = 10.$

$\prod_{p|60} p = 2 \cdot 3 \cdot 5 = 30.$

$\prod_{p|n} p^2$ = the product of the squares of all primes which divide n.

$$\prod_{p|10} p^2 = 2^2 \cdot 5^2 = 100.$$

$$\prod_{p|60} p^2 = 2^2 \cdot 3^2 \cdot 5^2 = 900.$$

$$\prod_{p^\alpha \| n} p^\alpha = n.$$

22 EXERCISES

1 Write out explicitly

$$\prod_{p^\alpha \| 700} p^\alpha; \quad \prod_{p|35} p^3; \quad \prod_{p|165} (p^2 - 1); \quad \prod_{d|18} d; \quad \prod_{p^\alpha \| 60} (\alpha + 1).$$

2 What value of α makes each of the following true?
   (a) $5^\alpha \| 375$ (b) $2^\alpha \| 320$ (c) $7^\alpha \| 686$

23 Theorem

If $n = \prod_{i=1}^{k} p_i^{\alpha_i}$ in standard form then $m|n$ if and only if $m = \prod_{i=1}^{k} p_i^{\beta_i}$, for some choice of β_i satisfying $0 \leq \beta_i \leq \alpha_i$ for $i = 1, 2, \ldots, k$. In words, m is a divisor of n if and only if every prime factor of m is also a prime factor of n and occurs no more often in the factorization of m than it does in n.

PROOF: If $m = \prod_{i=1}^{k} p_i^{\beta_i}$ with $0 \leq \beta_i \leq \alpha_i$, then $\frac{n}{m} = q = \prod_{i=1}^{k} p_i^{\alpha_i - \beta_i}$. Since $\alpha_i - \beta_i \geq 0$, q is an integer, and by definition $m|n$. Conversely, if $m|n$, then the unique prime factorization of n can be obtained by taking all the prime factors of m and $q = \frac{n}{m}$ combined. Thus no primes except $p_1, \ldots, p_k$ can divide m, and further, none of them can occur more often (with higher exponent) in m than n.

24 Definition

Two numbers are called *relatively prime* to one another if they have no common prime factors.

Note that every number is relatively prime to 1, but no number except 1 is relatively prime to 0.

25 EXERCISES

1. Show that two numbers are relatively prime if and only if they have no common factor greater than 1.
2. Prove that the number of divisors of a number n is given by
$$\prod_{p^\alpha \| n} (\alpha + 1).$$
3. Show that if m and n are relatively prime, then every divisor d of mn can be written in the form $d = d'd''$ where $d'|m$ and $d''|n$. Conversely, show that if $d'|m$ and $d''|n$, then $d'd''$ is a divisor of mn, and that every divisor of mn is obtained uniquely in this manner.
4. Let $d(n) =$ the number of divisors of n. Prove that if n and m are relatively prime then $d(n \cdot m) = d(n) \cdot d(m)$.
5. Show that the first sentence of Problem 3 and the first part of the second sentence are true even if m and n are not relatively prime, but that the second part of the second sentence is false.

Section 3 Euclidean Perfect Numbers

We return now to our search for perfect numbers. The following theorem is found in Euclid.

26 Theorem

If N is a number of the form $N = 2^{k-1}(2^k - 1)$ where $2^k - 1$ is prime, then N is perfect.

27 *Definition*

A perfect number given by the rule in Theorem 26 will be called a *Euclidean perfect number*.

According to Nicomachus (c. 100 A.D.), Euclid's rule gives all the even perfect numbers (cf. discussion at the end of Section 1). Iamblichus (c. 283–330 A.D.) repeated Nichomachus's claim and added to it that there is exactly one perfect number with a given number of digits and that the perfect numbers end alternately in 6 and 8. This of course implies that they are all even. These remarks of the ancients were all based on the knowledge of the first four perfect numbers $6 = 2(2^2 - 1)$, $28 = 2^2(2^3 - 1)$, $496 = 2^4(2^5 - 1)$, and $8{,}128 = 2^6(2^7 - 1)$. It is not known to what extent they could prove any of these claims or whether, the claims were purely empirical. Euclid did have some fairly advanced number-theoretic techniques. However, they were greatly hindered by lack of place notation for numbers, lack of symbolic notation for the arithmetic operations, and lack of algebra. The reader who would like to gain some insight into these handicaps might try to factor a moderate-sized number, say $2^9 - 1$, by use of only Roman numerals (or Greek or Hebrew, if you know them). (Ruffus stated in 1521 that $2^9 - 1$ is prime. Regius stated in 1536 that it is composite.) Leonardo of Pisa (Fibonacci) in his work *Liber Abaci* (1202) was the first strong European advocate of Arabic notation (created and discovered by the Hindus, brought to Europe by the Arabs) for numerals. It was not until about 1275 that Arabic notation became widely used among scholars of Europe. The earliest known use of it on gravestones was in 1371. The earliest use on coins was in Switzerland in 1424. For more details, consult Cajori [2, 3, 4].

28 Example

Show that $496 = 2^4(2^5 - 1)$ is perfect.

Since we need the divisors of 496, we first verify that $2^5 - 1 = 31$ is prime by testing that it has no divisors except 1 and itself. According to Theorem 23, the divisors of $496 = 2^4 \cdot 31$ are:

1	2	2^2	2^3	2^4
31	$2 \cdot 31$	$2^2 \cdot 31$	$2^3 \cdot 31$	$2^4 \cdot 31$

We wish to add all the divisors except 496 and see if their sum is 496. Adding the first row we get $1 + 2 + 2^2 + 2^3 + 2^4 = 31$. Adding the second row except for the last term we get

$$31 + 2 \cdot 31 + 2^2 \cdot 31 + 2^3 \cdot 31 = 31(1 + 2 + 2^2 + 2^3) = 31 \cdot 15$$

Adding these two rows of sums we get

$$31 + 15 \cdot 31 = 16 \cdot 31 = 2^4 \cdot 31 = 2^4(2^5 - 1) = 496.$$

Thus 496 is perfect.

The reader would do well to verify that 8128 is also perfect before reading the proof of Theorem 26.

Section 3   Euclidean Perfect Numbers   95

Notice how it was necessary to be able to add sums of the form $1 + 2 + 2^2 + \cdots + 2^k$ in the verification that 496 and 8128 are perfect. We shall need to add such sums in order to prove Theorem 26. Thus we prove the next lemma and corollary before beginning the proof of Euclid's theorem.

29   Lemma

The following is an identity:

$$x^k - 1 = (x - 1)(x^{k-1} + x^{k-2} + \cdots + x + 1).$$

PROOF: Multiply it out.

30   Corollary

The sum $1 + 2 + 2^2 + 2^3 + \cdots + 2^{k-1}$ is equal to $2^k - 1$.

PROOF OF THEOREM 26:   Let $N = 2^{k-1}(2^k - 1)$, with $(2^k - 1)$ a prime. We know from Theorem 23 (and Problem 2 of Exercises 25) that the $2k - 1$ proper divisors of N are $1, 2, 2^2, \ldots, 2^{k-1}, 1 \cdot (2^k - 1), 2 \cdot (2^k - 1), \ldots, 2^{k-2}(2^k - 1)$. By Corollary 30 the sum of the first half of the divisors from 1 to 2^{k-1} is $2^k - 1$ and the sum of the remaining divisors is $(1 + 2 + 2^2 + \cdots + 2^{k-2})(2^k - 1) = (2^{k-1} - 1)(2^k - 1)$. Thus the sum of all the proper divisors is $(2^k - 1) + (2^{k-1} - 1)(2^k - 1) = (1 + 2^{k-1} - 1)(2^k - 1) = 2^{k-1} \cdot (2^k - 1) = N$. Thus N is perfect. Theorem 26 is proved.

31   EXERCISES

1 With all your advantages of algebra and place notation can you find another perfect number?
2 Show that if k is even and $k \neq 2$, then $2^k - 1$ is not a prime.
3 Use Lemma 29 to show that if $a^k - 1$ is a prime for some $k > 1$ then $a = 2$.
4 Prove Corollary 30 by induction on k without using Lemma 29.
5 Prove Lemma 29 by induction on k.
6 If $N = 2^{k-1}(2^k - 1)$ and $2^k - 1$ is not prime, then N is abundant (cf. Problem 3 of Exercises 9).

Problems 2 and 3 Exercises 31 may be combined and strengthened to yield:

32   Theorem

If a and k are positive whole numbers with $k > 1$ and $a^k - 1$ is prime, then $a = 2$ and k is a prime.

PROOF: If $a = 1$, then $a^k - 1 = 0$ and is clearly not prime, so $a \geq 2$. From Lemma 29 we get $a^k - 1 = (a-1)(a^{k-1} + \cdots + 1)$. If $a - 1 > 1$, that is, if $a > 2$, then the above factors of $a^k - 1$ are proper and $a^k - 1$ is not prime. It follows that if $a^k - 1$ is prime, then $a = 2$. Suppose now that k is composite, say $k = m \cdot n$, with both $m > 1$ and $n > 1$. Then $a^k - 1 = (a^m)^n - 1$. By Lemma 29 with $x = a^m$, $(a^m)^n - 1 = ((a^m) - 1)((a^m)^{n-1} + (a^m)^{n-2} + \cdots + a^m + 1)$. Since $a > 1$, and $m > 1$ we see that $a^m - 1 \geq a^2 - 1 \geq 4 - 1 = 3 > 1$. Since $n > 1$ we see that $((a^m)^{n-1} + \cdots + 1) \geq a^m + 1 \geq 2$. Thus $a^k - 1$ is factorable if k is not a prime. The theorem is proved.

Thus the search for Euclidean perfect numbers reduces to a search for primes of the form $2^p - 1$ where p is a prime.

Marin Mersenne (1588–1648) was a Franciscan friar and spent most of his life in Parisian cloisters. His importance to the scientific and mathematical world of his day was due not so much to his own work, as to his energy, friendly enthusiasm, and extensive correspondence with intellectuals. He was a clearing house and disseminator of new scientific ideas, problems, and techniques. He encouraged Richeleau to establish the French Academy of Science in 1635. (Courtesy of the Smith Collection, Columbia University.)

33 *Notation and Definition*

A number of the form $2^p - 1$ is called a *Mersenne number* and is denoted by M_p.

These numbers are named after Marin Mersenne (1588–1648), who was certainly not the first person to think about them, but who put forth great effort to find primes of that form and to get his colleagues (many of whom were excellent mathematicians) to find them.

While we cannot yet answer the question of whether or not the perfect numbers alternately end in 6 or 8 we can prove the following:

34 Theorem

Every Euclidean perfect number ends in 6 or 28 when the numbers are written in base 10 notation.

This theorem is not proved here, but it follows from Theorem 42. At first sight this theorem appears to be quite different from our previous theorems in that it expresses not a property of the numbers themselves, but rather a property of the names of the numbers in a particular notational system (numerical language). But let us reflect for a minute and see if the property is really dependent on our notational system or if it is only the way we expressed the property which depends on our notational system. After all, it would be nice if we could communicate with the men from Mars, who may have 12 fingers and hence use base-12 notation, or with the computers, which use base 2 and base 8, or with people like the Babylonians, who used base 60. (This is why we today have 60-second minutes, 60-minutes hours, 360-degree circles, etc.)

To say that a number ends in 6 in the base-10 notation is another way of saying it has remainder 6 when it is divided by 10. Similarly, a number ends in 28 if and only if it has a remainder of 28 when divided by 100.

Many times we wish to prove something about the remainder of one number upon division by another without caring about what the quotient is. Thus odd or even is remainder 1 or 0 upon division by 2. (Or on and off if it is a light switch; in which case we certainly don't want to know how many times a switch has been flipped, but only if it is an odd or even number of times.) To tell time we use remainder upon division by 12 (24 in the army and much of Europe).

For day-of-the-week problems we like to know remainders upon division by 7.

For the above reasons, because the author thinks it is a beautiful subject, because we shall find it convenient later on, and also because it is basic to a large part of number theory, we digress for a moment to consider the arithmetic of remainders, or *theory of congruences*, as it is called.

The terminology, symbolism, and systematic development of the theory of congruences can be unambiguously attributed to the genius of Karl Friedrich Gauss (1775–1855). See Ore [12] for more details.

Karl Friedrich Gauss (1777–1855) was the son of an uneducated bricklayer, and he might have been lost to the mathematical world if his genius had not been recognized when he was very young. While still in grade school, he discovered and proved the formula $1 + 2 + \cdots + n = n(n+1)/2$. His first international fame was won by calculating the orbit of Ceres and enabling astronomers to rediscover this lost asteroid. He is famous not only for his pioneering work in the theory of numbers, which won him universal recognition as one of the greatest mathematicians of all time, but also for his work in geometry, analysis, astronomy, electricity, and magnetism. (Permission of the Granger Collection.)

35 Definition

Two numbers a and b are said to be *congruent for modulus m* if and only if $m|a - b$. Symbolically, we write $a \equiv b \bmod m$ to denote the fact that a and b are congruent for the modulus m.

It is not just chance that the symbols for congruence and equality are similar. As we shall see, there is a very strong analogy between congruence and equality. It is impossible to overemphasize the importance of good suggestive notation in helping one guess what statement to attempt to prove and in helping to find the proofs.

The next theorem shows explicitly the connection between congruences and remainders.

36 Theorem

Two numbers are congruent for a given modulus m if and only if they have the same remainder upon division by m.

PROOF: Let a and b be the two numbers. First, we suppose they have the same remainder, say r. Thus $a = qm + r$, $b = q'm + r$. In this case, $a - b = m(q - q')$, which is obviously divisible by m. Thus $a \equiv b \bmod m$. This proves that two numbers with the same remainder are congruent. We now show the other half of the theorem.

Suppose $a \equiv b \bmod m$. Let r and r' be the remainders of a and b, respectively, upon division by m. Thus $a = qm + r$ and $b = q'm + r'$, where $0 \leq r < m$ and $0 \leq r' < m$. Either $r \geq r'$ or $r' \geq r$. We argue the case where $r \geq r'$, and we will show that in this case we have $r = r'$. Since $a \equiv b \bmod m$, we know $m|a - b$. Now, $a - b = m(q - q') + r - r'$. Since $m|a - b$ and $m|m(q - q')$, by Theorem 11, it divides the difference, namely, $m|r - r'$. But $m > r \geq r - r' \geq 0$, and hence the only way m can divide $r - r'$ is if $r - r' = 0$. Similarly if $r' \geq r$ we can show that $r' - r = 0$. Thus $r = r'$ and a and b have the same remainders upon division by m. The proof is completed.

It is now easy to see that congruence has the following three properties (which equality also has).

37 Corollary

(i) For every number a, $a \equiv a \bmod m$.
(ii) For every two numbers a and b, $a \equiv b \bmod m$ if and only if $b \equiv a \bmod m$.

(iii) For any three numbers a, b, and c, if $a \equiv b$ and $b \equiv c$ mod m, then $a \equiv c$ mod m. (Things congruent to the same thing are congruent to each other.)

PROOF: All these statements follow since congruence is equality of remainders.

38 EXERCISES

1 Supply the part of the proof of Theorem 34 when $r' \geq r$.
2 Supply the proof of part 3 of Corollary 37.
3 Find four numbers congruent to 4 mod 7.
4 Find five numbers congruent to 2 mod 9. For each of these numbers, find the sum of its digits. If the sum has more than one digit, add its digits. Continue until you reach a one-digit number. Do you see a pattern? What is it? Can you prove it will always work?
5 Prove that a number is congruent to its last two digits for the modulus 4. Thus, 4 divides a number if and only if 4 divides the last two digits of the number.
6 Prove that $N \equiv r$ mod m if and only if $N = qm + r$ for some integer q.

The next theorem expresses the arithmetic theory of the addition, subtraction, and multiplication of congruences.

39 Theorem

If a, a', b, b', and m are integers such that $a \equiv a'$ mod m and $b \equiv b'$ mod m, then

(i) $a + b \equiv a' + b'$ mod m;
(ii) $a - b \equiv a' - b'$ mod m;
(iii) $a \cdot b \equiv a' \cdot b'$ mod m.

PROOF: Let a and a' have remainder r upon division by m, and let b and b' have remainder s. Then $a = qm + r$, $a' = q'm + r$, $b = tm + s$, and $b' = t'm + s$. Thus $a + b = qm + r + tm + s = m(q + t) + (r + s)$. It is now clear that $a + b \equiv r + s$ mod m. Putting primes on a, b, q, and t, we obtain $a' + b' \equiv r + s$ mod m. Thus $a + b \equiv a' + b'$ mod m. Replacing the plus sign by a minus sign in the above arguments, we obtain the second part of the theorem. A similar argument proves the third part.

Putting this theorem in language of more classic style, one might say, "If congruents be added to congruents their sums are congruents," etc.

We know that $a + b \equiv b + a$ mod m, since $a + b = b + a$ and thus have the same remainder. Also $(a + b) + c \equiv a + (b + c)$ mod m since they both have the same remainder, namely, the remainder of $a + b + c$.

40 Theorem

If p is a prime, then for any numbers a and b, $ab \equiv 0$ mod p if and only if $a \equiv 0$ mod p or $b \equiv 0$ mod p.

This theorem is merely a restatement of Theorem 17 in the present notation.

We could continue developing the analogy between congruence and equality, however our general aims and scope preclude this.

41 EXERCISES

1. Prove part (iii) of Theorem 39.
2. Prove Theorem 39 directly from the definition without using remainders.
3. Derive Theorem 36 from Theorem 39.
4. Prove that if $a + b \equiv a + c$ mod m, then $b \equiv c$ mod m.
5. Prove that if p is a prime, $a \not\equiv 0$ mod p, and $a \cdot b \equiv a \cdot c$ mod p, then $b \equiv c$ mod p.
6. Show that Problem 5 need not be true if p is replaced by a composite modulus.
7. Show that if k is odd, then either $k \equiv 1$ mod 4 or $k \equiv 3$ mod 4.

We restate and strengthen Theorem 34 using our new terminology.

42 Theorem

If $N = 2^{k-1}(2^k - 1)$ is a Euclidean perfect number, then $k = 2$ or k is odd, and either $N = 6$, $N \equiv 6$ mod 10, or $N \equiv 28$ mod 100, according as $k = 2$, $k \equiv 1$ mod 4, or $k \equiv 3$ mod 4, respectively.

PROOF: Let $N = 2^{k-1}(2^k - 1)$. From Theorem 32 we know that either $k = 2$ or k is odd. If k is 2, $N = 6$. We suppose now that k is odd; hence $k \equiv 1$ mod 4 or $k \equiv 3$ mod 4. We consider the even powers of 2 mod 10. Now, $2^4 \equiv 6$ mod 10 and $6^2 = 36 \equiv 6$ mod 10. Thus $6^n \equiv 6$ mod 10. It follows that $2^{4n} = (2^4)^n \equiv 6^n \equiv 6$ mod 10.

Suppose $k \equiv 1$ mod 4, then $k - 1 = 4n$, so $2^{k-1} = 2^{4n} \equiv 6$ mod 10, and $2^k - 1 \equiv 6 \cdot 2 - 1 \equiv 1$ mod 10. Thus $N = 2^{k-1}(2^k - 1) \equiv 6 \cdot 1 \equiv 6$ mod 10.

Suppose $k \equiv 3 \mod 4$, then $k - 1 = 4n + 2$, so $2^{k-1} \equiv 2^{4n+2} \equiv 2^{4n} \cdot 2^2 \equiv 6 \cdot 4 \equiv 4 \mod 10$. Also, $4|2^{k-1}$, and hence 4 divides the last two digits of 2^{k-1} (see Problem 5 of Exercises 38). It follows that the last digit of 2^{k-1} is 4 and also that 4 divides the last two digits. Since a number is congruent to its last two digits modulo 100, we get $2^{k-1} \equiv 4, 24, 44, 64,$ or $84 \mod 100$ and $2^k - 1 \equiv 2 \cdot 2^{k-1} - 1 \equiv 7, 47, 87, 27,$ or 67, respectively. Thus, $N = 2^{k-1}(2^k - 1) \equiv 4 \cdot 7, 24 \cdot 47, 44 \cdot 87, 64 \cdot 27,$ or $84 \cdot 67 \mod 100$. If we carry out the multiplication, we obtain in each case $N \equiv 28 \mod 100$, which establishes the theorem.

43 EXERCISES

1 Prove that if $N = 2^{p-1}(2^p - 1)$ is a Euclidean perfect number, and $N \neq 6$, then $N \equiv 1 \mod 3$.

2 Deduce from Problem 1 that if $N \neq 6$ then $N \equiv 4 \mod 6$.

3 Let $N = 2^{p-1}(2^p - 1)$ be a Euclidean perfect number with $p \neq 3$. Prove that $N \equiv 1 \mod 7$ or $N \equiv 6 \mod 7$ according as $p \equiv 1 \mod 3$ or $p \equiv 2 \mod 3$, respectively.

4 Figure out a similar statement for modulus 9. Prove it.

5 Show that $4^k \equiv 4 \mod 6$, $6^k \equiv 6 \mod 10$, $8^k \equiv 8 \mod 14$, and $10^k \equiv 10 \mod 18$ for any positive integral value of k.

6 Show that for odd numbers n, $(n + 1)^k \equiv n + 1 \mod 2n$ for any positive integral value of k.

Section 4 Primes and Their Distribution

We return to our search for perfect numbers. The best path open to us at the moment is to try and discover if there are more, hopefully infinitely many, primes of the form $2^p - 1$, that is, Mersenne primes. Before searching too far, it is sensible to ask what the prime numbers look like. Do they occur regularly, do they go on forever, or is there a largest prime, etc.?

One of the first significant tables was due to Rohnius in 1659, who gave factors of the numbers up to 24,000. The next major advance on this front, with much encouragement from the mathematical world, especially Lambert, was due to Felkel, a Viennese schoolmaster. Felkel, in 1776, computed factors of the numbers up to 408,000. His work was published by the Imperial Treasury of Austria, but as times got tough, and it was not a best seller, most of the copies were confiscated and the paper was used to make artillery cartridges for the war against the Turks. A truly colossal effort in this direction was done by Kulik (1773–1863), a mathematics professor at the University of Prague who produced a table (unpublished) up to 100,000,000. The best available table is due to D. N. Lehmer (1867–1938) of the U.S. Bureau of Standards, who published a factor table (with amazingly few errors) up to 10,006,721. The experimental evidence of these tables indicates that there is very little individual regularity among the primes but that there is no end to them. In fact, there will never be an end to them, as can be discovered by reading Euclid, who proved:

44 Theorem

There is an infinity of prime numbers.

PROOF: We shall show that there are infinitely many primes by showing that no finite list can contain all the primes. Let $p_1, p_2, \ldots, p_n$ be any list of primes. (Euclid took the first n.) Let $P = \prod_{i=1}^{n} p_i$. Consider $P + 1$; we have $P + 1 \equiv 1 \bmod p_i$, for $i = 1, 2, \ldots, n$. Thus, none of the p_i divide $P + 1$; hence, its prime factors are not in the list, which therefore is incomplete. The theorem is established.

We quote Rademacher and Toeplitz, who say [13, pp. 10–11]:

This part of Euclid is quite remarkable, and it would be hard to name its most admirable feature. The problem is only of theoretical interest. It can be proposed, for its own sake, only by a person who has a certain inner feeling for mathematical thought. This feeling for mathematics and appreciation of the beauty of mathematics was very evident in the ancient Greeks, and they have handed it down to later civilizations. Also, this problem is one that most people would completely overlook. Even when it is brought to our attention it appears to be trivial and superfluous, and its real difficulties are not immediately apparent. Finally, we must admire the ingenious and simple way in which Euclid proved the theorem. The most natural way to prove the theorem is not Euclid's. It would be more natural to try to find the next prime number following any given prime. This has been attempted but has always ended in failure because of the extreme irregularity of the formation of the primes.

Euclid's proof circumvents the lack of a law of formation for the sequence of primes by looking for *some* prime beyond instead of for the *next* prime after p. For example, his proof gives 2,311, not 13, as a prime past 11, and it gives 59 as one past 13. Frequently there are a great many primes between the one considered and the one given by the proof. This is not a sign of the weakness of the proof, but rather it is evidence of the ingenuity of the Greeks in that they did not try to do more than was required.

There is now some hope that there may be an infinity of Mersenne primes. Notice that $M_p = 2^p - 1 \equiv 4 - 1 \equiv 3 \bmod 4$. Is there an infinity of primes of this type? A Euclidean type argument answers this question.

45 Theorem

There is an infinity of primes p satisfying $p \equiv 3 \bmod 4$.

We shall need the following lemma.

46 Lemma

If $q_i \equiv 1 \bmod 4$ for $i = 1, 2, \ldots, n$, then $Q = \prod q_i \equiv 1 \bmod 4$.

The proof of the lemma is left as an exercise.

PROOF OF THEOREM 45: Let $p_1, p_2, \ldots, p_n$ be a list of primes satisfying $p_i \equiv 3 \bmod 4$, $i = 1, 2, \ldots, n$. Let $P = \prod_{i=1}^{n} p_i$. On the one hand, $4P - 1 \equiv -1 \bmod p_i$, so $p_i \nmid 4P - 1$, $i = 1, 2, \ldots, n$. On the other hand, $4P - 1 \equiv 3 \bmod 4$, and hence by Lemma 46, $4P - 1$ must have a prime factor p with $p \not\equiv 1 \bmod 4$. But clearly p cannot be even since $4P - 1$ is odd; thus $p \not\equiv 0, 2 \bmod 4$. Consequently, $p \equiv 3 \bmod 4$. Thus, the list $p_1, \ldots, p_n$ cannot contain all primes satisfying $p \equiv 3 \bmod 4$. The theorem is proved.

The first three such primes are 3, 7, and 11. By the above proof we know that $4 \cdot 3 \cdot 7 \cdot 11 - 1 = 923$ must have a new prime factor $p \equiv 3 \bmod 4$. In fact, $923 = 13 \cdot 71$ and $71 \equiv 3 \bmod 4$. We could now look at $4 \cdot 3 \cdot 7 \cdot 11 \cdot 71 - 1$ and get another, etc.

Since we have seen that there are infinitely many primes $p \equiv 3 \bmod 4$, we next inquire as to whether or not there are also infinitely many primes $p \equiv 1 \bmod 4$. The answer is affirmative.

47 Theorem

There are infinitely many primes $p \equiv 1 \bmod 4$.

Section 4   *Primes and Their Distribution*

We shall give this proof later, after Lemma 70 and some corollaries are proved.

It is possible, but difficult and beyond our reach in this short chapter, to prove that if a and b are two relatively prime numbers then there are an infinite number of primes $p \equiv a \bmod b$. This was first proved by Dirichlet in 1837.

48   EXERCISES

1 Prove Lemma 46.
2 Using Euclid's process find two more primes.
3 Using Euclid's process find two more primes which are congruent to 3 mod 4.
4 Verify that 2311 is the prime obtained after 11 by Euclid's method.
5 Prove that there are infinitely many primes $p \equiv 5 \bmod 6$.

While we have now settled the problem of whether or not there are infinitely many primes, we have not answered three questions which are both theoretically and practically important. The first: How far out in the number sequence is it necessary to go to find the 100th, or 1,000,000th, or, in general, the nth prime number? The second: Is there a method for finding all the primes less than 1,000 or 1,000,000 or, in general, some upper bound N? The third: Given a particular number, say, $2^{61} - 1$ or $2^{89} - 1$ or any other large number, how can we determine if it is prime? Let us discuss each question in turn.

Regarding the first, a second look at Euclid's proof of the infinitude of primes shows that if $p_1, p_2, \ldots, p_n$ are the first n primes, then there is a new prime, p, which divides $p_1 \cdot p_2 \cdots p_n + 1$. This observation enables us to prove:

49   Theorem

If p_n is the nth prime number, then $p_n \leq 2^{2^{(n-1)}}$.

PROOF: Note: $p_1 = 2 = 2^{2^0}$. We suppose inductively that the theorem is true for $p_1, p_2, \ldots, p_n$. We shall show that it is also true for p_{n+1}. By Euclid's argument, there is a prime p such that p is distinct from $p_1, p_2, \ldots, p_n$, and such that $p | p_1 p_2 \cdots p_n + 1$. Either p is the $(n+1)$st prime or the $(n+1)$st prime is between p_n and p. Thus, $p_{n+1} \leq p$ and $p \leq p_1 p_2 \cdots p_n + 1$, so we see that $p_{n+1} \leq p_1 p_2 \cdots p_n + 1$. By the

induction hypothesis
$$p_1 p_2 \cdots p_n + 1 \leq 2^{2^0} \cdot 2^{2^1} \cdots 2^{2^{n-1}} + 1 = 2^{(2^n-1)} + 1$$
$$\leq 2^{(2^n-1)} + 2^{(2^n-1)} = 2^{2^n}.$$

The theorem is proved.

Standing this result on its head we can say:

50 Corollary

The number of primes less than 2^{2^n} is at least $n + 1$.

PROOF: From the above theorem we see that $p_1, p_2, \ldots, p_{n+1} \leq 2^{2^n}$.

51 Corollary

If $k \geq 2$, the number of primes less than k is at least $\log_2 \log_2 (k)$.

We note that for $k = 2^{2^n}$, $\log_2 \log_2 k = n$ and this corollary is true by Corollary 50. If k is not of the form 2^{2^n} we shall work with the closest numbers which are of that form.

PROOF: There is an integer $n \geq 0$ such that $2^{2^n} \leq k < 2^{2^{n+1}}$. Thus, $n = \log_2 \log_2 2^{2^n} \leq \log_2 \log_2 k < \log_2 \log_2 2^{2^{n+1}} = n + 1$. Thus, $\log_2 \log_2 k < n + 1$, but there are $n + 1$ primes less than 2^{2^n}, whence less than k, which proves the corollary.

We denote by $\pi(k)$ the number of primes less than or equal to k. In this notation the two corollaries become $\pi(2^{2^n}) \geq n + 1$ and $\pi(k) > \log_2 \log_2 k$. It is possible to show that the ratio of $\pi(k)$ to $\frac{k}{\ln k}$ is approximately 1 for k large, and becomes closer and closer to one as k gets larger and larger, where $\ln k$ is the natural logarithm of k. While variants of this statement, known as the Prime Number Theorem, were made as early as 1796 by Legendre, it was not until the end of the last century that a proof was found by use of the calculus applied to the field of complex numbers. It is not unusual that in order to prove theorems about whole numbers it is necessary to go very far afield.

Even though the ratio of $\pi(k)$ and $\frac{k}{\ln k}$ becomes closer and closer to 1, the difference $\pi(k) - \frac{k}{\ln k}$, becomes larger and larger. These remarks also hold true for $\pi(k)$ and $\frac{k}{[(\ln k) - 1]}$, although their difference is smaller.

Section 4   *Primes and Their Distribution*   107

The chart below [10, 12, pp. 69 and 77] gives a comparison of $\pi(k)$, $\frac{k}{(\ln k) - 1}$, and $\int_2^k \frac{dx}{\ln x}$. The integral symbol $\int_2^k \frac{dx}{\ln x}$ denotes the area under the graph of the curve $y = \frac{1}{\ln x}$ between $x = 2$ and $x = k$. All but the last two values of $\pi(k)$ were obtained by actual count. The last three values were computed by Meisell, N. P. Bertelsen, and D. H. Lehmer [10] using recursive formulas for $\pi(k)$ in terms of its value for smaller numbers. The values of $\int_2^k \frac{dx}{\ln x}$ are accurate, at least through the first eight digits (Table 1).

Table 1

k	$\pi(k)$	$\frac{k}{(\ln k) - 1}$	$\int_2^k \frac{dx}{\ln x}$
1,000	168	169	178
10,000	1,229	1,218	1,246
100,000	9,592	9,512	9,630
1,000,000	78,498	78,030	78,628
10,000,000	664,579	661,459	664,918
100,000,000	5,761,455	5,740,380	5,762,209
1,000,000,000	50,847,478	50,701,542	50,849,235
10,000,000,000	455,052,511	454,011,971	455,055,613

Regarding the second question, on the method of listing the primes, we have made very little theoretical advance since the Greeks; however, with the advent of electronic computing machines, we have made colossal practical advances. We can carry their methods much farther and much faster than they could have hoped. The basic theoretic technique for a systematic computation of all primes below a given limit, say N, is the sieve (or crib) of Eratosthenes (276–194 B.C.).

Before describing the sieve of Eratosthenes we need the following lemma.

52   Lemma

If N is composite, then N has a prime factor p such that $p \leq \sqrt{N}$.

PROOF: Since N is composite it may be written as a product $N = r \cdot s$, where $1 < r < N$ and $1 < s < N$. One of r and s is less than or equal to $\sqrt{N}$, for if not $r > \sqrt{N}$ and $s > \sqrt{N}$ whence $N = r \cdot s > \sqrt{N} \cdot \sqrt{N} = N$. But $N > N$ is clearly a contradiction. Thus one of r and s is less than $\sqrt{N}$. We suppose $r < \sqrt{N}$. Since $r \neq 1$, r has at least one prime factor. Let p be any prime factor of r. Then $p \leq r$ whence $p \leq \sqrt{N}$. Further, $p|N$ since $p|r$ and $r|N$. The lemma is proved.

In order to construct the sieve of Eratosthenes up to N, we first make a list of all the numbers from 2 to N. Since 2 is the first prime we begin by striking out all multiples of two, except 2 itself. The next number which remains, 3, is a prime, but no other multiple of 3 can be prime, so we cross out 3^2 and every third number counting those already crossed out. (Why can we begin with 3^2?) The next number remaining is the next prime, namely, 5; we now repeat the process crossing out 5^2 and the higher multiples of 5, every fifth number counting those already crossed out, and so on, until we reach the point where the next number which is not crossed out is greater than $\sqrt{N}$. The numbers which remain are the primes. This is so since, according to Lemma 52, every composite number which is less than p^2 is also a multiple of a prime less than p; hence any composite will have been crossed out as a multiple of its smallest prime factor.

In practice one crosses off the multiples of 2 by not writing them.

53 Examples of the Sieve of Eratosthenes up to 100

The multiples of 3 are crossed out by—; the multiples of 5 are crossed out by /; the multiples of 7 are crossed out by \. Since $11^2 > 100$ we stop after crossing out the multiples of 3, 5, and 7.

2, 3, 5, 7, 9̶, 11, 13, 1̶5̶, 17, 19, 2̶1̶, 23, 2̶5̶, 2̶7̶, 29, 31, 3̶3̶, 3̶5̶, 37, 3̶9̶, 41, 43, 4̶5̶, 47, 4̶9̶, 5̶1̶, 53, 5̶5̶, 5̶7̶, 59, 61, 6̶3̶, 65, 67, 6̶9̶, 71, 73, 7̶5̶, 7̶7̶, 79, 8̶1̶, 83, 8̶5̶, 8̶7̶, 89, 9̶1̶, 9̶3̶, 95, 97, 9̶9̶

54 EXERCISES

1 Find all the primes less than 200.

2 Compute $\pi(100)$ and $\dfrac{100}{(\ln 100) - 1}$.

3 Compute $\pi(200)$ and $\dfrac{200}{(\ln 200) - 1}$.

Section 5 Factorization Techniques

This photograph, taken in 1933, shows from left to right, Dr. R. C. Burt of the Burt Scientific Laboratories, Pasadena, Cal., Dr. Derrick Norman Lehmer, then Professor of Mathematics at the University of California, and Derrick Henry Lehmer, then National Research Fellow in Mathematics and currently Professor of Mathematics at the University of California.

An end view of the congruence machine showing one series of different sized gears with holes in the rim under each cog. Some are stopped up and some are open. At the right are two small prisms which reflect the light after it passes through one series of gears. It is then sent back again up the series on the other side.

(The following passage and all the photographs are from an article that appeared in the March 12, 1933 issue of the Carnegie Institute of Washington, News Service Bulletin, School Edition, Vol. III, Number 3.)

"This machine, constructed at the laboratories of the Robert Burt Company, by the inventor, D. H. Lehmer, is the first attempt ever made to apply the magic of the photoelectric cell to this problem of the study of remote numbers.

It is proposed to use it, first of all, in clearing up certain outstanding factorizations of numbers of the form of $2^n \pm 1$ and of $10^n \pm 1$, numbers which have so far baffled the efforts of mathematicians."

Section 5 Factorization Techniques

The third question, on methods of factorization of a large number, has a vast and varied literature, especially if one includes methods which prove the existence of a factor without necessarily finding it.

Given a number N, we can find out if it has any nontrivial (that is, neither 1 nor N) factors by dividing it by all the primes between 2 and $\sqrt{N}$, as we saw from Lemma 52. This can be an arduous process and presupposes a list of primes up to $\sqrt{N}$ which may not exist if N is really large.

The first real improvement in the method of finding factors was due to Pierre de Fermat (1601–1665), a jurist, parliamentarian, and amateur mathematician. He was the outstanding mathematician of his day and is often called the father of modern number theory. The following theorem and technique are derived from an undated letter he wrote probably to Mersenne in 1643.

Pierre de Fermat's life (1601–1665) centered around Toulouse, where he became a prominent lawyer. His scientific and mathematical achievements were many and varied, but his great love was number theory. Although he published his results, he was uncommunicative about his methods. Most of what little we know about his work has been found in his notes and letters. He once wrote in the margin of a book that he had discovered a most remarkable proof of the fact that $x^n + y^n = z^n$ has no solution in positive whole numbers if $n > 2$. To this day, no one has been able to prove or disprove this statement, now known as Fermat's last theorem. The many attempts to do so have probably stimulated more mathematical discoveries than any other single problem! (Courtesy of the Smith Collection, Columbia University.)

55 Theorem

An odd number can be written as a product of two factors in exactly as many ways as it can be expressed as the difference of two squares of nonnegative numbers.

PROOF: Let N be an odd number. Suppose it can be factored as $N = q \cdot r$, with $q \geq r$. Since N is odd both q and r are odd so that $\frac{q+r}{2}$ and $\frac{q-r}{2}$ are nonnegative whole numbers. Furthermore, $N = q \cdot r = \left(\frac{q+r}{2}\right)^2 - \left(\frac{q-r}{2}\right)^2$. It is also easy to see that distinct factorization gave distinct values for $q + r$ and $q - r$. For if $q' + r' = q + r$ and $q' - r' = q - r$, then adding, we get $2q' = 2q$. Whence $q = q'$ and $r = r'$.

Now suppose $N = k^2 - l^2$; then $N = (k + l)(k - l)$ is a factorization of N. Again distinct values of k and l lead to distinct values of the factors.

Let us return to the problem of perfect numbers to find a good candidate on which to apply this method. We know that to find a Euclidean perfect number we need only find a Mersenne prime $M_n = 2^n - 1$. We also saw that we need only consider the case in which n is a prime. The Mersenne primes corresponding to the perfect numbers 6, 28, 496, and 8128 known to antiquity come from $n = 2, 3, 5, 7$. The next likely candidate is $n = 11$.

Before applying this method to M_{11}, we note that for any number N, $N = N \cdot 1$ leads us to $N = \left(\frac{N+1}{2}\right)^2 - \left(\frac{N-1}{2}\right)^2$ that is, $k = \frac{N+1}{2}$. If $k > \frac{N+1}{2}$ there is no way of writing $N = k^2 - l^2$ since in that case

$$k^2 - l^2 \geq k^2 - (k-1)^2 = 2k - 1 > 2\left(\frac{N+1}{2}\right) - 1 = N.$$

On the other hand, $k^2 \geq N$, whence $k \geq \sqrt{N}$. Thus, we should consider k in the range $\sqrt{N} \leq k < \frac{N+1}{2}$ and see if $k^2 - N$ is a square. The search would be facilitated if we could tell easily from looking at a number whether or not it is a perfect square. To this end we prove:

56 Theorem

If N is a perfect square, then the last two digits of N are one of the following combinations: 00, 01, 04, 09, 16, 21, 24, 25, 29, 36, 41, 44, 49, 56, 61, 64, 69, 76, 81, 84, 89, 96.

PROOF: We first note that every number is congruent to its last two digits modulo 100. Thus, if N is any number with last digits a_1a_0, then $N \equiv a_1a_0$ mod 100 so $N^2 \equiv (a_1a_0)^2$ mod 100. It is therefore sufficient to look at the squares of the number between $0 = 00$ and 99 inclusive. Also $x^2 \equiv (50 + x)^2$ mod 100 since $(50 + x)^2 = 2{,}500 + 100x + x^2 \equiv x^2$ mod 100. Thus it suffices to look at the squares of the first fifty numbers. Next, we note that $x^2 \equiv (50 - x)$ mod 100 since $(50 - x)^2 = 2500 - 100x + x^2 \equiv x^2$ mod 100. Thus, it suffices to consider the final digits of x^2 for $x = 0, 1, 2, \ldots, 25$. One could reduce further, but at this point it is probably just as easy to compute these 26 squares and see that the last digits are as stated. The student should make this computation (or look in a table) to finish the proof.

57 EXERCISES

1 Prove that if $a \equiv b$ mod c, then $a^2 \equiv b^2$ mod c.
2 Is it true that if $a^2 \equiv b^2$ mod c, then $a \equiv \pm b$ mod c. Prove, if true; give a counterexample if false.
3 For what values of c is it true that $a^2 \equiv b^2$ mod c implies $a \equiv \pm b$ mod c?
4 Show that $a \equiv b$ mod 100 if and only if a and b have the same last two digits.

One final remark before we test $M_{11} = 2^{11} - 1$: we shall use the identity $(k + 1)^2 - N = k^2 - N + 2k + 1$.

Our search for factors of N will proceed as follows: we shall take the smallest possible value of $k \geq \sqrt{N}$, then look successively at $k^2 - N$, $(k + 1)^2 - N$, ... until we find a value $m \geq \sqrt{N}$ such that $m^2 - N$ is a square. If this occurs with $m < \dfrac{N + 1}{2}$, then we shall know N is not prime and we shall also have a factorization; if this does not occur, then we shall know that N is a prime.

We return to $N = 2^{11} - 1 = 2047$. Either by tables, by the square root algorithm, or by noting that $\sqrt{20} \approx 4.5$ we find $45^2 < 2047 < 46^2$. Thus, we consider values of k in the range $46 \leq k \leq \dfrac{2048}{2} = 1024$. All congruences below are modulo 100.

$46^2 - N = 69$
not a square.

$47^2 - N = 46^2 - N + 2 \cdot 46 + 1 = 69 + 92 + 1 = 162 \equiv 62$
not a square by Theorem 50.

$48^2 - N = 162 + 2 \cdot 47 + 1 = 257 \equiv 57$
not a square by Theorem 50.

$49^2 - N = 257 + 2 \cdot 48 + 1 = 354 \equiv 54$
not a square by Theorem 50.

$50^2 - N = 354 + 2 \cdot 49 + 1 = 453 \equiv 53$
not a square by Theorem 50.

$51^2 - N = 453 + 2 \cdot 50 + 1 = 554 \equiv 54$
not a square by Theorem 50.

$52^2 - N = 554 + 2 \cdot 51 + 1 = 657 \equiv 57$
not a square by Theorem 50.

$53^2 - N = 657 + 2 \cdot 52 + 1 = 762 \equiv 62$
not a square by Theorem 50.

$54^2 - N = 762 + 2 \cdot 53 + 1 = 869 \equiv 69$
not a square by computation or Problem 4 of Exercises 60.

$55^2 - N = 869 + 2 \cdot 54 + 1 = 978 \equiv 78$
not a square by Theorem 50.

$56^2 - N = 978 + 2 \cdot 53 + 1 = 1089 = (33)^2$
a square.

Thus, $N = 56^2 - 33^2 = (56 + 33)(56 - 33) = 89 \cdot 23$. $M_{11} = 2047 = 89 \cdot 23$. The reader will note that if he had chosen to divide by all the primes less than $\sqrt{N}$, that is, less than 45 to see if any divided N, he would have had to perform several long divisions to discover that 23 was a factor. While divisibility by 2, 3, 5, is easily tested, division by 7, 11, and 13 is a little more difficult to test and by 17, 19, and 23 is best done by the division algorithm (long division).

We note that despite great arithmetic effort the number M_{11} was thought to be prime and hence $2^{10} \cdot M_{11}$ thought to be perfect up until the time of Fermat's factorization. In fact, Bovillus (1470–1553) along with many others claimed that both $2^9 - 1 = 512 - 1 = 511 = 7 \cdot 73$ and $2^{11} - 1 = 2047 = 89 \cdot 23$ are prime. The first recorded statement (known to the authors) in which this is corrected appears in book by Hudalrichus Regius, *Utriusque Arithmetices* (Strasburg, 1536), in which he correctly gives (without proof) the fifth perfect number as $33{,}550{,}336 = 2^{12} \cdot (2^{13} - 1)$.

Thus by use of Fermat's method we can solve in half a page a problem which skilled mathematicians as late as 1603 could not solve. In that year Cataldi published a table of all primes less than 5150. His method of testing primality was precisely that of dividing N by all primes not exceeding $\sqrt{N}$. He *proved* that the fifth and sixth Euclidean perfect numbers are 33,550,336 and 8,589,869,056 arising from $M_{13} = 8191$ and $M_{17} = 131,071$. He *claimed* that M_p is prime for $p = 2, 3, 5, 7, 13, 17, 19, 23, 29, 31$, and 37; as we shall see, not all his claims are valid.

We can see from the fifth and sixth perfect numbers, that the Euclidean perfect numbers do not alternately end in 6 and 8, and also that there is not one with every number of digits. In fact, considering the size of the fifth Euclidean perfect number it is not surprising that the ancients had not found it.

58 EXERCISES

1 Show that if N is odd and $3|N$ then $N = \left(\dfrac{N+9}{6}\right)^2 - \left(\dfrac{N-9}{6}\right)^2$, where $\dfrac{N \pm 9}{6}$ is an integer.

2 Deduce from Problem 1 that if N is odd and not prime then $N = k^2 - l^2$ for some choice of k with $\sqrt{N} \leq k \leq \dfrac{N+9}{6}$.

3 Show that if $2 \nmid N$, $3 \nmid N$, $5 \nmid N$, $7 \nmid N$, and $N \neq 11$, then N is composite if and only if $N = k^2 - l^2$ for some k, where $\sqrt{N} \leq k \leq \dfrac{N + 121}{22}$.

4 Fermat stated his method in connection with the factorization of 2,027,651,281. Show that this number is not prime.

5 Is 10,122,197 a prime?

6 Is 125,249 a prime?

In using Fermat's method we could easily see that many values of $k^2 - N$ could not be squares by noting which congruence types modulo 100 could be squares. It is possible to use this type of test for other moduli. Of course, if the test is to be practical, the modulus must be chosen so that it is easy to determine the congruence type of large numbers with respect to the given modulus. Thus for a number written in base 10 notation, it is easy to find its congruence type

Section 5 Factorization Techniques

modulo 100, since this is given by the last two digits. The modulus 9 also works nicely for base 10 notation. We begin with a theorem many of you have seen in a different form.

59 Theorem (Casting Out Nines)
If a number is written in its base 10 notation, then it is congruent to the sum of its digits modulo 9. In particular, 9 divides a number if and only if it divides the sum of its digits.

PROOF: Let $a_k a_{k-1} \cdots a_2 a_1 a_0$ be the base-10 notation for N; that is, $N = a_k 10^k + a_{k-1} 10^{k-1} + \cdots + a_1 10 + a_0$, where $0 \leq a_i \leq 9$.
We note that $10 \equiv 1 \bmod 9$, whence for every integer n, $10^n \equiv 1 \bmod 9$. Thus, $N = a_k 10^k + a_{k-1} 10^{k-1} + \cdots + a_1 \cdot 10 + a_0 \equiv a_k + a_{k-1} + \cdots + a_1 + a_0 \bmod 9$. The theorem is established.

60 EXERCISES
1 Prove that $N \equiv a_0 - a_1 + a_2 - \cdots + (-1)^k a_k \bmod 11$. Whence $11 | N$ if and only if 11 divides the alternating sum of the digits.
2 Use the fact that $7 \cdot 11 \cdot 13 = 1{,}001$ to devise a method for finding the remainder of N upon dividing by 7, 11, or 13.
3 Prove that if N is a perfect square, $N \equiv 0, 1, 4, 7 \bmod 9$.
4 Use Problem 3 to show that 869 is not a perfect square. See our proof that $2^{11} - 1$ is not prime.
5 What congruence types can be perfect squares for the modulus 11?
6 Prove that N is congruent to the sum of its digits modulo 3.
7 Analyze the puzzle about the sum of digits being 9 which is given in the first paragraph of this chapter.

The study, initiated by Gauss, of which numbers can be squares for a given modulus has led to a great many remarkable and interesting theorems. This theory is called the *theory of quadratic residues* and is developed in any of the number theory books in the bibliography.
Soon we shall prove some other divisibility tests, but first we shall need to do some groundwork in the theory of congruences.

61 *Definition*
A collection of numbers $a_1, a_2, \ldots, a_n$ is called a *complete residue system* for the modulus m if every integer is congruent to one and only one of the numbers $a_1, a_2, \ldots, a_n$.

We recall that given two positive numbers N and d, the division algorithm (long-division process) yields a quotient q and a remainder r upon division of N by d, for which we have $N = qd + r$ and $0 \leq r < d$.

62 Theorem

If $m > 1$, then the numbers $0, 1, 2, \ldots, m - 1$ form a complete residue system for the modulus m.

PROOF: By the division algorithm every number has a remainder which is one of the numbers $0, 1, 2, \ldots, m - 1$. Thus, every number is congruent to at least one of the numbers $0, 1, \ldots, m - 1$; it suffices to show that no two of these numbers are congruent. For, if some number N is congruent to two of them, say, $N \equiv j \bmod m$ and $N \equiv k \bmod m$, then by Corollary 37 Part 3 (two things congruent to the same thing are congruent to each other), $j \equiv k \bmod m$. But, no two of $0, 1, 2, \ldots, m - 1$ are congruent, since these numbers are their own remainders on division by m, and two numbers are congruent if and only if they have the same remainder, the theorem is proved.

63 Theorem

If a and m are relatively prime, then $ab \equiv ac \bmod m$ if and only if $b \equiv c \bmod m$; that is, division by a is possible modulo m.

We shall use the following lemma to prove Theorem 63.

64 Lemma

If a and m are relatively prime and $ad \equiv 0 \bmod m$, then $d \equiv 0 \bmod m$.

PROOF: Let $m = p_1^{\alpha_1} p_2^{\alpha_2} \cdots p_k^{\alpha_k}$ where $p_1, p_2, \ldots, p_k$ are distinct primes and $\alpha_i \geq 1$. Since $m|ad$, $ad = p_1^{\gamma_1} p_2^{\gamma_2} \cdots p_k^{\gamma_k} q_1^{\beta_1} q_2^{\beta_2} \cdots q_l^{\beta_l}$, where $\gamma_i \geq \alpha_i$, $q_1, q_2, \ldots, q_l$ are distinct primes, and $\beta_i \geq 1$. Since none of $p_1, p_2, \ldots, p_k$ occur in the prime factorization of a and since the prime factorization of ad is simply all the prime factors of a and d together, we conclude that $d = p_1^{\gamma_1} p_2^{\gamma_2} \cdots p_k^{\gamma_k} q_1^{\delta_1} q_2^{\delta_2} \cdots q_l^{\delta_l}$ where $\beta_i \geq \delta_i \geq 0$. It follows that $m|d$ and hence $d \equiv 0 \bmod m$. The lemma is established.

PROOF OF THEOREM 63: If $ab \equiv ac \bmod m$, then $a(b - c) \equiv 0 \bmod m$. Applying Lemma 64, we see that $b - c \equiv 0 \bmod m$. It follows that $b \equiv c \bmod m$. Conversely, if $b \equiv c \bmod m$, then we know from Theorem 39 that $ab \equiv ac \bmod m$.

Note that if m is a prime, then every number is either divisible by m, ($\equiv 0 \bmod m$) or relatively prime to m. That is to say in the system of arithmetic modulo a prime, cancellation is possible except for zero.

65 Corollary

If p is a prime and $a \not\equiv 0 \bmod p$, then $ab \equiv ac \bmod p$ if and only if $b \equiv c$.

Now we are prepared to prove another of Fermat's divisibility theorems.

66 Theorem

If p is a prime, then for every number a, $a^p \equiv a \bmod p$.

PROOF: If $a \equiv 0 \bmod p$, the statement is clearly true. If $a \not\equiv 0 \bmod p$, then a and p are relatively prime. From Theorem 62 and Corollary 65, we see that none of the numbers $0 \cdot a, 1 \cdot a, 2a, 3a, \ldots, (p-1)a$ is congruent to another, for if $0 \leq i \leq j < p$ and $ia \equiv ja \bmod p$, then $i \equiv j \bmod p$ by Corollary 65, whence $i = j$ by Theorem 62. Since $0 \cdot a \equiv 0 \bmod p$, each of $a, 2a, \ldots, (p-1)a$ is congruent to one of $1, 2, 3, \ldots, (p-1)$ because $0, 1, 2, \ldots, (p-1)$ is a complete residue system. On the other hand, since no two of the numbers in the sequence $a, 2a, 3a, \ldots, (p-1)a$ are congruent to one another, they must each be congruent to distinct numbers in the sequence $1, 2, \ldots, (p-1)$. Since each sequence has the same number of terms, the two sequences can be paired in such a way that each pair has one number from each sequence and the numbers are congruent. Hence, $a \cdot 2a \cdot 3a \cdots (p-1)a \equiv 1 \cdot 2 \cdot 3 \cdots (p-1) \bmod p$. Thus, $a^{p-1}(p-1)! \equiv (p-1)! \bmod p$. From Corollary 65 it follows that $a^{p-1} \equiv 1 \bmod p$. Multiplying by a, we obtain the theorem.

67 Corollary

If p is a prime and $a \not\equiv 0 \bmod p$, then $a^{p-1} \equiv 1 \bmod p$.

68 REMARK

The above really amounts to saying that if $a \not\equiv 0 \bmod p$, then the reciprocal of a can be defined, for modulus p; in fact, $a^{-1} = a^{p-2} \bmod p$, since then $a \cdot a^{-1} \equiv a^{-1} \cdot a \equiv 1 \bmod p$.

69 Corollary

If p is a prime, then the equation $ax \equiv b \bmod p$ is solvable whenever $a \not\equiv 0 \bmod p$. One solution is $x = a^{p-2}b$. All solutions are congruent to one another.

PROOF: Clearly, $x = a^{p-2}b$ is a solution. Let x and y be any solutions. Thus $ax \equiv b \equiv ay$, so, by Corollary 65, $x \equiv y$ mod p.

Considerations of time and space prevent us from developing a complete theory of algebra for a given modulus. Such a development can be found in a more systematic number theory book, for example, Hardy and Wright [9].

We are now in a position to prove Theorem 47, which states that there are infinitely many primes $p \equiv 1$ mod 4. We shall need two lemmas.

70 Lemma

If a is any integer and m is any positive odd integer, then $a^m + 1 \equiv 0$ mod $a + 1$.

PROOF: Since m is a positive odd integer

$$a^m + 1 = (a + 1)(a^{m-1} - a^{m-2} + a^{m-3} - \cdots + a^2 - a + 1).$$

71 Lemma

If $b \equiv 0$ mod c and $c \equiv 0$ mod d, then $b \equiv 0$ mod d.

PROOF: We see from the hypothesis that $d|c$ and $c|b$. It follows that $d|b$ and whence $b \equiv 0$ mod d. See Exercises 5, Problem 6.

PROOF OF THEOREM 47: To show that there are infinitely many primes $p \equiv 1$ mod 4, it is sufficient to show that given any integer $n \geq 2$, there is such a prime greater than n. (Why?) We do this by showing that $(n!)^2 + 1$ has such a factor. No prime which divides $n!$ can divide $(n!)^2 + 1$; hence all the prime factors of $(n!)^2 + 1$ are greater than n. Let p be a prime factor of $(n!)^2 + 1$. Since $(n!)^2 + 1$ is odd for $n \geq 2$, it follows that p is odd. If $p \equiv 1$ mod 4, we are finished; if not, since p is odd, $p \equiv 3$ mod 4. Thus, for some nonnegative integer k, $p = 4k + 3$. By Lemma 70 with $a = (n!)^2$ and $m = 2k + 1$, we have

$$(n!)^{2(2k+1)} + 1 \equiv 0 \text{ mod } (n!)^2 + 1.$$

We also have $(n!)^2 + 1 \equiv 0$ mod p. By Lemma 71, we have

$$(n!)^{4k+2} + 1 = (n!)^{2(2k+1)} + 1 \equiv 0 \text{ mod } p.$$

Multiplying by $n!$, we obtain

$$(n!)^{4k+3} + n! = (n!)^p + n! \equiv 0 \text{ mod } p.$$

On the other hand, by Fermat's theorem (Theorem 66), $(n!)^p - n! \equiv 0$ mod p. Subtracting the last two equations we obtain $2n! \equiv 0$ mod p. Since p

is odd we get $n! \equiv 0 \mod p$ which is a contradiction to $(n!)^2 + 1 \equiv 0 \mod p$. Thus, $p \equiv 3 \mod 4$ is contradictory, and it must be the case that $p \equiv 1 \mod 4$ since p is an odd prime. Theorem 47 is proved.

We return to our search for Mersenne primes.

72 Theorem

If $2p + 1$ is prime, then either $2p + 1 | M_p$ or $2p + 1 | M_p + 2$.

PROOF: From Corollary 67 we see that $2^{2p} \equiv 1 \mod 2p + 1$. That is, $2^{2p} - 1 \equiv 0 \mod 2p + 1$. But $2^{2p} - 1 = (2^p - 1)(2^p + 1) = M_p \cdot (M_p + 2)$, so $M_p(M_p + 2) \equiv 0 \mod 2p + 1$. Since $2p + 1$ is prime, the conclusion follows from Theorem 40.

Since $M_p + 2 = 2^p + 1$, the above theorem says that if $2p + 1$ is prime, $2^p \equiv \pm 1 \mod 2p + 1$. If $2^p \equiv 1 \mod 2p + 1$, then we know that M_p is composite.

Suppose we test the next unknown Mersenne number, M_p, for which we have not settled the question of primality and for which $2p + 1$ is a prime. This is the case for $p = 23$ and $2p + 1 = 47$. We wish to determine if $2^{23} \equiv 1 \mod 47$ or $2^{23} \equiv -1 \mod 47$. We use $2^{23} = (2^5)^4 \cdot 2^3$ and $2^5 = 32 \equiv -15 \mod 47$. Thus $(2^5)^2 \equiv 225 \equiv -10 \mod 47$ and $(2^5)^4 \equiv (-10)^2 \equiv 100 \equiv 6 \mod 47$. Combining these results $2^{23} \equiv 6 \cdot 8 \equiv 1 \mod 47$. Thus, $M_{23} = 2^{23} - 1 \equiv 0 \mod 47$. We obtain:

73 Theorem

M_{23} is composite. In fact $47 | 2^{23} - 1$.

Note that the above theorem disproves Cataldi's claim that M_{23} is prime. We now know that M_p is prime for $p = 2, 3, 5, 7, 13, 17$ and composite for $p = 11, 23$. The case $p = 19$ is now the smallest case which we have not decided.

74 EXERCISES

1 Are the following complete residue systems for the given modulus?
   (a) $-1, 0, 1 \mod 3$
   (b) $-3, -2, -1, 0, 1, 2, 3 \mod 6$
   (c) $-3, -2, -1, 0, 1, 2, 3 \mod 7$
   (d) $-3, -2, -1, 0, 1, 2, 3 \mod 8$

2 Verify computationally whether or not $0 \cdot 5, 1 \cdot 5, 2 \cdot 5, 3 \cdot 5, 4 \cdot 5, \ldots, 10 \cdot 5$ is a complete residue system modulo 11.

3 Are $0, 2, 2^2, 2^3, 2^4$ a complete residue system modulo 5? How do you know?

4 Let $a_1, a_2, \ldots, a_k$ be all the positive integers less than n which are relatively prime to n. Prove that every number which is relatively prime to n is congruent to exactly one of these numbers.

5 Keeping the notation of Problem 4 and supposing that a is relatively prime to n, show that $aa_1, aa_2, \ldots, aa_k$ are all relatively prime to n and that no two of them are congruent.

6 Following Euler we let $\phi(n)$ denote the number of positive integers among $1, 2, \ldots, n$ which are relatively prime to n. Show (Euler's theorem) that if a is relatively prime to n, then $a^{\phi(n)} \equiv 1 \bmod n$. (Hint: Compare with Theorem 66 and Corollary 67.)

7 Prove that if $2k + 1$ is a prime, and $a \not\equiv 0 \bmod 2k + 1$ then $a^k \equiv 1 \bmod 2k + 1$ or $a^k \equiv -1 \bmod 2k + 1$.

8 Show that M_{83} is composite. (Hint: $2^{83} = 4((2^9)^3)^3$.)

9 Show that $2^k \equiv 1 \bmod 2k + 1$ for the choices $k = 8, 20, 36$ which make $2k + 1$ prime.

10 Show that $2^k \equiv -1 \bmod 2k + 1$ for the choices $k = 1, 5, 9$ which make $2k + 1$ prime.

11 Show that $2^k \equiv -1 \bmod 2k + 1$ for the choices $k = 2, 6, 18$ which make $2k + 1$ prime.

12 Show that $2^k \equiv 1 \bmod 2k + 1$ for the choices $k = 3, 11, 15$ which make $2k + 1$ prime.

Problems 9 to 12 above, together with Problem 8 above and Theorem 73, lead us to make the following conjecture, first proved by Euler.

75 Theorem

If $2k + 1$ is prime, then $2^k \equiv 1 \bmod 2k + 1$ for $k \equiv 0, 3 \bmod 4$; and $2^k \equiv -1 \bmod 2k + 1$ for $k \equiv 1, 2 \bmod 4$.

The usual methods of proof of the above theorem are somewhat involved, and usually follow a discussion of the solvability of the equation $x^2 \equiv a \bmod p$ and how to tell if there is any solution (square root) for a given number a. The interested reader is urged to consult the sections on quadratic residues of a standard number theory book for further reference, for example, Hardy and Wright, Niven and Zuckerman, Shanks, or Sierpinski [9, 11, 14, 15].

The above theorem has the corollary:

76 Corollary
If both p and $2p + 1$ are primes and $p \equiv 3 \bmod 4$, then M_p is composite; in fact, $2p + 1 | M_p$.

Fermat was also able to prove that in testing whether or not a given M_p is prime, one could restrict one's attention to certain types of prime divisors. In particular, Fermat proved:

77 Theorem
If p is an odd prime, then every prime divisor q of M_p satisfies $q \equiv 1 \bmod 2p$.

We also know that if M_p is composite it has a prime factor less than $\sqrt{M_p}$, thus to test the primality of M_p, we need only to look at primes $q \equiv 1 \bmod 2p$ with $q < \sqrt{M_p}$. We shall not prove this theorem.

We also state without proof:

78 Theorem
If p is an odd prime, then every divisor, q, of M_p satisfies $q \equiv \pm 1 \bmod 8$.

The above two theorems considerably simplify the work of testing whether or not M_p is a prime. For instance, by Theorem 77, we can restrict our attention to the prime numbers among $2p + 1, 2(2p) + 1, 3(2p) + 1, \ldots$, which are less than $\sqrt{M_p}$. Among these primes we need consider only those which are congruent to ± 1 for the modulus 8 by Theorem 78. We apply these methods to M_{19}.

79 Example
Test M_{19} for primality.

SOLUTION: We restrict our attention to the numbers (Theorem 77): $38 + 1, 2 \cdot 38 + 1, 3 \cdot 38 + 1, \ldots, 19 \cdot 38 + 1$, since $20 \cdot 38 + 1$

$= 761$ and $(761)^2 > M_{19} = 524{,}287$. Of these 19 numbers, we need consider only those which are congruent to $\pm 1 \bmod 8$. In particular, $k \cdot 38 + 1 \equiv 1 \bmod 8$ for $k \equiv 0 \bmod 4$, that is, for $k = 4, 8, 12, 16$, and $k \cdot 38 + 1 \equiv -1 \bmod 8$ for $k \equiv 1 \bmod 4$, that is, for $k = 1, 5, 9, 13, 17$. Of these nine possible divisors, we need consider only those which are prime.

If $k \equiv 1 \bmod 3$, then $k \cdot 38 + 1 \equiv 1 \cdot 2 + 1 \equiv 0 \bmod 3$ and thus $k \cdot 38 + 1$ is not prime. Also, if $k \equiv 3 \bmod 5$, then $k \cdot 38 + 1 \equiv 3 \cdot 3 + 1 \equiv 0 \bmod 5$ so $k \cdot 38 + 1$ is not prime. Four possibilities remain: $k = 5, 9, 12, 17$ which yield for $k \cdot 38 + 1$ the four numbers 191, 343, 457, and 647. But 7 divides $343 = 7^3$, hence 343 is not prime and needn't be considered. We know that either M_{19} is prime or one of the three prime numbers 191, 457, or 647 must divide it. Now $M_{19} = 524{,}287$; thus we can simply divide out these three by long division or compute $2^{19} \equiv 2(2^9)^2$ using each of the three numbers as the modulus and find that $2^{19} - 1 \not\equiv 0 \bmod 191, 457,$ or 647; thus M_{19} is prime.

This tells us that Cataldi's claim about M_{19} was right and that $2^{18} \cdot M_{19} = 137{,}438{,}691{,}328$ is the seventh perfect number.

80   EXERCISES

1 Verify that $2^{19} - 1 \not\equiv 0 \bmod 191, 457,$ or 647.
2 Use Corollary 76 to show that M_{11} is composite.
3 Show that M_{83} is composite.
4 Verify Cataldi's claim that M_{13} is prime.
5 Do the same for M_{17}.

Euler developed these methods and applied them to M_{31}. He discovered that none of the prime numbers p which are less than $\sqrt{M_{31}}$ and which satisfy both $p \equiv 1 \bmod 62$ and $p \equiv \pm 1 \bmod 8$ divides M_{31}. Thus M_{31} is prime yielding the perfect number $2^{30} \cdot M_{31} = 2{,}305{,}843{,}008{,}139{,}952{,}128$. This number was the largest perfect number known in its time (1772). The size of this number led Barlow [1] to claim in his number-theory book, published in 1811, that this perfect number "is the greatest that will ever be discovered, for as they are merely curious without being useful it is not likely that any person will attempt to find one beyond it." This type of claim, like those which suggest closing the patent office, is made from time to time by people who underestimate the power of human curiosity, whether it be to find the next perfect number, send a man to the moon, or investigate how many people can fit in a telephone booth.

We record below discoveries made since that time. In 1886, more than 100 years after M_{31} was discovered to be prime, Seelhoff and Pervusin, independently, proved that M_{61} is prime and hence that

$$2^{60} \cdot M_{61} = (1{,}152{,}921{,}504{,}606{,}846{,}976)(2{,}305{,}843{,}009{,}213{,}693{,}951)$$

is a perfect number. This number, when multiplied out, has 36 digits. In the year 1912, Powers and Cunningham, independently and by different methods, showed that M_{89} is prime, and hence $2^{88} \cdot M_{89}$, a number of 48 digits, is a perfect number. Thanks to the work of H. S. Uhler, who showed in a paper published in 1952 that M_p is composite for $p = 157; 167; 193; 199; 227;$ and 229, we can now say that if $p \leq 257$, M_p is prime for precisely the values $p = 2, 3, 5, 7, 13, 17, 19, 31, 61, 89, 107,$ and 127.

In 1952 with the aid of the SWAC computer, Robinson showed that M_p is prime for $p = 521; 607; 1279; 2203;$ and 2281. The perfect number $2^{2280}M_{2281}$ has 1,372 digits. By use of highly efficient tests for primality due to Lucas and D. H. Lehmer (Sierpinski [15, Chapter X, Sections 2, 3]) and with the aid of the most advanced computers in the world, we can now say that for $p \leq 12000$, M_p is prime precisely for the values above plus $p = 3217; 4253; 4423; 9689; 9941;$ and 11213. Thus the twentieth Euclidean perfect number is $2^{4422}M_{4423}$, a number of 2,663 digits; the twenty-third discovered by Donald Gillies in 1963 is $2^{11212}M_{11213}$, a number of 6,957 digits. Gillies' work was done on the ILLIAC computer at the University of Illinois ["Three new Mersenne primes and the statistical theory," *Mathematics of Computation*, **V18** (1964)]. The University of Illinois Mathematics Department uses advertising on their postal meter to spread the word on the largest Mersenne prime known. (B. Tuckerman announced in June 1971 that M_{19937} is the next Mersenne prime.)

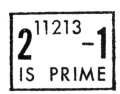

Much larger values of p are known for which M_p is composite. For example, from Corollary 76 and the fact that both $p = 16{,}188{,}302{,}111$ and $2p + 1 = 32{,}376{,}604{,}223$ are primes, we see that $M_{16,188,302,111}$ is composite. We note here that it would be a truly monumental chore to divide this number, which has about 4,873,- million digits, by $2p + 1$ to find out that $2p + 1$ is really a factor.

Shanks [14, p. 29] makes the point nicely. For this choice of p:

M_p is a number which, if written out in decimal, would be nearly five billion digits long. Each such number would more than fill the telephone books of all five boroughs of New York City. Imagine then, if Cataldi were alive today, and if he set himself the task of proving these M_p composite—by *his* methods! Can't you see the picture—the ONR contract—the thousands of graduate assistants gainfully employed—the Beneficial Suggestion Committee, etc.? But we are digressing.

Section 6   Non-Euclidean Perfect Numbers

At this point in our search for more perfect numbers by finding more Mersenne primes we are at the limit of both modern theory and modern technology. As computers improve and as (hopefully) better techniques for testing Mersenne primes are developed, more Mersenne primes will be found. Perhaps someday we shall be able to decide if there is an infinity of Mersenne primes; at the moment there seems to be little hope.

There are, however, other approaches to the problem of finding perfect numbers which we have not yet tried. For example, are the Euclidean perfect numbers, namely, $2^{p-1}M_p$ for M_p a prime, the only even perfect numbers? Are there odd perfect numbers? Many? Infinitely many?

Since the notion of a perfect number involves the sum of the divisors of a number, we begin by studying the sums of divisors of numbers. Since the phrase "the sum of the divisors of n" will be used many times, we define the following notation.

81   *Notation*

We denote by $\sigma(n)$ (σ for sum) the sum of all the divisors of the number n.

Section 6   Non-Euclidean Perfect Numbers   125

82   Examples

1   $\sigma(1) = 1$.
2   If p is a prime, then $\sigma(p) = 1 + p$.
3   If $n = p \cdot q$ where p and q are distinct primes, then

$$\sigma(n) = 1 + p + q + p \cdot q.$$

Why?

4   If n is perfect, then $\sigma(n) = 2n$, and conversely.

83   EXERCISES

1   Compute $\sigma(27)$.
2   Compute $\sigma(5^2)$.
3   Compute $\sigma(2^5)$.
4   Compute $\sigma(2^5 \cdot 5^2)$.
5   Do you notice any patterns in Problems 1 to 4? If not, compute some more.

84   Lemma

If p is a prime and α is a positive integer, then $\sigma(p^\alpha) = \dfrac{p^{\alpha+1} - 1}{p - 1}$.

PROOF: The divisors of p^α are simply $1, p, p^2, \ldots, p^\alpha$ (cf. Theorem 23). While from Lemma 29 we see that $1 + p + p^2 + \cdots + p^\alpha = \dfrac{p^{\alpha+1} - 1}{p - 1}$.

85   Corollary

$\sigma(2^\alpha) = 2^{\alpha+1} - 1$ for every positive integer α.

86   Lemma

If $p \nmid n$, then $\sigma(p^\alpha n) = \sigma(p^\alpha)\sigma(n)$.

PROOF: We observe (Theorem 23) that $d | p^\alpha n$ if and only if $d = p^\beta m$ where $0 \leq \beta \leq \alpha$ and $m | n$. Let $m_0 = 1, m_1, m_2, \ldots, m_k = n$ denote all the divisors of n in increasing order; then the divisors of $p^\alpha n$ are

$$\begin{aligned}
(0) \quad & m_0, \quad m_1, \quad m_2, \quad \ldots, m_k, \\
(1) \quad & pm_0, \quad pm_1, \quad pm_2, \quad \ldots, pm_k, \\
& \cdots\cdots\cdots\cdots\cdots\cdots\cdots \\
(\beta) \quad & p^\beta m_0, \; p^\beta m_1, \; p^\beta m_2, \; \ldots, p^\beta m_k, \\
& \cdots\cdots\cdots\cdots\cdots\cdots\cdots \\
(\alpha) \quad & p^\alpha m_0, \; p^\alpha m_1, \; p^\alpha m_2, \; \ldots, p^\alpha m_k.
\end{aligned}$$

The sum of the numbers in the βth row is

$$p^\beta(m_0 + m_1 + \cdots + m_k) = p^\beta \sigma(n).$$

Adding these row sums together, we obtain

$$\sigma(p^\alpha n) = \sigma(n) + p\sigma(n) + p^2\sigma(n) + \cdots + p^\alpha \sigma(n)$$
$$= (1 + p + \cdots + p^\alpha)\sigma(n) = \sigma(p^\alpha)\sigma(n),$$

which proves the lemma.

87 Theorem

Let $n = p_1^{\alpha_1} \cdots p_k^{\alpha_k}$ with $p_1, \ldots, p_k$ distinct primes. Then

$$\sigma(n) = \sigma(p_1^{\alpha_1}) \cdot \sigma(p_2^{\alpha_2}) \cdots \sigma(p_k^{\alpha_k}) = \prod_{p^\alpha \| n} \frac{p^{\alpha+1} - 1}{p - 1}.$$

PROOF: Applying the preceding theorem successively, we obtain

$$\sigma(p_1^{\alpha_1}(p_2^{\alpha_2} \cdots p_k^{\alpha_k})) = \sigma(p_1^{\alpha_1}) \cdot \sigma(p_2^{\alpha_2}(p_3^{\alpha_3} \cdots p_k^{\alpha_k}))$$
$$= \sigma(p_1^{\alpha_1}) \cdot \sigma(p_2^{\alpha_2}) \cdot (p_3^{\alpha_3}(p_4^{\alpha_4} \cdots p_k^{\alpha_k})) = \cdots$$
$$= \sigma(p_1^{\alpha_1})\sigma(p_2^{\alpha_2}) \cdots \sigma(p_k^{\alpha_k}).$$

From Lemma 84, we get

$$\sigma(n) = \frac{p_1^{\alpha_1+1} - 1}{p_1 - 1} \cdot \frac{p_2^{\alpha_2+1} - 1}{p_2 - 1} \cdots \frac{p_k^{\alpha_k+1} - 1}{p_k - 1}.$$

The theorem is proved.

88 Theorem

If m and n are relatively prime, then $\sigma(m \cdot n) = \sigma(m) \cdot \sigma(n)$.

PROOF: Let $m = p_1^{\alpha_1} p_2^{\alpha_2} \cdots p_k^{\alpha_k}$ and $n = q_1^{\beta_1} q_2^{\beta_2} \cdots q_l^{\beta_l}$ be representations in standard form. Since m and n are relatively prime, the p's and the q's are all distinct. Thus $m \cdot n = p_1^{\alpha_1} \cdots p_k^{\alpha_k} \cdot q_1^{\beta_1} \cdots q_l^{\beta_l}$. By the preceding theorem $\sigma(m \cdot n) = \sigma(p_1^{\alpha_1}) \cdots \sigma(p_k^{\alpha_k})\sigma(q_1^{\beta_1}) \cdots \sigma(q_l^{\beta_l}) = \sigma(m) \cdot \sigma(n)$, which proves the desired result.

89 EXERCISES

1 Compute $\sigma(3^2 \cdot 5^4 \cdot 7^2)$.
2 Let $\sigma_2(n)$ be the sum of the squares of the divisors of n.
   Evaluate $\sigma_2(p^\alpha)$ where p is a prime and α a positive integer.
3 Find and prove an analog of Theorem 87 for $\sigma_2(n)$.
4 Prove the analog of Theorem 88 for $\sigma_2(n)$.

Section 6 Non-Euclidean Perfect Numbers

5. Let $\sigma_k(n)$ be the sum of the kth powers of the divisors of n. Prove the analogs of Theorems 87 and 88 for $\sigma_k(n)$.

6. One could define a perfect number as one for which the *sum* of all the divisors is equal to the number *added* to itself. What of numbers such that the *product* of the divisors is equal to the number *multiplied* by itself? Prove that for any number n, $\prod_{d|n} d = n^2$ if and only if $n = p^3$ or $n = pq$, where p and q are primes.

7. Characterize all numbers n such that $\prod_{d|n} d = n^3$.

8. Find some numbers such that $\prod_{d|n} = n^k$. Can you find them all?

9. Show that if n is a positive integer, then $\sigma_2(n) = \sigma(n) \prod_{p^\alpha \| n} \frac{p^\alpha + 1}{p + 1}$.

10. Let n be a perfect number. Show that $\frac{\sigma_2(n)}{n} = 2 \prod_{p^\alpha \| n} \frac{p^{\alpha+1} + 1}{p + 1}$.

11. Show that for any number q there is a number $n > 1$ such that
$$1 \leq \frac{\sigma(n)}{n} \leq 1 + \frac{1}{q}.$$
This is often stated as "1 is the infimum or inferior limit of the numbers $\frac{\sigma(n)}{n}$."

*12. Show that for any number q there is a number n such that $\frac{\sigma(n)}{n} > q$. This is often stated as "the numbers $\frac{\sigma(n)}{n}$ are unbounded above."

We can now prove the theorem conjectured by the ancient Greeks and first proved by Euler, namely, that the Euclidean perfect numbers are the only even perfect numbers. We recall that in our new notation, a number N is perfect if and only if $\sigma(N) = 2N$.

90 Theorem

An even number N is perfect if and only if $N = 2^{p-1}(2^p - 1)$, where $2^p - 1$ is a (Mersenne) prime.

PROOF: We have already shown that such a number is perfect. To show the converse let N be any even number. Then $N = 2^q n$ where $q \geq 1$ and n is odd.

$\sigma(N) = \sigma(2^q) \cdot \sigma(n) = (2^{q+1} - 1)\sigma(n)$. If N is perfect we must have $2N = (2^{q+1} - 1)\sigma(n)$; that is, $2^{q+1} \cdot n = (2^{q+1} - 1)\sigma(n)$. Thus

$$\frac{n}{\sigma(n)} = \frac{2^{q+1} - 1}{2^{q+1}}.$$

Since $2^{q+1} - 1$ and 2^{q+1} are relatively prime, we see that $n = (2^{q+1} - 1)d$ and $\sigma(n) = 2^{q+1}d$ for some integer d. We now show that d must be 1. If $d > 1$, then $1, d$, and $d(2^{q+1} - 1)$ are all distinct divisors of n whence $\sigma(n) \geq 1 + d + d(2^{q+1} - 1) = 1 + d \cdot 2^{q+1} > d \cdot 2^{q+1} = \sigma(n)$. But $\sigma(n) > \sigma(n)$ is impossible, whence $d > 1$ is impossible.

It follows that $d = 1$, $n = 2^{q+1} - 1$ and $\sigma(2^{q+1} - 1) = 2^{q+1}$. This last equality is possible only if $2^{q+1} - 1$ is a prime. The theorem is proved.

We have now shown that the problem of finding even perfect numbers is precisely the problem of finding Mersenne primes, which as we have seen can be a very difficult task, but one which has led and continues to lead to many fruitful mathematical discoveries.

We now turn our attention to the case of odd perfect numbers.

We wish to find an odd number such that $\sigma(n) = 2n$. Let $n = p_1^{\alpha_1} \cdots p_k^{\alpha_k}$ where $\alpha_1, \alpha_2, \ldots, \alpha_k \geq 1$ and $p_1, \ldots, p_k$ are distinct odd primes.

The condition of perfection imposes some strong restrictions on the choices of the α_i, as is shown in the following:

91 Theorem

If $n = p_1^{\alpha_1} \cdots p_k^{\alpha_k}$ is an odd perfect number then $n = p^\alpha \cdot m^2$, where p is a prime and $p \equiv 1 \equiv \alpha \mod 4$. In other words, for all primes p_i, except one, α_i is even, and the one exceptional prime, p, satisfies $p \equiv 1 \mod 4$, its exponent α also satisfies $\alpha \equiv 1 \mod 4$.

PROOF: Since $2n = \sigma(n)$, we get $2n = \sigma(p_1^{\alpha_1}) \cdots \sigma(p_k^{\alpha_k}) = (1 + p_1 + p_1^2 + \cdots + p_1^{\alpha_1})(1 + p_2 + \cdots + p_2^{\alpha_2}) \cdots (1 + p_k + \cdots + p_k^{\alpha_k})$.

Since n is odd, $n \equiv 1, 3 \mod 4$ so $2n \equiv 2, 6 \mod 4$, but $2 \equiv 6 \mod 4$, thus $2n \equiv 2 \mod 4$. Thus $\sigma(n) = 2n$ is divisible by 2 but not by 4. It follows that the product $\sigma(p_1^{\alpha_1})\sigma(p_2^{\alpha_2}) \cdots \sigma(p_r^{\alpha_r})$ is divisible by 2 but not by 4. Thus all but one of the $\sigma(p_i^{\alpha_i})$ are odd, and the one which is even is not divisible by 4. If $p_i \equiv 3 \mod 4$, then $p_i \equiv -1 \mod 4$, so that $\sigma(p_i^{\alpha_i}) = 1 + p_i + p_i^2 + \cdots + p_i^{\alpha_i} \equiv 1 + (-1) + (-1)^2 + (-1)^3 + \cdots + (-1)^{\alpha_i} \mod 4$, a sum which alternates $+1$'s and -1's. If α_i is odd, the last term is -1 and $\sigma(p_i^{\alpha_i}) \equiv 0 \mod 4$, which is impossible since we have already seen that no term is divisible by 4. If α_i is even, then $\sigma(p_i^{\alpha_i}) = 1, 1 + p_i + p_i^2 + \cdots + p_i^{\alpha_i} \equiv 1 + (-1) + 1 + (-1) + \cdots + 1 \equiv 1 \bmod 4$. If $p_i \equiv 1 \bmod 4$, then

$$\sigma(p_i^{\alpha_i}) = 1 + p_i^1 + p_i^2 + \cdots + p_i^{\alpha_i} \equiv 1 + 1^1 + 1^2 + \cdots + 1^{\alpha_i} \equiv \alpha_i + 1.$$

Since exactly one of these terms must be $\equiv 2 \bmod 4$, we have exactly one $\alpha \equiv 1 \bmod 4$. For the remaining terms $\sigma(p_i^{\alpha_i}) \equiv 1, 3 \bmod 4$ and hence $\alpha_i \equiv 0, 2 \bmod 4$. Thus α_i is even for all but one prime p with exponent α, and as we have seen $p \equiv \alpha \equiv 1 \bmod 4$ for this exceptional prime. The proof is completed.

92 Corollary

If n is odd and perfect, then $n \equiv 1 \bmod 4$.

PROOF: From the theorem $n = p^\alpha m^2$. Since $p \equiv 1 \bmod 4$, $p^\alpha \equiv 1 \bmod 4$. Since m^2 is odd it follows that m is odd, that is, $m \equiv 1, 3 \bmod 4$. Squaring, we see that $m^2 \equiv 1 \bmod 4$. Thus $n = p^\alpha m^2 \equiv 1 \cdot 1 \equiv 1 \bmod 4$. The corollary is established.

93 EXERCISES

1 Show that if m is odd, then $m^2 \equiv 1 \bmod 8$.
2 Show that if $n = p^\alpha m^2$ is odd and perfect, then $n \equiv p \bmod 8$.
3 Show that if n is odd and $\sigma(n) = 6n$, then $n = p^\alpha m^2$ where $p \equiv 1 \equiv \alpha \bmod 4$.

Let us notice in the proof of Theorem 91 that all we have used from the relation $2n = \sigma(n)$ is that n is odd and that $2n \equiv 2 \bmod 4$. The same analysis could be made for the numbers n which satisfy the conditions $6n = \sigma(n)$, n odd, or $22n = \sigma(n)$, n odd.

We now make use of the fact that $\sigma(n)$ must be exactly twice n for a perfect number. The theorems we prove are not the best known results, but their proofs illustrate the techniques needed to obtain better results.

The equation $2n = \sigma(n)$ can be rewritten as $2 = \dfrac{\sigma(n)}{n}$. If $n = p_0^{\alpha_0} p_1^{\alpha_1} \cdots p_k^{\alpha_k}$, then

$$\sigma(n) = \frac{p_0^{\alpha_0+1} - 1}{p_0 - 1} \cdot \frac{p_1^{\alpha_1+1} - 1}{p_1 - 1} \cdots \frac{p_k^{\alpha_k+1} - 1}{p_k - 1}.$$

Thus

$$\frac{\sigma(n)}{n} = \frac{1}{p_0^{\alpha_0} \cdot p_1^{\alpha_1} \cdots p_k^{\alpha_k}} \cdot \left(\frac{p_0^{\alpha_0+1} - 1}{p_0 - 1} \cdot \frac{p_1^{\alpha_1+1} - 1}{p_1 - 1} \cdots \frac{p_k^{\alpha_k+1} - 1}{p_k - 1} \right)$$

$$= \frac{p_0^{\alpha_0+1} - 1}{p_0^{\alpha_0+1} - p_0^{\alpha_0}} \cdot \frac{p_1^{\alpha_1+1} - 1}{p_1^{\alpha_1+1} - p_1^{\alpha_1}} \cdots \frac{p_k^{\alpha_k+1} - 1}{p_k^{\alpha_k+1} - p_k^{\alpha_k}}.$$

This discussion may be summarized in the following lemma:

94 Lemma

If n is odd and perfect and $n = p_0^{\alpha_0} \cdot p_1^{\alpha_1} \cdots p_k^{\alpha_k}$, where $p_0, p_1, \ldots, p_k$ are distinct primes, then

$$2 = \frac{p_0^{\alpha_0+1} - 1}{p_0^{\alpha_0+1} - p_0^{\alpha_0}} \cdot \frac{p_1^{\alpha_1+1} - 1}{p_1^{\alpha_1+1} - p_1^{\alpha_1}} \cdots \frac{p_k^{\alpha_k+1} - 1}{p_k^{\alpha_k+1} - p_k^{\alpha_k}}.$$

We shall eventually show that if n is perfect, then it must have several factors. This will be true since, for p large, the numbers $(p^{\alpha+1} - 1)/(p^{\alpha+1} - p^{\alpha})$ are only a little bit larger than 1 and hence lots of them are needed. On the other hand, if too many of the p's are small, the product will be more than 2. In order to make these methods precise, we need simpler expressions which give us information on the size of $(p^{\alpha+1} - 1)/(p^{\alpha+1} - p^{\alpha})$. The first theorem of this type which we prove is that an odd perfect number can't have too many small factors; to do this, we will need simple estimates for $(p^{\alpha+1} - 1)/(p^{\alpha+1} - p^{\alpha})$ which are slightly smaller.

95 Lemma

If $n = p_0^{\alpha_0} p_1^{\alpha_1} \cdots p_k^{\alpha_k}$ is an odd perfect number with $p_0, p_1, \ldots, p_k$ distinct primes and numbered so that $p_0 \equiv \alpha_0 \equiv 1 \bmod 4$, then

$$2 \geq \frac{p_0 + 1}{p_0^2} \cdot \frac{p_1^2 + p_1 + 1}{p_1^2} \cdots \frac{p_k^2 + p_k + 1}{p_k^2}.$$

PROOF: Let p be prime and α be an integer; then

$$\frac{p^{\alpha+1} - 1}{p^{\alpha+1} - p^{\alpha}} = \frac{1 - 1/p^{\alpha+1}}{1 - 1/p}.$$

If $\alpha \geq 1$, then $1 - 1/p^{\alpha+1} \geq 1 - 1/p^2$ since $1/p^{\alpha+1} \leq 1/p^2$ in this case. If $\alpha \geq 2$, then $1 - 1/p^{\alpha+1} \geq 1 - 1/p^3$ since $\alpha + 1 \geq 3$. Thus, since $\alpha_0 \geq 1$, we get

$$\frac{p_0^{\alpha_0+1} - 1}{p_0^{\alpha_0+1} - p_0^{\alpha_0}} = \frac{1 - 1/p_0^{\alpha_0+1}}{1 - 1/p_0} \geq \frac{1 - 1/p_0^2}{1 - 1/p_0}.$$

Section 6 Non-Euclidean Perfect Numbers

But
$$\frac{1-1/p_0^2}{1-1/p_0} = \frac{p_0^2-1}{p_0^2-p_0} = \frac{(p_0+1)(p_0-1)}{p_0(p_0-1)} = \frac{p_0+1}{p_0}.$$

Thus
$$\frac{p_0^{\alpha_0+1}-1}{p_0^{\alpha_0+1}-p_0^{\alpha_0}} \geq \frac{p_0+1}{p_0}.$$

Similarly, since $\alpha_i \geq 2$ for $i \neq 0$, we get

$$\frac{p_i^{\alpha_i+1}-1}{p_i^{\alpha_i+1}-p_i^{\alpha_i}} = \frac{1-1/p_i^{\alpha_i+1}}{1-1/p_i} \geq \frac{1-1/p_i^3}{1-1/p_i} = \frac{p_i^3-1}{p_i^1(p_i-1)} = \frac{p_i^2+p_i+1}{p_i^2}.$$

Lemma 95 can now be obtained from Lemma 94 by replacing $(p_0^{\alpha_0+1}-1)/(p_0^{\alpha+1}-p_0^\alpha)$ by the possibly smaller $(p_0+1)/p_0$ and $(p_i^{\alpha_i+1}-1)/(p_i^{\alpha_i+1}-p_i^{\alpha_i})$ by the possibly smaller $(p_i^2+p_i+1)/p_i^2$.

96 Theorem

If n is odd and perfect, then $n \not\equiv 0 \bmod 105$; that is, not all of 3, 5, and 7 can be prime factors of n.

PROOF: We suppose that 3, 5, and 7 divide n.

Case 1. $p_0 = 5$ We may suppose that $p_1 = 3$ and $p_2 = 7$. For $i \geq 2$, $(p_i^2+p_i+1)/p_i^2 \geq 1$. Thus from Lemma 95 we get

$$2 \geq \frac{5+1}{5} \cdot \frac{3^2+3+1}{3^2} \cdot \frac{7^2+7+1}{7^2} \cdot 1$$
$$= \tfrac{6}{5} \cdot \tfrac{13}{9} \cdot \tfrac{57}{49} = \tfrac{494}{245} > 2.$$

This is impossible, so Case 1 cannot arise.

Case 2. $p_0 \neq 5$ We may suppose $p_1 = 3$, $p_2 = 5$, $p_3 = 7$.
Since $(P+1)/P \geq 1$, we obtain from Lemma 95 the inequality

$$2 \geq 1 \cdot \frac{3^2+3+1}{9} \cdot \frac{5^2+5+1}{25} \cdot \frac{7^2+7+1}{49} \cdot 1$$
$$= \tfrac{13}{9} \cdot \tfrac{31}{25} \cdot \tfrac{57}{49} = \tfrac{7,657}{3,675} > 2.$$

This is also impossible.

Since one of these cases must arise if $n \equiv 0 \bmod 105$, it is impossible that $n \equiv 0 \bmod 105$. The proof is complete.

Before we can prove a theorem to the effect that an odd perfect number has many factors, we need an "opposite number" to Lemma 95.

97 Lemma

If $n = p_0^{\alpha_0} p_1^{\alpha_1} \cdots p_k^{\alpha_k}$ is an odd perfect number with $p_0, p_1, \ldots, p_k$ distinct primes, then

$$2 \leq \frac{p_0}{p_0 - 1} \cdot \frac{p_1}{p_1 - 1} \cdots \frac{p_k}{p_k - 1}.$$

PROOF: For any prime p and positive integer α, we see that

$$\frac{p^{\alpha+1} - 1}{p^{\alpha+1} - p^\alpha} < \frac{p^{\alpha+1}}{p^{\alpha+1} - p^\alpha} = \frac{p}{p - 1}.$$

Replacing $(p_i^{\alpha_i+1} - 1)/(p_i^{\alpha_i+1} - p_i^{\alpha_i})$ by the larger $p_i/(p_i - 1)$, $i = 0, 1, \ldots, k$, in Lemma 94, we obtain Lemma 97.

98 Theorem

If n is odd and perfect, then n has at least three distinct prime factors.

PROOF: We suppose n is odd, perfect, and has at most two distinct prime factors. We shall show that this supposition leads to a contradiction and is, therefore, false. In this manner, we will prove the theorem.

Case 1. n has only one prime factor.

In this case, $n = P^a$. Since $P \equiv 1 \bmod 4$, we know $P \geq 5$; using Lemma 97, we derive the contradictory inequality

$$2 < \frac{P}{P - 1} = 1 + \frac{1}{P - 1} \leq 1 + \frac{1}{5 - 1} = \frac{5}{4}.$$

Thus n cannot have only one prime factor.

Case 2. n has exactly two prime factors.

In this case, $n = P^a p^\alpha$, where α is even and $a \equiv 1 \equiv P \bmod 4$. Thus $P \geq 5$ and $p \geq 3$. We again apply Lemma 97, which yields

$$2 < \frac{P}{P - 1} \cdot \frac{p}{p - 1} \leq \frac{5}{4} \cdot \frac{3}{2} = \frac{15}{8}.$$

This is contradictory, so Case 2 cannot arise.

Section 6 Non-Euclidean Perfect Numbers

Since neither Case 1 nor Case 2 can arise, it follows that n must have at least three prime factors. The theorem is proved.

We next prove a slightly stronger result, namely, that an odd perfect number has at least four distinct prime factors. The proof is somewhat intricate, and its purpose here is to illustrate for the more curious reader the style and technique used in the proofs of the more advanced theorems.

The less serious or more hurried reader might well be advised to skip this proof.

99   Theorem

If n is odd and perfect, then n has at least four distinct prime factors.

PROOF: We suppose that n has exactly three prime factors, and we show that this is impossible. Let $n = q^\gamma r^\delta s^\epsilon$, where q, r, and s are distinct odd primes satisfying $q < r < s$. The proof will be split into several cases.

Case 1. $q \geq 5$.

In this case $r \geq 7$ since $r > q = 5$ and r is prime. Also $s \geq 11$. Applying Lemma 97, we obtain

$$2 < \frac{q}{q-1} \cdot \frac{r}{r-1} \cdot \frac{s}{s-1} \leq \frac{5}{4} \cdot \frac{7}{6} \cdot \frac{11}{10} = \frac{77}{48} < 2.$$

This is a contradiction.

In the remaining cases $q = 3$.

Case 2. $q = 3, r \geq 7$.

Again we apply Lemma 97 and get the contradiction,

$$2 < \frac{q}{q-1} \cdot \frac{r}{r-1} \cdot \frac{s}{s-1} \leq \frac{3}{2} \cdot \frac{7}{6} \cdot \frac{11}{10} = \frac{77}{40} < 2.$$

This leads to the last case.

Case 3. $q = 3, r = 5$.

This case itself will be broken into several parts.

Subcase 3.1. $s \geq 17$.

In this subcase, our old standby, Lemma 97, yields

$$2 \leq \tfrac{3}{2} \cdot \tfrac{5}{4} \cdot \tfrac{17}{16} = \tfrac{255}{128} < 2,$$

a contradiction.

There are only two remaining possibilities: $s = 11$ and $s = 13$. We must consider each separately. A quick check will show that Lemma 97 fails us in these two remaining subcases. We shall find it necessary to use Lemma 94, which is more accurate than Lemma 97, in which there has been a definite loss of information.

Before beginning the next subcases, we pause to prove a lemma.

100   Lemma

If $b \equiv 3 \bmod 5$, then $b^k - 1 \equiv 0 \bmod 5$ if and only if $k \equiv 0 \bmod 4$.

PROOF: Because of the rules of congruence we may take $b = 3$. Let $k = 4l + r$, where $0 \leq r \leq 3$. By Fermat's theorem (Corollary 67), $3^4 \equiv 1 \bmod 5$. By simple computation, none of 3^1, 3^2, or 3^3 is congruent to $1 \bmod 5$, while $3^0 = 1 \equiv 1 \bmod 5$. Thus $b^k = b^{4l+r} = (b^4)^l b^r \equiv 1^l b^r \equiv b^r \bmod 5$. Thus $b^k \equiv 1 \bmod 5$ if and only if $b^r \equiv 1 \bmod 5$; that is, if and only if $r = 0$. The lemma is established.

Returning to the proof of Theorem 99, we consider

Subcase 3.2. $q = 3$, $r = 5$, $s = 13$.

Let $n = 3^{2\beta} 5^\delta 13^\epsilon = P^\alpha m^2$ where we know that either $\delta \equiv 1 \bmod 4$ and $\epsilon \equiv 0 \bmod 2$ or $\delta \equiv 0 \bmod 2$ and $\epsilon \equiv 1 \bmod 4$ according as $P = 5$ or $P = 13$, respectively. However, $\epsilon \not\equiv 3 \bmod 4$, so $\epsilon + 1 \not\equiv 0 \bmod 4$, a fact we shall use in a moment.

We now apply Lemma 94 and get,

$$2 = \frac{3^{2\beta} - 1}{3^{2\beta} \cdot 2} \cdot \frac{5^{\delta+1} - 1}{5^\delta \cdot 4} \cdot \frac{13^{\epsilon+1} - 1}{13^\epsilon \cdot 12}.$$

Since the fraction on the right-hand side is equal to 2, it follows that the denominator is half the numerator. In particular, 5 must divide the numerator. Thus (Theorem 17) 5 must divide one of $3^{2\beta} - 1$, $5^{\delta+1} - 1$ or $13^{\epsilon+1} - 1$. Clearly, $5^{\delta+1} - 1 \equiv -1 \bmod 5$, while we see from Lemma 100 that $13^{\epsilon+1} - 1 \not\equiv 0 \bmod 5$ since $\epsilon + 1 \not\equiv 0 \bmod 4$. Thus $3^{2\beta} - 1 \equiv 0 \bmod 5$. But $2\beta + 1 \not\equiv 0 \bmod 4$ either, and by Lemma 100, $3^{2\beta+1} \not\equiv 1 \bmod 5$. The last two sentences show that this subcase leads to a contradiction.

There remains only the most difficult subcase.

Subcase 3.3. $q = 3$, $r = 5$, $s = 11$.

Section 6   Non-Euclidean Perfect Numbers

In this case for $n = P^\alpha m^2$ there is no choice for P since only one of the primes involved is congruent to 1 mod 4. Thus $n = 5^\alpha \cdot 3^{2\beta} \cdot 11^{2\zeta}$, where $\alpha \equiv 1$ mod 4 and $\beta, \zeta \geq 1$.

We start with the equality

101
$$2 = \frac{5^{\alpha+1} - 1}{5^\alpha \cdot 4} \cdot \frac{3^{2\beta+1} - 1}{3^{2\beta} \cdot 2} \cdot \frac{11^{2\zeta+1} - 1}{11^{2\zeta} \cdot 10}.$$

Since $\dfrac{3^{2\beta+1} - 1}{3^{2\beta}} < \dfrac{3^{2\beta+1}}{3^{2\beta}} = 3$ and $\dfrac{11^{2\zeta+1} - 1}{11^{2\zeta}} < \dfrac{11^{2\zeta+1}}{11^{2\zeta}} = 11.$

We obtain from Formula 101

$$2 \leq \frac{5^{\alpha+1} - 1}{5^\alpha \cdot 4} \cdot \frac{3}{2} \cdot \frac{11}{10} \quad \text{or} \quad 5^\alpha \cdot 160 \leq 5^\alpha \cdot 5 \cdot 3 \cdot 11 - 3 \cdot 11.$$

Simplifying, we get $33 \leq 5^{\alpha+1}$. Thus $\alpha > 1$. But $\alpha \equiv 1$ mod 4, so $\alpha \geq 5$.

Now using the inequalities $\dfrac{5^{\alpha+1} - 1}{5^\alpha} < 5$ and $\dfrac{11^{2\zeta+1} - 1}{11^{2\zeta}} < 11$, we obtain from Equation 101,

$$2 < \frac{5}{4} \cdot \frac{11}{10} \cdot \frac{3^{2\beta+1} - 1}{3^{2\beta} \cdot 2} = \frac{11}{16} \cdot \frac{3^{2\beta+1} - 1}{3^{2\beta}}.$$

Simplifying, we get

$$32 \cdot 3^{2\beta} < 33 \cdot 3^{2\beta} - 11$$

or $11 < 3^{2\beta}$. Thus $2\beta > 2$, and since 2β is an even integer, $2\beta \geq 4$, or $\beta \geq 2$.

In the same manner in which Lemma 95 was derived from Lemma 94 we shall now derive Formula 102 below from Formula 101. First we note that for any whole number $\eta \leq \alpha$,

$$\frac{p^{\alpha+1} - 1}{p^{\alpha+1} - p^\alpha} = \frac{1 - 1/p^{\alpha+1}}{1 - 1/p} \geq \frac{1 - 1/p^{\eta+1}}{1 - 1/p} = \frac{p^{\eta+1} - 1}{p^{\eta+1} - p^\eta}.$$

In particular, since $\alpha \geq 5$

$$\frac{5^{\alpha+1} - 1}{5^\alpha \cdot 4} = \frac{5^{\alpha+1} - 1}{5^{\alpha+1} - 5^\alpha} \geq \frac{5^6 - 1}{5^6 - 5^5} = \frac{5^6 - 1}{5^5 \cdot 4},$$

and since $2\beta \geq 4$ and $2\zeta \geq 2$,

$$\frac{3^{2\beta+1} - 1}{3^{2\beta} \cdot 2} \geq \frac{3^5 - 1}{3^4 \cdot 2} \quad \text{and} \quad \frac{11^{2\zeta+1} - 1}{11^{2\zeta} \cdot 10} \geq \frac{11^3 - 1}{11^2 \cdot 10}.$$

Using these inequalities in Formula 101, we get

102
$$2 \geq \frac{5^6 - 1}{5^5 \cdot 4} \cdot \frac{3^5 - 1}{3^4 \cdot 2} \cdot \frac{11^3 - 1}{11^2 \cdot 10}.$$

With a little arithmetic calculation, we deduce from Formula 102, that

$$2 \geq \frac{15624}{3125 \cdot 4} \cdot \frac{242}{81 \cdot 2} \cdot \frac{1330}{121 \cdot 10}.$$

Reducing, we obtain $2 \geq \frac{434}{3125} \cdot \frac{1}{9} \cdot \frac{133}{1} = \frac{57722}{28125} > 2$. This is a contradiction.

Thus the assumption that an odd number having three distinct prime divisors could be perfect leads to a contradiction since any such number must fall into one of the above cases. (The reader should check that we have not skipped any possibilities.) The theorem is thus established.

By using the above techniques, together with more powerful techniques, it can be shown that if an odd perfect number exists, it must have at least six distinct prime factors, and Kanold has shown in 1957 that it must be greater than 100,000,000,000,000,000,000. For more information and detailed references, consult Sierpinski [15, Chapter IV, Section 5]. It is also possible to find more stringent conditions on the exponents; that is, the a and α's. Despite all this effort no one has either found an odd perfect number or proved that one cannot exist. It is felt by some that no odd perfect number exists.

Even though we don't know if any odd perfect numbers exist, it is still possible to prove meaningful things about them.

103 EXERCISES

1. Show that if n is odd and perfect and $n \not\equiv 0 \mod 3$, then n has at least seven distinct prime factors.
2. Can you now find an odd abundant number (cf. Problem 3 of Exercises 9)?
3. Derive a result on the number of distinct prime factors of a perfect number which is divisible by neither 3 nor 5.
4. Let d be the least positive integer for which $11^d \equiv 1 \mod 25$. If $a > 0$ and $11^a \equiv 1 \mod 25$, then $d|a$.
5. If $11^a \equiv 1 \mod 25$ then $10|a$.
6. Using Problem 5 and Formula 101, give another proof of subcase 3.3.

Section 7 Extensions and Generalizations

The theory we have developed in our discussion of perfect numbers gives us powerful techniques for handling other problems that arise in mathematics and its application. We mention briefly a few related problems.

The first application is to *multiply perfect numbers*. These numbers were first named and investigated in depth by the seventeenth century French mathematicians.

104 *Definition*

A number n is called $k$ply perfect (sub-$k$ple) if and only if it satisfies $\sigma(n) = kn$.

In this notation, what we have been calling perfect numbers are doubly perfect (or sub-double) numbers.

Several triply perfect numbers have been found. For example,

$\sigma(120) = 3 \cdot 120$ (Recarde, 1557)

$\sigma(672) = 3 \cdot 672$ (Fermat, c. 1637)

$\sigma(2^9 \cdot 3 \cdot 11 \cdot 31) = 3 \cdot (2^9 \cdot 3 \cdot 11 \cdot 13)$ (Jumeau, Prior of St. Croix, April, 1638)

$\sigma(1{,}476{,}304{,}896) = \sigma(2^{13} \cdot 3 \cdot 11 \cdot 43 \cdot 127) = 3 \cdot (2^{13} \cdot 3 \cdot 11 \cdot 43 \cdot 127)$ (Descartes, December, 1638)

105 EXERCISES

1 Prove that if k is an odd integer, then every odd $k$ply perfect number is a perfect square.

2 Prove Fermat's rule that if $\dfrac{2^{n+3} - 1}{2^n + 1}$ is an integer and is prime, then $3 \cdot 2^{n+2} \cdot \dfrac{2^{n+3} - 1}{2^n + 1}$ is triply perfect. (Note that not all the examples are of this type. Descartes thought that there were no more of this type and that this was an ad hoc rule in view of the examples Fermat knew when he (Fermat) made it. To the author's knowledge this question is still unsettled.)

One of the great skeptics of the seventeenth century, René Descartes (1596–1650) devoted his life to destroying old ideas and creating new ones. The mathematical achievement for which he is chiefly remembered is his contribution toward the development of coordinate geometry (see Chapter 4). However, he also made important discoveries in number theory and algebra, and is best known as a philosopher and scientist. (Courtesy of the Smith Collection, Columbia University.)

Descartes in 1638 gave the following quadruply perfect (subquadruple) numbers:

$$30240 = 2^5 \cdot 3^2 \cdot 5 \cdot 7.$$
$$32760 = 2^3 \cdot 3^2 \cdot 5 \cdot 7 \cdot 13.$$
$$23569920 = 2^9 \cdot 3^3 \cdot 5 \cdot 11 \cdot 31.$$
$$142990848 = 2^9 \cdot 3^2 \cdot 7 \cdot 11 \cdot 13 \cdot 31.$$
$$66433720320 = 2^{13} \cdot 3^3 \cdot 5 \cdot 11 \cdot 43 \cdot 127.$$
$$403031236608 = 2^{13} \cdot 3^2 \cdot 7 \cdot 11 \cdot 13 \cdot 43 \cdot 127.$$

He also gave one quintuply perfect number:

$$14{,}182{,}439{,}040 = 2^7 \cdot 3^4 \cdot 5 \cdot 7 \cdot 11^2 \cdot 17 \cdot 19.$$

106 EXERCISES

1. Prove the following rule of Descartes: If n satisfies $\sigma(n) = 3n$ and $3 \nmid n$, then $3n$ satisfies $\sigma(3n) = 4 \cdot (3n)$.
2. If $\sigma(n) = 3n$, $3\|n$, but $5 \nmid n$, then for $m = 45n$, $\sigma(m) = 4m$.
3. Apply rule 2 to some of the above triply perfect numbers.
4. Verify that the numbers given by Descartes are quadruply perfect.
5. Verify Descartes' quintuply perfect number.

For further history and discussion, the reader is referred to Dickson's *History* [7] or Sierpinski [15, Chap. IV, Section 5].

Certain pairs of numbers, called *amicable pairs*, have also enjoyed great favor among numerologists, astrologers, and mystics throughout the ages.

107 *Definition*

A pair of numbers, m and n, are called an *amicable pair* if and only if the sum of the proper divisors of each of them is equal to the other. In our present notation $\sigma(m) = \sigma(n) = m + n$.

The oldest, and only classically known, amicable pair is $220 = 2^2 \cdot 5 \cdot 11$ and $284 = 2^2 \cdot 71$.

We quote from Dickson [7] the following two passages.

According to Iamblichus, "Certain men steeped in mistaken opinion thought that the perfect number was called love by the Pythagoreans on account of the union of different elements and affinity which exists in it; for they call certain other numbers, on the contrary, amicable numbers, adopting virtues and social qualities to numbers, as 284 and 220, for the parts of each have the power to generate the other, according to the rule of friendship, as Pythagoras affirmed. When asked what is a friend, he replied, 'another I,' which is shown in these numbers. Aristotle so defined a friend in his *Ethics*."

Among Jacob's presents to Esau were 200 she-goats and 20 he-goats, 200 ewes and 20 rams (Genesis, XXXII, 14). Abraham Azulai (1570–1643), in commenting on this passage from the Bible, remarked that he had found written in the name of Rau Nachshon (ninth century A.D.): Our ancestor Jacob prepared his present in a wise way. This number 220 (of goats) is a hidden secret, being one of a pair of numbers such that the parts of it are equal to the other one 284, and conversely. And Jacob had this in mind; this has been tried by the ancients in securing the love of kings and dignitaries.

A further illustration is from the works of the great Arab scholar Ibn Khaldun (1332–1406) (Dickson [7]).

Let us mention that the practice of the art of talismans has also made us recognize the marvelous virtues of amicable (or sympathetic) numbers. These numbers are 220 and 284. One calls them amicable because the aliquot parts of one when added give a sum equal to the other. Persons who occupy themselves with talismans assure that these numbers have a particular influence in establishing union and friendship between two individuals. One prepares a horoscope theme for each individual, the first under the sign of Venus while this planet is in its house or in its exaltation and while it presents in regard to the moon an aspect of love and benevolence. In the second theme the ascendant snould be in the seventh sign. On each one of these themes one inscribes one of the numbers just indicated, but giving the strongest number to the person whose friendship one wishes to gain, the beloved person. I don't know if by the strongest number one wishes to designate the greatest one or the one which has the greatest number of aliquot parts. There results a bond so close between the two persons that they cannot be separated. The author of the Ghaia and other great masters in this art declare that they have seen this confirmed by experience.

The knowledge of amicable numbers was brought to Europe by the Arabs.

108 EXERCISES

1. Prove the rule of Abu-l-Hasan Thabit ben Karrah (Tabit Ibn Karra, 836–901): Let $k_n = 3 \cdot 2^n - 1$. If all of k_n, k_{n-1}, and $9 \cdot 2^{2n-1} - 1 = l_n$ are primes, then $M = 2^n k_{n-1} k_n$ and $N = 2^n l_n$ are an amicable pair.
2. Use this rule to find another pair of amicable numbers.
3. Verify that 220 and 284 are an amicable pair.

There are at least 391 known pairs of amicable numbers. See, for example, E. B. Escott, "Amicable Numbers," *Scripta Mathematics*, **12**(1946), pp. 61–72. Dickson [7, pp. 45–46] also gives a short list due to Euler. The most recently discovered pair, 79,750 and 88,730, was found by H. L. Rolf in 1965. Reference to this, as well as to other interesting new facts about amicable pairs, can be found in [8].

Letting $\sigma^*(n) = \sigma(n) - n$, we see that for an amicable pair $\sigma^*(m) = n$ and $\sigma^*(n) = m$. Can you find longer chains? For example, numbers n_1, n_2, and n_3 for which $\sigma^*(n_1) = n_2$, $\sigma^*(n_2) = n_3$, and $\sigma^*(n_3) = n_1$.

Section 7 *Extensions and Generalizations* 141

It was conjectured by Catalan that starting with any number n if we take $\sigma^*(n)$ then $\sigma^*(\sigma^*(n))$, then $\sigma^*(\sigma^*(\sigma^*(n)))$ that eventually we get a cyclic pattern. For a discussion of this problem and many other interesting unsolved problems the reader should consult Sierpinski [15, p. 166, and 16].

We end with a passage of the play *Sapientia* by Roswitha (Hrotsvit c. 930 A.D.), the outstanding playwright of feudal Europe, taken from *The Plays of Roswitha*, translated by Christopher St. John. Sapientia and her three daughters are brought before the Emperor Hadrian for an interview on account of their subversive Christian activities.

HADRIAN. And how old are they?

SAPIENTIA. What do you say, children? Shall I puzzle his dull brain with some problems in arithmetic?

FAITH. Do, mother. It will give us joy to hear you.

SAPIENTIA. As you wish to know the ages of my children, O Emperor, Charity has lived a diminished evenly even number of years; Hope a number also diminished, but evenly uneven; and Faith an augmented number, unevenly even.

HADRIAN. Your answer leaves me in ignorance.

SAPIENTIA. That is not surprising, since not one number, but many, come under this definition.

HADRIAN. Explain more clearly, otherwise how can I follow you?

SAPIENTIA. Charity has now completed two olympiads, Hope two lustres, and Faith three olympiads.

HADRIAN. I am curious to know why the number "8," which is two olympiads, and the number "10," which is two lustres, are called "diminished"; also why the number "12," which is made up of three olympiads, is said to be "augmented."

SAPIENTIA. Every number is said to be "diminished" the parts of which when added together give a sum which is less than the number of which they are parts. Such a number is 8. For the half of 8 is 4, the quarter of 8 is 2, and the eighth of 8 is 1; and these added together give 7. It is the same with 10. Its half is 5, its fifth part 2, its tenth part 1, and these added together give 8. On the other hand, a number is said to be "augmented" when its parts added together exceed it. Such, for instance, is 12. Its half is 6, its third 4, its fourth 3, its sixth 2, its twelfth 1, and the sum of these figures 16. And in accordance with the principle which decrees that between all excesses shall rule the exquisite proportion of the mean, that number is called "perfect" the sum of the parts of which is equal to its whole. Such a number is 6, whose parts—a third, a half, and a sixth—added together come to 6. For the same reason 28, 496, and 8000 are called "perfect."

HADRIAN. And what of the other numbers?

SAPIENTIA. They are all either augmented or diminished.

HADRIAN. And that "evenly even" number of which you spoke?

SAPIENTIA. That is one which can be divided into two equal parts, and these parts again into two equal parts, and so on in succession until we come to indivisible unity: 8 and 16 and all numbers obtained by doubling them are examples.

HADRIAN. Continue. We have not yet heard of the "evenly uneven" number.

SAPIENTIA. One which can be divided by two, but the parts of which after that are indivisible: 10 in such a number, and all others obtained by doubling odd numbers. They differ from the "evenly even" numbers because in them only the minor term can be divided, whereas in the "evenly even" the major term is also capable of division. In the first type, too, all the parts are evenly even in name and in quantity, whereas in the second type when the division is even the quotient is uneven, and vice versa.

HADRIAN. I am not familiar with these terms and divisors and quotients alike mean nothing to me . . . . And the "unevenly even" numbers?

SAPIENTIA. They, like the "evenly even" can be halved, not only once, but sometimes twice, thrice, and even four times, but not down to indivisible unity.

HADRIAN. Little did I think that a simple question as to the ages of these children could give rise to such an intricate and unprofitable dissertation.

SAPIENTIA. It would be unprofitable if it did not lead us to appreciate the wisdom of our Creator, and the wondrous knowledge of the Author of the world, Who in the beginning created the world out of nothing, and set everything in number, measure and weight, and then, in time and the age of man, formulated a science which reveals fresh wonders the more we study it.

109 EXERCISES

1 Write a mathematical definition of "evenly even."

2 How could we use the notion of congruences to distinguish between evenly uneven and unevenly even?

3 Are the perfect numbers of the play abundant, deficient, or perfect?

4 Calculate $\sigma^*(100)$, $\sigma^*\sigma^*(100)$, $\sigma^*\sigma^*\sigma^*(100)$ until a pattern arises. What is it?

5 Do the same with $12496 = 2^4 \cdot 11 \cdot 71$.

References

1. Peter Barlow, *An Elementary Investigation of the Theory of Numbers*, J. Johnson and Co., London, 1811.
2. Florian Cajori, *A History of Mathematical Notations*, Vol. I, The Open Court Publishing Company, La Salle, Ill., 1928.
3. ———, *A History of Mathematics*, The Macmillan Company, New York, 1931.
4. ———, *A History of Elementary Mathematics*, The Macmillan Company, New York, 1893 (reprinted 1963).
5. Tobias Dantzig, *Number the Language of Science*, Doubleday & Company, Inc., Garden City, N.Y., 1954.
6. H. Davenport, *The Higher Arithmetic*, Hutchinson House, New York, 1952.
7. L. E. Dickson, *History of the Theory of Numbers*, Vols. I, II, and III, Carnegie Institute of Washington Publication No. 256, 1918, 1920 (reprinted Chelsea Publishing Company, New York, 1952).
8. A. A. Giola and A. M. Vaidya, "Amicable numbers with opposite parity" *Amer. Math. Monthly*, (8) **74** (October, 1967), pp. 969–973.
9. G. H. Hardy and E. M. Wright, *An Introduction to the Theory of Numbers* (corrected 4th ed.), Oxford University Press, London, 1962.
10. D. H. Lehmer, "On the exact number of primes less than a given limit," *Ill. J. Math.*, (3) **3** (September, 1959), pp. 381–388.
11. I. Niven and H. S. Zuckerman, *An Introduction to the Theory of Numbers*, John Wiley & Sons, Inc., New York, 1960.
12. Oystein Ore, *Number Theory and Its History*, McGraw-Hill Book Company, New York, 1948.
13. H. Rademacher and O. Toeplitz, *The Enjoyment of Mathematics* (trans. H. Zuckerman), Princeton University Press, Princeton, 1957.
14. Daniel Shanks, *Solved and Unsolved Problems in Number Theory*, Vol. I, Spartan Books, Washington, 1962.
15. Wacław Sierpiński, *Elementary Theory of Numbers* (trans. Hulanicki), Panstwowe Wydawnictwo Naukowe, Warsaw, 1964.
16. ———, *A Selection of Problems in the Theory of Numbers* (trans. A. Sharma), Pergamon Press, New York, 1964.
17. Sherman Stein, *Mathematics, the Man-made Universe*, W. H. Freeman and Company, San Francisco, 1963.

EXCURSIONS INTO MATHEMATICS

CHAPTER   3   **WHAT IS AREA?**

BY ANATOLE BECK

Section 1 Introduction

Section 1 Introduction

We live today in a world in which measuring is a process we take for granted. Not only do we measure lengths, masses, accelerations, and voltages, but we even measure intelligence, learning, and consistency of observation. It must appear to us that whatever is quantitative, in the sense of being perceptible as "more" or "less," can be assigned a number which answers the question, "How much?" So it is natural that we accept easily the idea that area is a measurable quantity.

To get some idea of what it must have been like before the measurement of area was an accepted technique, we must turn today to the social sciences. There we measure such things as intelligence, but without much genuine faith in the measurements, and we find other quantities such as aggression, satisfaction, and fear which we perceive vaguely as quantitative but for which we have no concept of measurement.

Even when measurements do exist, they are not, in general, additive. We know something about measuring intelligence, for example, by the IQ. It is not the case, however, that if we have a job for a genius, three morons will do equally well. The intelligence of the combination is not the sum of the individual intelligences.

Thus it was a great advance in human knowledge when the Egyptians broke through to a quantitative measurement of area. We, who have lived all our lives with area as a simple concept, must strain a little to imagine a world in which it was an elusive idea. When the Greeks put geometry on a deductive basis, they imagined that they did not bring to the subject any preconceptions. But one of their basic ideas was the existence of an additive measurement of area.

In our courses in high school geometry, we make this same supposition, so it may come as something of a surprise to learn that if we are to have the subject on a sound footing, then we must find justification for believing in additive areas, especially in the case of curvilinear figures.

We usually think of our high-school geometry as Euclidean, and perhaps we imagine that it was all known to the Greeks. Certainly, our modern concept of area was not. For example, let us consider the following quotations from Boyer [1]:

The question "What is the area of a circle" would have had no meaning to the Greek geometers. But the query "What is the ratio of the areas of two circles" would have been a legitimate one.

The Pythagoreans, among the Greeks, were driven by philosophical and mystical considerations to attempt to reduce every idea in the universe to *number*. They developed a method for comparing rectilinear areas, but even they stumbled on the problem of measurement. Quoting Boyer again,

> One very important result of the Pythagorean search for unity in nature and geometry was the theory of application of areas. This . . . became fundamental in Greek geometry, in which it later led to the method of exhaustion. . . . The method of application enabled them to say of a figure bounded by straight lines that it was greater than, equivalent to, or less than a second figure. Such a superposition of one area upon another constitutes a first step in the attempt to make exactly definable the notion of area, in which a given unit of area is said to be contained in a second area a given number of times. . . . Greek mathematicians did not speak of the *area of one figure*, but of the *ratio of two surfaces*, a definition which could not . . . be made precise. . . .

With the method of applications, it was possible to compare the areas of rectilinear figures (for example, the Pythagorean Theorem), but this left unsettled the areas of circles, of ellipses, and of various curvilinear and irregular "figures." The Greeks attempted to settle these problems by "exhausting" the circle of its area, removing successively bits of the "area" until none remained. Of course, it never happened that none remained. Writing of the Greek view, Boyer says,

> The area of a circle could not literally be exhausted by applying rectilinear configurations to it inasmuch as curvilinear magnitudes were fundamentally different. The area of a circle was not to be compared with that of a square, for area was not a numerical concept. . . . The number π would have had no meaning to the Greeks.

It may be that you yourself had some qualms about this problem when you studied about area in elementary geometry. If so, you were very likely told not to worry about it, that everything worked. In fact, it is true that everything does work, if the figures are at all reasonable, but it is only in relatively recent times that a theory has been developed which puts the entire question of area on sound mathematical ground. Note that as late as 1604, the question was not settled. The methods we shall use in this chapter were beginning to be developed, but they were far from being established. In that year, Luca Valerio published a book on geometry which sought to deal with this question. Boyer says of his method:

Then Valerio assumed without proof that if the difference between the inscribed and the circumscribed figures is smaller than any given area, this will be true also of the difference between the curve and either of these figures.

It was not until the time of the development of the infinitesimal calculus that the Problem of Area began to get a mathematically satisfactory solution, and it was not until the work of Lobachevsky (axiomatization of mathematical systems), Lebesgue, and Carathéodory (measure theory) that there appeared an approach to this problem which could be made to hold water without recourse to the techniques of the differential and integral calculus.

Henri Lebesgue (1875–1941) showed early promise in mathematics; his Paris thesis (1902) on "Integral, Length, Area," remains a classic. His concept of measure for very complicated collections of points, and the consequences of this concept, took hold quickly. "Lebesgue measure" and the "Lebesgue integral" soon became basic subject in the mathematical curriculum. Although he did valuable work in other fields, his fame rests on his contribution to measure theory. He was appointed to the Sorbonne in 1910 and remained in Paris throughout his life. (Permission of the Granger Collection.)

Constantin Carathéodory (1873–1950) was born in Berlin of a distinguished Greek family. He was educated in Belgium, gave up a promising career as an engineer in Egypt to study mathematics in Berlin and Göttingen (then at the height of its mathematical glory) and taught at various German Universities. He returned to Greece to help build a new university before settling in Munich (1924–1950). The theory of measure, which had been shown by Lebesgue to be of fundamental mathematical importance, was given a final elegant axiomatization (comparable to Hilbert's axiomatization of geometry) by Carathéodory in the 1930's. (This photograph shows him in his Munich study in 1932.)

Actually, the Problem of Area is so closely tied to the problem of Integration that much of the work in this chapter is similar to work done at some point of Integration Theory. However, it will not be necessary for you to learn about limits in order to handle this material.

Once we call into question the mathematical basis of the measurement of curvilinear areas, we soon see that the measurement of rectilinear areas also needs some shoring up, by modern standards.

We can state our problem as follows: We feel intuitively that we can assign to each geometric "figure" a number which represents

Section 1 Introduction 151

its "size" in two dimensions. The numbers assigned to the figures will have certain properties, and it will be our task to show that there actually does exist a grand assignment scheme which gives such a measurement for each of many figures. We shall also see that some "figures" cannot be assigned an area by this method. In Illustration 1, you see a "puff-ball," which has no area, and the last section in this chapter consists of the construction of a curve whose interior has no area in this sense.

Illustration 1

We must be very careful of arguments that are merely plausible, rather than rigorously correct. There is a well-known example (cf. for example, Kreyszig [5], pp. 115 ff.) of a process of mismeasurement of the surface area of a simple circular cylinder. You no doubt know that the area of a cylinder is the product of its height by the perimeter of the base. Yet for any number greater than this product, a plausible argument can be made that *that* number is in fact the correct area. Certain of the introductory theorems seem obvious on their face, and yet are quite difficult to prove.

We now ask what are the fundamental properties of these numbers which we want to assign. Often we find, in mathematics, that it is possible to build a whole theory from a few basic requirements, and so it is in this case. In elementary geometry, we conceived of area as having these properties:

A1: Every geometric "figure" has an area, and this area is a non-negative real number.

A2: If one "figure" is contained in another, then the area of the contained "figure" is no larger than the area of the containing "figure."

A3: If a "figure" is made up in the proper way as a combination o two other "figures," then the area of the large figure is the sum of the areas of the two smaller ones. (See Illustration 2.) †

A4: The area of a rectangle is the product of its length and its width.

In laying down these rules, we did not bother too much with what was a "figure," and what was not. Polygons of the sort that one was used to seeing were clearly figures, as were circles and ellipses. It was not then important to consider whether an object like the one in Illustration 1, which is to be thought of as consisting of infinitely

Illustration 2

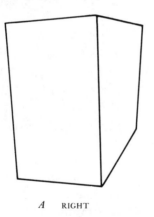
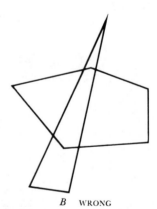

A RIGHT *B* WRONG

† The Euclidean statement of this property, which was held to be self-evident, was that "the whole is the sum of its parts." Actually, this apparently bland fact is a deep assertion. Life abounds in situations where the whole is more, or less, than the sum of its parts, in some sense. Thus, a team, or an army, is more than that many men, or even those very men, considered as an unorganized collection, or set. On the other hand, the mixing of a quart of water with a quart of alcohol does not give two quarts of liquid.

many lines, but not making up the whole interior of a circle, was or was not a figure. In A2, we never questioned the possibility that one figure might be contained in another, and that they might have the *same* area. As we shall see, a careful study of these questions will lead to some interesting answers.

Most important of these is the question of whether there is a way of assigning areas to all "figures" in a way consistent with A1, A2, A3, and A4. It is natural to feel that experience with cutting and measuring cloth, paper, etc. makes the concept of area in some sense equivalent to the concept of mass. Every figure surely can be assigned an area in a way consistent with A1, A2, A3, and A4, or so it seems. But do we have solid mathematical evidence to support this intuition?

Section 2 Rectangles and Grid Figures

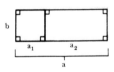

Illustration 3

Let us now retrace the steps by which we define the areas of certain plane figures and see how this leads to a concept of area. We start with rectangles. Is it easy to see that the definition of the area of a rectangle given in A4 actually satisfies A1, A2, and A3?

A1 is satisfied, if we think only of rectangles as "figures," since each product of length by width is a nonnegative real number. A3 is likewise easily satisfied (see Illustration 3) since a rectangle can be cut into two rectangles in only one way, that is by cutting it with a line parallel to one of the sides. Let the side divided have length a, and let the divisions be a_1 and a_2, respectively. Let the undivided side have length b. Then the small rectangles have areas $a_1 \cdot b$ and $a_2 \cdot b$, respectively, while the large rectangle has area $(a_1 + a_2) \cdot b = a_1 \cdot b + a_2 \cdot b$. Thus, A3 is shown. As for A2, we would like to know that a rectangle can be put into another rectangle only if the length of the inside rectangle is less than the length of the outer one, and the widths also (see Illustration 4). However, this is not always the case (see Illustration 5). In this case, the proof that the inside rectangle has the smaller area is quite difficult, and cannot be accomplished by the use of rectangles alone.

There are two ways around this minor embarrassment. One is to seek other building blocks for our "figures" than rectangles, and

Illustration 4

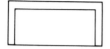

the other is to consider only rectangles that behave themselves, as in Illustration 4. We shall use both methods to see where they lead us.

Taking the second method first, we choose a fixed line in the plane, and work only with rectangles whose sides are parallel or perpendicular to that line. For such rectangles Illustration 4 is a perfectly good picture, and will give a proof of A2. We denote by the letter ℜ all the rectangles of this sort, and we state that A4 assigns to each rectangle in ℜ an area so that A1, A2, and A3 hold for these "figures."

Illustration 5

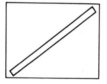

1   EXERCISES

1 Suppose A4 required that the area of a rectangle be the product of its length by twice its width. Would A1, A2, and A3 hold for rectangles in ℜ?

2 Suppose the area were the sum of the length and the width. Would A1, A2, and A3 hold?

3 Suppose A4 set the area of a rectangle to be the product of the length by the square of the width. Which of A1, A2, and A3 would hold?

4 Suppose that five rectangles are arranged as in Illustration 6, so as to form a single large rectangle. Show that the sum of the five areas given by A4 is the area of the large rectangle given by A4. That is, show that

$$a \cdot f + c \cdot g + b \cdot k + d \cdot i + e \cdot h = (a + b) \cdot (f + g).$$

Our next step is to extend the class of figures with which we shall deal. Suppose some of the rectangles of ℜ are combined (see Illustration 7) to form another "figure." Is it possible to extend the concept of area to that figure?

If we want to say that the area of the figure in Illustration 7 is the sum of the areas of the four rectangles that make it up, then we have to know which they are. Of course, it might be that it does not matter how we got the figure (see Illustration 8), and that no matter how the same figure is made up of rectangles from ℜ, the sum of the areas will be the same. That is actually the case, but we shall have to prove it.

An interesting technique used in situations like this is what is known as "common refinement." In this case, a common refinement

Section 2   Rectangles and Grid Figures   155

Illustration 6

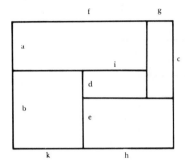

Illustration 7

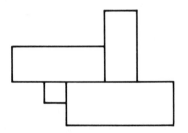   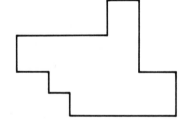

Illustration 8

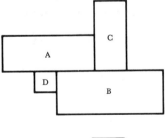   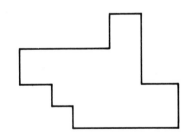

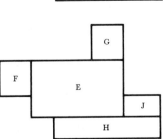

Illustration 9

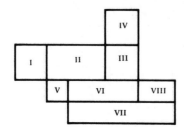

Illustration 10

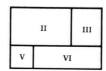 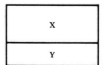

involves a resolution of the figure in question as in Illustration 9, where each line which occurs in either decomposition in Illustration 8 occurs. To show that the decompositions in Illustration 8 are equal, we shall show that both are equal to the one in Illustration 9, with which each has something in common.

The argument we now wish to make runs as follows:

$$A + B + C + D = (I + II) + (VI + VII + VIII)$$
$$+ (III + IV) + V$$
$$= (II + III + V + VI) + I + IV$$
$$+ VII + VIII$$
$$= E + F + G + H + J,$$

where we are using the same letters here to represent both the rectangles and their areas. Can you see that we have not proved anything yet? For instance, we do not know that

$$II + III + V + VI = E.$$

But we shall now prove these also. Let us combine II and III to form X, and V and VI to form Y, as in Illustration 10. II + III = X, as we already know from our discussion of A3, and V + VI = Y,

Illustration 11

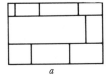

a

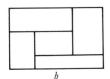

b

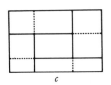

c

Illustration 12

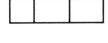

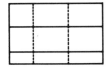

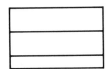

by the same reasoning, so that $(II + III) + (V + VI) = X + Y = E$, again by the same reasoning. And in a similar way, we know that

$$I + II = A, \quad III + IV = C,$$

and

$$VI + VII + VIII = (VI + VIII) + VII = B.$$

All these last arguments, in fact, deal with a single problem, which we can summarize in the following statement:

2 Proposition

If any rectangle be decomposed into (finitely many) smaller rectangles, all with their sides parallel or perpendicular to a given direction, then the sum of the areas of the small rectangles (as defined in A4) is the area of the large rectangle.

In proving this assertion, it would be useful if all rectangles, like E in Illustration 8 or the rectangle in Illustration 11(a) were cut first horizontally and then vertically, or *vice versa*. This is not the case, however, as we see in the rectangle in Illustration 11(b) and so we have to be a bit subtler. If we add the dotted lines, then we have the kind of decomposition we want. In this case, every line in the interior of the rectangle goes all the way through, and the rectangle will be said to be *sectioned*. If we can prove the proposition for the case when the rectangle is sectioned, then we may be able to use the result in the way indicated in Illustration 11(c). How do we prove the theorem for sectioned rectangles? We use the method of the previous argument.

When we have a sectioned rectangle, as in Illustration 12, we see that each horizontal row is made up of a number of rectangles of equal height, laid end to end. The area of the long rectangle thus formed is $b(a_1 + a_2 + a_3 + \cdots + a_n)$, where $a_1, a_2, \ldots, a_n$ are the lengths of the rectangles of the row. Of course, $b(a_1 + \cdots + a_n) = b \cdot a_1 + \cdots + b \cdot a_n$, so that the areas of the small rectangles making up a row is equal to that of a single rectangle comprising the same row, and the length of that rectangle is the length of the large rectangle, while its height is the height of the row. Thus we see that since the argument above is valid for each row, then the sum of all the little rectangles is the same in area as the row rectangles. But a similar argument shows that the row rectangles add up to the large rectangle in area, so that the little rectangles add up to the big one in area. Now we are ready to tackle the main proposition.

Let the rectangle A be decomposed into the rectangles A_1, A_2, A_3, . . . , A_n, as shown in Illustration 13. Let each side of each small rectangle be extended (as shown) to section the large rectangle. Now call the new rectangles B_1, . . . , B_m. It is not possible for any of the sides of the rectangles A_1, . . . , A_n to intersect any of the rectangles B_1, . . . , B_m because any such intersecting side would then have to be a part of a sectioning line, and no sectioning line passes through any of the B rectangles. So we see that each B rectangle is wholly contained in one of the rectangles A_1, . . . , A_n, and in only one, since the A rectangles have no interior points in common. Each line which sections the large rectangle A either sections A_1 or does not intersect its interior, and the same is true concerning A_2, A_3, . . . , A_n. Thus, each of the rectangles A_1, . . . , A_n is also sectioned by these

Illustration 13

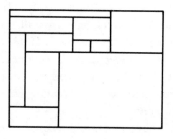

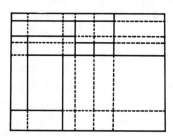

lines. Each of the rectangles $A_1, \ldots, A_n$ therefore has the same area as the sum of the B rectangles it contains. Thus, the sum of the B rectangles is, on the one hand, equal to the area of the large rectangle and, on the other hand, equal to the sum of the areas of the rectangles $A_1, \ldots, A_n$. This proves Proposition 2.

The method we have used is known as the method of "common refinement," and we shall have occasion to use it again in this chapter. Now that we have this technique, we define a category of figures which are made up of finitely many rectangles, each with its sides parallel or perpendicular to a given direction, as in Illustration 14. We call them *grid figures* or *right-angle figures* and this class of figures is denoted by $\mathcal{G}$. For technical reasons, we consider the empty set as belonging to $\mathcal{G}$. The empty set consists of 0 rectangles, and when we assign areas to the figures of $\mathcal{G}$, we shall assign the empty set the area 0.

Illustration 14

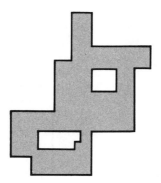

3 EXERCISES

1 Show that if a figure A belongs to $\mathcal{G}$ and A is represented as being composed of rectangles in two different ways, then the sum of the areas of the rectangles is the same for the two ways.

2 Show informally that if we assign to each figure in $\mathcal{G}$ the number indicated in Problem 1, then that assignment of "areas" to the figures in $\mathcal{G}$ will satisfy A1, A2, A3 and A4.

Illustration 15

Illustration 16

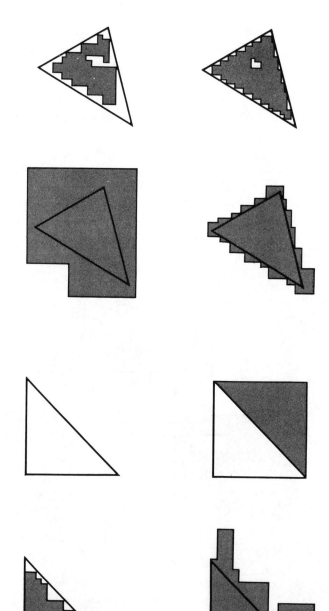

Section 3 Triangles

Illustration 17

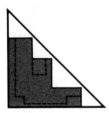

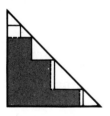

Illustration 18

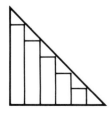

Having taken this approach, we now find that $\mathcal{G}$ represents only a small proportion of all the figures that we are interested in measuring for area. For instance, it does not include any triangles, nor does it include any rectangles whose sides are skew to the original direction in which all the figures of $\mathcal{G}$ are oriented. So our next program is to measure triangles for area.

If we are to assign an area to a triangle, then by A2, it must be at least as large as the area of any figure from $\mathcal{G}$ lying within it, and no larger than any figure from $\mathcal{G}$ within which it lies (see Illustration 15).

We shall start with the isosceles right triangle in the position shown in Illustration 16. We let the legs have length 1. If we knew for some reason that the area of the triangle were definable, and that it were necessarily the same as the area of the complementary triangle, then it would follow by A3, if we knew that A3 held, that the area would be half the product of the legs. Both of these things seemed "obvious" to Euclid. The first, that every proper sort of figure has an area, we have discussed. The second arose from his assumption that congruent figures have equal areas, itself a nontrivial assertion. In this treatment, we do not assume either of these things, so we shall have to look for different ways of showing this same result.

For any figure out of $\mathcal{G}$ which is contained in the interior of the triangle in question, we can choose another figure from $\mathcal{G}$ which contains it and is shaped like a staircase (see Illustration 17). The area of the triangle must then be at least as big as the area of the staircase figure. Let us now choose such a figure whose area we can easily calculate. Which one? Choose a number $n \geq 3$, and divide the base of the triangle into n pieces. Illustration 18 shows $n = 6$. Then build the "staircase" corresponding to this division, in the way shown. Note that there are 5 rectangles in the picture, and that in general there will be $n - 1$ rectangles. In the picture, each rectangle has width $\frac{1}{6}$, while in general the widths will be $\frac{1}{n}$. The tallest rectangle falls $\frac{1}{6}$ short of the full leg of the triangle, so that the tallest rectangle is $\frac{5}{6}$ high. The next tallest falls $\frac{1}{6}$ short of the height of the tallest, and is thus $\frac{4}{6}$ tall. In general, the heights of the rectangles, in order, is $\frac{5}{6}, \frac{4}{6}, \frac{3}{6}, \frac{2}{6}, \frac{1}{6}$. In the case of n divisions, the rectangles have heights $1 - \frac{1}{n}, 1 - \frac{2}{n}, \ldots, \frac{2}{n}, \frac{1}{n}$. Thus, their areas are $\frac{n-1}{n} \cdot \frac{1}{n}, \frac{n-2}{n} \cdot \frac{1}{n}, \ldots, \frac{2}{n} \cdot \frac{1}{n}, \frac{1}{n} \cdot \frac{1}{n}$. Adding these and factoring out the common factor $\frac{1}{n^2}$, we have $\frac{1}{n^2}((n-1) + (n-2) + \cdots + 2 + 1)$. We now need the well-known formula that $1 + 2 + \cdots + (n-2) + (n-1) = \frac{n(n-1)}{2}$. Applying this, we have the area of the staircase figure as $\frac{1}{n^2}\left(\frac{n(n-1)}{2}\right) = \frac{1}{2}\left(1 - \frac{1}{n}\right)$. So we see that the area of the triangle must be at least $\frac{1}{2}\left(1 - \frac{1}{n}\right)$ for every integer $n \geq 3$. Now, we form a similar staircase figure *around* the triangle, as shown in Illustration 19. In this case, there are n rectangles with widths $\frac{1}{n}$ each and heights $\frac{1}{n}, \frac{2}{n}, \ldots, \frac{n-1}{n}, \frac{n}{n}$, and a similar analysis will show that the area of the triangle cannot be more than $\frac{1}{n} \cdot \frac{1}{n} + \frac{1}{n} \cdot \frac{2}{n} + \cdots + \frac{1}{n} \cdot \frac{n-1}{n} + \frac{1}{n} \cdot \frac{n}{n} = \frac{1}{2}\left(1 + \frac{1}{n}\right)$. Thus, the area of this triangle must be $\frac{1}{2}$, since $\frac{1}{2}$ is the only number lying between $\frac{1}{2}\left(1 - \frac{1}{n}\right)$ and $\frac{1}{2}\left(1 + \frac{1}{n}\right)$ for all possible values of n.†

Of course, it is possible that in some other way, we might be able to show that the area of the triangle would have to be something else, say $\frac{2}{5}$, so that we would know that there is no way to define an area for the triangle *which is consistent with what we have already done*. Actually, this cannot happen, as we shall prove later. Taking the area of the triangle as $\frac{1}{2}$ *is* consistent with the earlier results.

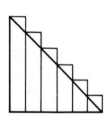

Illustration 19

---

† This is actually a sophisticated point in the argument. If the student sees a difficulty here, that difficulty arises from delicate considerations in the theory of real numbers and their representations (see Chapter 6). For purposes of continuing with the geometric argument, we shall not plumb the depths of this question.

4 EXERCISES

1 Use the same sort of analysis to derive a presumptive area for a right triangle whose base is 1 and whose altitude is 2.

2 Do the same for the right triangle both of whose legs are 2.

3 Do the same for the right triangle of base 4 and altitude 3.

We can now use the same techniques (see Exercises 4) to make an analysis of a right triangle with legs of length a and b parallel and perpendicular to the given direction. This would give a proof of Theorem 5, below.

5 Theorem

Let a right triangle have one leg parallel to the selected direction, and let its legs be of length a and b. If this triangle can be assigned an area consistent with A1, A2, A3, and A4, then that area must be $\frac{1}{2}a \cdot b$.

Illustration 20

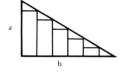

PROOF: Let the staircase figures be drawn as before, with the base divided into n equal pieces, as in Illustration 20. Then the area must lie between

$$\frac{a \cdot b}{2}\left(1 - \frac{1}{n}\right) \text{ and } \frac{a \cdot b}{2}\left(1 + \frac{1}{n}\right),$$

for every integer $n \geq 3$. Thus, it must be $\frac{1}{2}a \cdot b$.

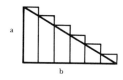

Following that line of reasoning, we could investigate the same question for a triangle of the sort shown in Illustration 21 and observe that the area would have to be $\frac{1}{2}b_1 h + \frac{1}{2}b_2 h = \frac{1}{2}(b_1 + b_2)h$.

6 Theorem

Let a triangle be chosen with one side parallel to the selected direction. Assume that the altitude on that side lies within the triangle, and that the side and altitude have lengths b and h, respectively. If this triangle has an area consistent with the foregoing analysis, then that area must be $\frac{1}{2}bh$.

Illustration 21

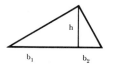

PROOF: Proved above.

In Illustration 22, we see a triangle with a horizontal base, and we note that the right triangles AFC and ABE are similar, since they share the angle at A. Thus, we see that $c/h = b/k$, and $\frac{1}{2}bh = \frac{1}{2}ck$, so that $\frac{1}{2}ck$ is also the area of the triangle. Have we now proved that the area of any triangle is equal to one-half the product of any side by

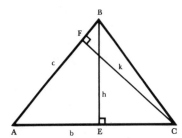

Illustration 22

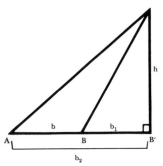

Illustration 23

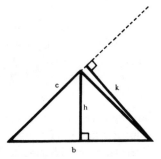

Illustration 24

the altitude on that side? Not yet. We have only shown that if one side of the triangle is horizontal, and each altitude falls within the triangle, then there can be no other area for the triangle consistent with A1, A2, A3, and A4. Note that we have not shown that this number does, in fact, give a consistent area, but only that no other number is possible.

By now, we have spoken a few times of the fact that a certain area could only be a certain number, and then commented that such a proof did not show that the figure had an area. If a certain number did not exist, for example, the smallest positive real number, we still might be able to show that *if it did exist*, it would have to have some particular value. As an example, try the following exercises.

7   EXERCISES

1 Show that if $x > 0$ and $\sqrt{x} \neq x$, then either $\sqrt{x} < x$ or $x^2 < x$.
2 Show that if x is the least positive real number, then $x^2 = x$.
3 Show that if there is a least positive real number, it must be 1.
4 Is there a least positive real number?
*5 Prove that if there is a least positive real number, it must be 2.

Our next step is to remove some of the restrictions in Theorem 6. In Illustration 23 we see that the area, if there is to be an area, of the triangle ABC must, by A3, add up with the area of triangle $BB'C$ to the area of $AB'C$. Since these last two are $\frac{1}{2}b_1h$ and $\frac{1}{2}b_2h$, respectively, we must have

$$\text{Area}(\triangle ABC) + \tfrac{1}{2}b_1h = \tfrac{1}{2}b_2h,$$

so that $\text{Area}(\triangle ABC) = \tfrac{1}{2}b_2h - \tfrac{1}{2}b_1h = \tfrac{1}{2}bh$.

Similarly, the equation $bh = ck$ does not depend on the altitude falling inside the triangle, as we see in Illustration 24. Thus, we see that for any triangle, if it has a horizontal side (that is, parallel to the direction we originally chose), then its area is equal to one-half the product of the length of any side by the length of the altitude on that side. Let's make that a theorem.

8   Theorem

Let a triangle be chosen which has a side parallel to the selected direction. If this triangle can be assigned an area consistent with the preceding theory, then that area is equal to the product of the length of any side by the length of the altitude on that side.

PROOF: Proved above.

One can easily see that the same sort of reasoning applies if the triangle has a vertical side (that is, one perpendicular to the chosen direction). If the triangle does *not* have a horizontal side, then perhaps it can be divided into two triangles with horizontal sides, as in

Illustration 25

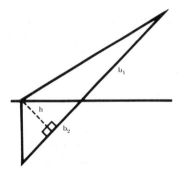

Illustration 25. The areas of the two triangles are $\frac{1}{2}b_1h$ and $\frac{1}{2}b_2h$ respectively, so that the area of the whole triangle, by A3, must be $\frac{1}{2}h(b_1 + b_2)$, that is, half the product of that side by the altitude on that side. By the methods used before, it now follows that for any triangle, the only possible area, if there is one at all, is half the product of a side by its altitude, and it doesn't matter which side.

9   EXERCISES

1 Can every triangle be cut into two triangles by a horizontal line?
2 Prove the area formula for *all* triangles.

Section   4   Polygons

As a matter of fact, in what is to follow, it is really not so convenient to work with rectangles, and we would do better to work with triangles. The reason for this is that every figure which can be constructed out of rectangles can also be constructed out of triangles, but not conversely. So we shall define the "area" of any triangle as half the product of any side by the altitude on that side, and as we see from the theory above, it does not matter which side we choose. If a "figure" can be decomposed into triangles, in some satisfactory way, then we define the area of the figure as the sum of the areas of the component triangles. That is all well and good if we have some sort of reason to know that this process will always give the same answer, no matter how the figure is cut, but how can we know that?

We return to our previous method of "common refinement." The lines of division do not necessarily cut the figure into triangles, but possibly the areas into which it does cut it can themselves be cut into triangles.

Thus, in Illustration 27, we see that each partition cuts the figure into three triangles, while the "common refinement" cuts it into 8 areas, which then are cut further to give 9 triangles. We would like to say that the areas of the triangles combine in this way:

$$A + B + C = (I + II + III) + (IV + V + VI + VII)$$
$$+ (VIII + IX)$$
$$= (I + IV) + (II + V + VI + VIII)$$
$$+ (III + VII + IX)$$
$$= D + E + F.$$

What we need is a theorem that says that if a triangle is cut up into triangles, in any way, then the area assigned to the large triangle by our formula (one-half the product of a side by the altitude on that side) is the sum of the areas assigned to the smaller triangles by the

Illustration 26

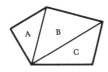 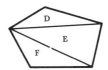

Illustration 27

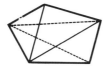 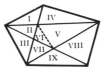 

Illustration 28

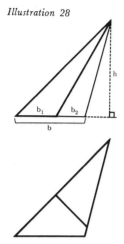

same formula. In many cases, when we want to cut up something into many pieces, and be sure that everything adds up, we find it sufficient to know that we can cut it into two pieces, and then we cut each piece into two pieces, and so on until we have all the pieces. So when we have some cutting to do, we often talk in terms of cutting things in two (notice A2, for instance). In a triangle, this works beautifully, as long as the triangle is cut into two *triangles*, as in Illustration 28. That can only be done in one way, essentially, namely by a line joining a vertex with a point of the opposite side. Then the areas of the small triangles are $\frac{1}{2}b_1 h$ and $\frac{1}{2}b_2 h$ respectively, and their sum is $\frac{1}{2}(b_1 + b_2)h = \frac{1}{2}bh$. Thus, we have a proof in this case.

If we go to three, we can sometimes do it in the same way, and sometimes not (see Illustration 29).

Even in the case where we cannot add up the bases to give a base, we can sometimes write $C + D = B$, as in Illustration 30, and $A + C + D = A + B =$ the whole triangle, using the principle for 2 again, but using it twice.

Illustration 29

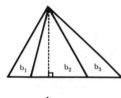
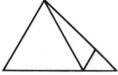

On the other hand, we can cut up a triangle into three triangles as in Illustration 31, so that no two will form a triangle together. If we extend one of the lines, however, as in Illustration 32, then we have $M + N = C$, $A + M = E$, and $B + N = F$. Therefore, we write

$$A + B + C = A + B + M + N$$
$$= (A + M) + (B + N)$$
$$= E + F = \text{the whole triangle.}$$

Illustration 30

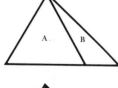
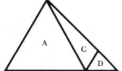

Similarly, we can locate four triangles in a triangle so that no two form a triangle. Another cut will help us however, as we see in Illustrations 33 and 34. Then $I + II + III + IV = A + C = E + F$, using the theorem for two again, and $A + B + C + D = E + F + B + D = G + H =$ the area of the large triangle. So we score again. The general process is to include in the "cutting up" process all the lines joining some chosen vertex with all the vertices of triangles, and then make triangles out of whatever quadrilaterals this might produce. Another example is seen in Illustration 35.

Let us pursue this example. If we think of the original triangle as the "big" triangle, and imagine it cut originally into "little"

Section 4   Polygons   169

Illustration 31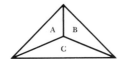

Illustration 32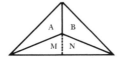

Illustration 33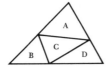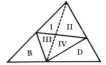

Illustration 34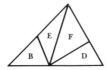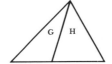

Illustration 35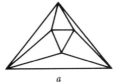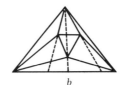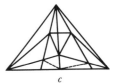

                                                        a   *b*   *c*

triangles, then this process cuts each of the "little" triangles into "tiny" triangles. The next step is to show that

(i) the "big" triangle has for its area the sum of the areas of the "tiny" triangles;

(ii) each "little" triangle has for its area the sum of the areas of the "tiny" triangles it contains.

Illustration 36

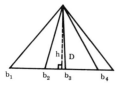

It will then follow that the area of the "big" triangle equals the sum of the areas of the "little" triangles, since each of these numbers is equal to the sum of the areas of the "tiny" triangles. Of course, this last statement also involves showing that the sum of the areas of triangles making up a triangle add up to the same area. But perhaps we shall find that these triangles are placed in some especially simple way (remember that that was the case for rectangles when we made the common refinement). In the long, thin triangles in Illustration 36 which emanate from the top vertex, the tiny triangles are so arranged that all the triangles from some point upward form a triangle. Thus, in Illustration 37, $I + II = A$, $A + III = B$, $B + IV = C$, $C + V = D$, where D is the whole triangle. Then $D = C + V = B + IV + V = A + III + IV + V = I + II + III + IV + V$. Thus, the tiny triangles in D add up to D in area, using only the fact that we know about the case when two triangles combine to make another triangle. And the same is presumably true for *each* of the long triangles. Now, we know that the sum of the areas of the long triangles is the sum of the areas of the tiny triangles they contain, on the one hand, and on the other hand, they add to $\frac{1}{2}b_1h + \frac{1}{2}b_2h + \frac{1}{2}b_3h + \frac{1}{2}b_4h = \frac{1}{2}bh$, the area of the large triangle.

Let us look once again at the triangulation in Illustration 35, and let us select one of the triangles in the triangulation, i.e., what we call a "little" triangle. To illustrate, we shall choose the triangle which is adjacent to the base of the big triangle. We see that this "little" triangle contains six tiny triangles, and we would like to know that the presumptive areas for the six tiny triangles add up to the presumptive area for the little triangle.

We note first that when we draw the lines from the top vertex to the vertices of this little triangle, one of the lines cuts the little triangle into two triangles. Not every little triangle is cut in this way, but some are. The two triangles then have the property that their presumptive areas add up to the presumptive area of the little triangle. Also, these triangles are made up of whole "tiny" triangles. Let us now examine one of these two triangles—in this case, the left-hand one (see Illustrations 38 and 39). This particular triangle is made up of three tiny triangles, and the presumptive areas add up in the proper way, as shown in Illustration 40.

We now see that we have built up enough of a mechanism for showing that the desired theorem holds in the triangulation shown in Illustration 35. For we now show that the presumptive areas of

Illustration 37

Illustration 38

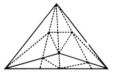 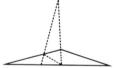

Illustration 39

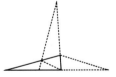

Illustration 40

Illustration 41

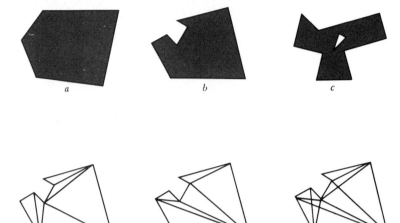

Illustration 42

Illustration 43

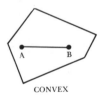

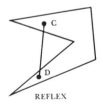

the tiny triangles add up, on the one hand, to the presumptive area of the big triangle and, on the other hand, to the sum of the presumptive areas of the little triangles.

Now suppose we were faced with another big triangle divided into little triangles in another way. Would the same techniques work there? You could try a few cases, and in fact, you should. You will find that an adaptation of these methods will work in each case. However, neither your experiments nor my statement constitute a proof. We shall have a proof, later on, but it will be difficult. Before we do, however, let us see how this gives us a concept of area. Suppose now that we are given a polygon. A polygon may be *convex* or *reflex*, and we shall define these words later. However, to give you some idea of their meaning, the polygon in Illustration 41(a) is convex; the one in 41(b) is reflex. The figure in 41(c) is a polygonal region. If we could divide any of these figures into triangles, we might be able to obtain an area for the figure by adding the areas of the triangles. It would help if the resulting number were always the same, no matter how we cut the figure into triangles. That is, indeed the case, as we shall see by our method of common refinements. (See Illustration 42.)

If a figure can be triangulated in two ways, then we can make a triangulation into "tiny" triangles. The sum of the areas of the

Illustration 44

Illustration 45

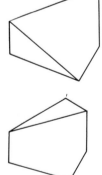

Illustration 46

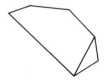

"tiny" triangles will then be the sum of the areas of the triangles in each decomposition, that is, the "area" of the figure.

10   EXERCISES

*1 Can every convex polygon be divided into triangles?
*2 Can every polygon be divided into triangles?

This now leaves us with some questions:

(1) How does one *prove* the theorem on triangulation of triangles?
(2) Which areas can be triangulated?

To answer these questions, we shall use a technique called *induction*. We shall have a few examples of this method, which consists of putting the theorem to be proved into a number of separate categories, one for each of the integers 1, 2, 3, ... Then we show the theorem in the case of one of the integers, and show that each succeeding case can be made to follow on the assumption that all the cases up to that point are known to be true. Then all the cases are known to be true. Does this sound complicated? Well, it is, but maybe an example of how we use the principle will help. First, we need a definition:

11   *Definition*

A polygon is said to be *convex* if every line joining two points inside it also lies inside. Otherwise, it is called *reflex*. (See Illustration 43.)

12   Theorem

Every convex polygon can be triangulated.

PROOF: We start with the trivial observation that every polygon of three sides is a triangle, and is thus triangulated as it stands, in a certain funny sense. Every convex polygon of four sides is cut by either diagonal into two triangles (Illustration 44). Now, how about pentagons? It is quite sufficient to note that we can cut any convex pentagon into a convex quadrilateral and a triangle (Illustration 45) and that we can cut any convex hexagon into a convex pentagon and a triangle. Thus, since we can triangulate a convex quadrilateral, we can also triangulate a convex pentagon, and in the same way, we can also use the fact concerning the pentagon to show the same thing for the hexagon. See how the truth of the theorem is passed on, from one figure to another, in a manner of speaking, from the triangle to the quadrilateral, to the pentagon, to the hexagon, and from there in turn to the heptagon to the octagon, and so on. That is, it seems we can now prove these two assertions:

(i) Every convex polygon of 3 sides can be triangulated.
(ii) If n is any number so chosen that every convex polygon with n sides can be triangulated, then $n + 1$ is also such a number.

A mathematician would say that the statements (i) and (ii) justify the assertion that:

Every convex polygon with three or more sides can be triangulated.

He gives as his reason "the principle of induction." This principle of induction is a basic property of the numbers 1, 2, 3, 4, ... Intuitively, it is based on the fact that there cannot be a *least* positive integer m for which one can have a nontriangulable convex polygon of m sides. Three is clearly not that number, and if the number were 14,297, for instance, then we would know that every convex polygon of 14,296 sides was triangulable. Thus, by (ii), we would know that every convex polygon of 14,297 sides was also triangulable, contrary to our hypothesis. So there is no such m.

Note that we have not yet shown the truth of assertion (ii). We have only discussed the fact that (i) and (ii) together would prove the theorem. We proceed now to a proof of (ii).

If we have the statement that every convex polygon of n sides is triangulable and if we are given a convex polygon of $n + 1$ sides, we cut the $(n + 1)$-gon into a triangle and an n-gon by connecting the outside vertices of two adjacent edges. We then use our hypothesis to triangulate the n-gon. With the new triangle, that gives a triangulation of the $(n + 1)$-gon. Thus, we have proved this theorem.

It is worthwhile to examine this method. Each integer "passes the theorem along" to the next integer, and we know that the integer 3 passes it to 4. Thus, we see that it must reach 5, 6, 7, 8, etc. The "etc." is deceptive, and the notion of time is also (as, for instance, *when* does it reach n?), but we deal with it on the basis that the time factor is not relevant, and the process is not defined in time.

A statement about the mathematical universe does not *become* true. It is or it isn't. In this case, all the convex n-gons are triangulable and always have been. Later on, we shall have a much more sophisticated induction to deal with.

Now, we take up the case of the reflex polygons. To show that one of these is triangulable, we shall divide it into convex polygons,

Section 4   Polygons   175

each of which is triangulable by the previous theorem. Let us now consider this present method.

Each line in the plane divides the plane into two parts, and both of these parts are convex, in the sense that the line segment joining any two points in a half-plane lies entirely in that half-plane. If we now add a second line, the two lines together cut the whole plane into three or four pieces, according as they are or are not parallel. In either case, the regions thus created are convex, and there are no more than four of them. Three lines cut the plane into as few as four regions (if they are parallel) or as many as seven (extend the sides of a triangle). And again all the regions formed are convex. In fact, n lines divide the plane into convex regions, and there are at least $n + 1$ and no more than $\frac{1}{2}(n(n + 1)) + 1$ of these regions. To verify this assertion, we need two lemmas:

13   Lemma

If K is a convex region, and a line l intersects K, then l cuts K into two convex sets.

14   Lemma

The number r of convex regions into which n lines divide the plane satisfies $n + 1 \leq r \leq \frac{1}{2}(n(n + 1)) + 1$.

PROOFS: We shall not prove the former lemma, and shall only sketch the proof of the latter. The student is to make up this deficiency.

Once more, we proceed by the inductive method. If $n = 1$, then we know that one line divides the plane into 2 regions, and that $1 + 1 = 2 = \frac{1}{2}(1 \cdot 2) + 1$, so that the lemma holds for $n = 1$. Now, if the lemma could fail to hold, let m be the smallest number for which it could fail and let $l_1, \ldots, l_m$ be any m lines in the plane. Then the lines $l_1, \ldots, l_{m-1}$ divide the plane into no fewer than m and no more than $\frac{1}{2}((m-1)m) + 1$ convex pieces. We know that l_m can intersect l_1 in at most one point, and it can intersect l_2 in at most one point, etc. If l_m intersects each line $l_1, \ldots, l_{m-1}$ once, and all points of intersection are distinct, then there are $m - 1$ points of intersection on line l_m. If l_m is parallel to some of the other lines, or if some of the points of intersection coincide, then there might be fewer. Let p be the number of points of intersection. Then these points divide l_m into $p + 1$ pieces, and $1 \leq p + 1 \leq m$. Each of the $p + 1$ pieces cuts through one of the regions into which $l_1, \ldots, l_{m-1}$ divide the plane and makes two new regions. Thus, if $l_1, \ldots, l_{m-1}$ divide the plane into r^* regions, then $l_1, \ldots, l_m$ divide it into $r^* + p + 1$.

Since m is the smallest number for which the lemma could fail to hold (we call this phrase the *hypothesis of the induction*), we know that $m - 1 \leq r^* \leq \frac{1}{2}((m - 1)m) + 1$. Since $r = r^* + p + 1$, and $1 \leq p + 1 \leq m + 1$, we have $(m - 1) \leq r^*$ and $1 \leq p + 1$, so that $m \leq r^* + p + 1 = r$. Similarly, $r^* \leq \frac{1}{2}((m - 1)m) + 1$ and $p + 1 \leq m + 1$, so that $r = r^* + p + 1 \leq \frac{1}{2}((m - 1)m) + 1 + m + 1 = \frac{1}{2}(m(m + 1)) + 1$. That is, $m \leq r \leq \frac{1}{2}(m(m + 1)) + 1$. Thus the lemma is true for m also. But this is nonsense, since it was assumed that the lemma was false for m. Thus, we conclude that there is no such m, and the lemma holds for all values of n. (These last three sentences are usually compressed into the assertion that "the theorem is thus true by induction.")

Our next task is to turn this machinery to the problem of triangulating the reflex polygon:

15 Theorem

Every reflex polygon can be triangulated.

PROOF: Let $s_1, \ldots, s_n$ be the sides of a reflex polygon, and let $l_1, \ldots, l_m$ be the lines on which they lie (see Illustration 47). Note that we write $l_1, \ldots, l_m$ instead of $l_1, \ldots, l_n$, since two sides might lie on the same straight line, as is shown in Illustration 48. Thus $m \leq n$. The lines $l_1, \ldots, l_m$ divide the plane into at most $\frac{1}{2}(m(m + 1)) + 1$ convex regions, and none of these is cut by any of the lines $l_1, \ldots, l_m$, because of the way in which we defined them. It follows that none of the regions is cut by any side, and that each is thus either entirely inside the polygon or entirely outside it. Those that lie inside make up the whole interior of the polygon, and since they are all convex, each can be triangulated. Thus, the original reflex polygon can be triangulated also, which was to be proved. (In Latin, *quod erat demonstrandum*.)

Illustration 47

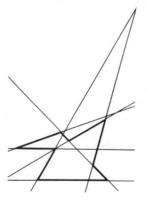
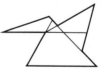

Illustration 48

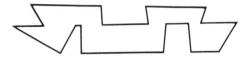

16 EXERCISES

In a polygon, any line joining two vertices which is not a side is called a *diagonal*. If it lies entirely inside the polygon, it is called an *interior diagonal*.

1. Show that in a convex polygon, all the diagonals are interior diagonals.
2. Show that a convex polygon of n sides can be cut by diagonals into $n-2$ triangles.

*3. Show that every reflex polygon has an interior diagonal.

*4. Of the r regions into which n given lines divide the plane, let b be the number of regions which are bounded. (A region is called *bounded* if it can be included within some rectangle.) Show that

$$0 \leq b \leq \tfrac{1}{2}(n-1)(n-2)$$

5. Show that for every positive number n, we can find n lines which divide the plane into $n+1$ regions.

*6. Show that for every positive number n, there are n lines which divide the plane into $\tfrac{1}{2}[n(n+1)] + 1$ regions.

*7. Show that for every positive number n, there are n lines which divide the plane so that $\tfrac{1}{2}(n-1)(n-2)$ of the resulting regions are bounded.

8. Show that if the plane is cut by n lines, not all parallel, then the number of unbounded regions is always $2n$.

Section 5 Polygonal Regions

Illustration 49

Now that we have triangulated all polygons, we shall go on to more complicated things of the same sort. A *polygonal region* is any figure which is bounded by finitely many straight line segments—as, for example, the figures in Illustration 49. A polygon, convex or reflex, is a polygonal region. A polygonal region may come in one piece or several pieces. Its boundary might include a polygon whose interior does not lie in the region, or does so only in part. Each such polygon is called a *hole* of the region. A polygonal region with no holes is made up of finitely many polygons.

To triangulate polygonal regions, we can use the same technique we used for the reflex polygons or we can invent a new one. Suppose, for example, that we had a region with one hole. Then Illustration 50 shows how we might break it into two reflex polygons. Each of these can be triangulated, so the polygonal area can be triangulated also. Suppose it has two "holes" (Illustration 51). We might try cutting it into two reflex polygons, but this is a method which we might find hard to describe in a later case when we had more holes. So instead, we shall cut it into a number of regions, *each of which has fewer holes* than the original one (Illustration 52).

Illustration 50

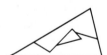
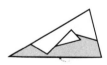
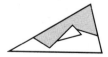

To prove this, we use a new tool, a new kind of induction, which we call a *course-of-values* induction. You remember that the ordinary mode of induction works like this:

(i) The statement is true for $n = k$.
(ii) Whenever the statement is true for a particular number $m - 1$, then it is true for m as well.

THEREFORE, the statement is true for all $n \geq k$.

In the course-of-values induction, we change (ii) to (ii'):

(ii') Whenever the statement is true for *all* values of n satisfying $k \leq n < m$, then it is true for m also.

Thus, the mode of the course-of-values induction is as follows:

(i) The statement is true for $n = k$.
(ii') Whenever the statement is true for all values of n satisfying $k \leq n < m$, then it is true for m also.

THEREFORE, the statement is true for all values of $n \geq k$.

Let R be any polygonal region with n holes, where $n > 1$.

Section 5   Polygonal Regions

Illustration 51

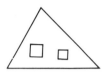

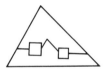

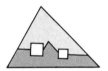

Illustration 52

Let P be some point *inside* one of the holes, and let l be a line through P, as in Illustration 53. Then l cuts R into a finite number of regions (how many?), each of which lies entirely on one side of l. Let R_1 be any of these subregions. Then R_1 may or may not have holes, but every hole of R_1 is also a hole of R. (Prove this!) Furthermore, the hole containing p is not a hole in R_1, since that hole is "cut open" by l. Thus, R_1 has fewer holes than R. *By the hypothesis of the induction*, R_1 can be triangulated. In fact, each of the regions that l cuts R into can be triangulated. Thus, R can be triangulated.

Now, let us be a little more careful about what we mean. First, we must see that each line l which cuts a polygonal region with s sides (we count inner and outer edges as sides) cuts it into at most $\frac{s}{2} + 1$ subregions. (Confirm this assertion!) Then we observe that if R is a polygonal region with n holes, then R is certainly triangulable if $n = 1$, since l would then cut R into polygons (no more than $\frac{s}{2} + 1$ of them), each of which would be triangulable.

If it were *not* the case that every polygonal region can be triangulated then let n_0 *be the smallest number of holes* a nontriangulable region could have, and let R be a polygonal region with n_0 holes and s sides which cannot be triangulated. Note that $n_0 > 1$. Let P and l be chosen as before, and assume that l cuts R into k regions R_1, $R_2, \ldots, R_k$. Note that $1 < k \leq \frac{s}{2} + 1$. Each of these k regions has fewer than n_0 holes (maybe none), and is polygonal. Thus, each of them can be triangulated. But this gives us a triangulation of R, contrary to the hypothesis that R is not triangulable. Thus, every such polygonal region is triangulable. This demonstrates the method known as *course-of-values induction*, where the truth of the assertion in the nth case rests not merely on the $(n-1)$st case but on all the cases up to the $(n-1)$st.

The above argument is now a proof of the following theorem:

Illustration 53

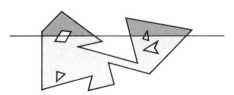

17 Theorem

Every polygonal region can be decomposed into a finite collection of triangles.

PROOF: Can you see that this statement has been proved?

18 EXERCISES

Now that we have this new tool, we continue Exercises 16 with the following:

1 Show that every reflex polygon of n sides can be cut by interior diagonals into $n-2$ triangles.
*2 Show that every reflex polygon has as interior diagonal which, together with two of the sides of the polygon, forms a triangle whose interior lies inside the polygon.

Now that we have set up the machinery of induction, let us apply it to the problem of the triangulation of triangles. Remember that one of our earliest propositions was this:

19 Theorem

If any triangle be decomposed into finitely many triangles, and if each of the triangles mentioned is assigned for its area half the product of one of its sides by the altitude on that side, then the area of the large triangle is equal to the sum of the areas of the smaller triangles.

We "proved" this assertion using an example, and we asserted that a similar situation would obtain in any case we attempted to deal with. Now we shall give a formal proof by induction. In any induction, we must choose the "parameter of the induction," that is, the number n of the induction. We might naturally choose, for example, the number of the smaller triangles, or the number of vertex points, and attempt to work from there. Perhaps a proof can be found based on these parameters. However, we shall choose for the parameter of our induction *the number of points lying inside the large triangle which are vertices of a small triangle*. By "inside" we mean really inside, not on the boundary. The reason for our choice is that the induction step, going to m from the smaller values, is very easy. By putting a line through any vertex of the large triangle and any interior vertex, we

Illustration 54

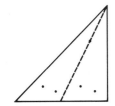

divide the triangle into two triangles, each of which has fewer interior vertices than the large triangle (Illustration 54). There are still some details to be attended to in this induction step, but before we get around to them, let us see that the entire induction does not work unless we can prove the theorem in the case where the number of interior vertices is 0. To prove *this*, we need *another* induction, and so we shall have to prove it first as a lemma before we start on the theorem.

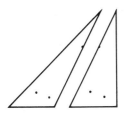

20 Lemma

If a triangle be divided into lesser triangles in such a way that no vertex of any of these lesser triangles lies in the interior of the large triangle, and if each triangle be assigned for its area half the product of some side by the altitude on that side, then the sum of the areas of the lesser triangles equals the area of the large triangle.

Illustration 55

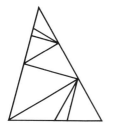

To prove this lemma, we use another induction, and in this case, we take for the parameter of our induction another number depending on the vertices. We see from our picture that the total number of vertices might or might not be a good parameter. In fact, we choose another unusual parameter, *the number of vertices on the edge having the fewest vertices*. In Illustration 55, that number is 2 and the edge in question is the base (for reasons of convenience, we do not count the vertices of the large triangle). A line drawn from the opposite vertex will divide the large triangle into two triangles in such a way that each of them has no interior vertices in the triangulation, and so that the parameter of induction is decreased. Again, the sticky point is showing the lemma in the case where the parameter of induction is 0 and, believe it or not, this requires yet another lemma, and yet another induction.

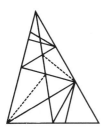

21 Lemma

Let a triangle be decomposed into lesser triangles in such a manner that none of the vertices of the lesser triangles lies in the interior of the large triangle. Suppose further that one of the edges of the large triangle contains no vertices of the lesser triangles except its end points (which are vertices of the large triangle). If we assign to each triangle mentioned above for its area a number equal to half the product of a side by the altitude on that side, then the sum of the areas of the lesser triangles is equal to the area of the large triangle.

PROOF: We have already shown this in the case where the number of lesser triangles is 2. We now proceed by induction on the number n of lesser triangles. Thus, suppose we know the lemma to be true for all values of n up to and including some number $m - 1$. We wish to show the lemma for m little triangles, based on the knowledge that it is true for $2, 3, \ldots, m - 1$ little triangles. Let us label as A and B the end points of a side of the large triangle containing no vertex of a little triangle, and let C be the third vertex of the large triangle. Then $\overline{AB}$ is also a side of a little triangle, and let D be its third vertex. We know that D does not lie inside $\triangle ABC$ by the hypothesis, and so it must lie on $\overline{AC}$ or $\overline{BC}$. Label the large triangle so that $\overline{AC}$ is the side it lies on. Then the line $\overline{BD}$ divides $\triangle ABC$ into 2 triangles, $\triangle ABD$ and $\triangle BCD$. Since $\triangle ABD$ is one of the little triangles, we have the other $m - 1$ little triangles comprising $\triangle BCD$. Thus, the area of $\triangle BCD$ is equal to the sum of the areas of the $m - 1$ little triangles *by the hypothesis of the induction*. Since the area of $\triangle BCD$ plus the area of $\triangle ABD$ adds up to the area of $\triangle ABC$, we have that the sum of the areas of the $m - 1$ triangles, plus the area of the mth, adds up to the area of the large triangle. Thus, the lemma holds for m little triangles as well. *By the induction*, the lemma is now true for all cases, *quod erat demonstrandum* (Q.E.D.).

We now proceed to prove Lemma 20.

PROOF OF LEMMA 20: Let the large triangle be divided as required into lesser triangles, with no vertex of any of the lesser triangles lying inside the large triangle. For each side of the large triangle, count the number of vertices of small triangles lying on it, exclusive of the end points themselves. Let n denote the least of these numbers, and n will be the parameter of our new induction. If $n = 0$, we know the theorem to be true by Lemma 21. Now we shall prove our induction step. We assume the lemma is known in the cases where the parameter of induction equals $0, 1, \ldots, m - 1$. Let $\triangle$ be a large triangle which is decomposed into lesser triangles $\triangle_1, \triangle_2, \ldots, \triangle_k$ where the smallest number of vertices of $\triangle_1, \triangle_2, \ldots, \triangle_k$ lying on a side of $\triangle$ (exclusive of the end points) is m. Label that side of $\triangle$ as $\overline{AB}$ and let the third vertex of $\triangle$ be denoted as C. Choose any vertex of a small triangle lying on $\overline{AB}$ and denote it as D. Draw the line $\overline{CD}$. For each of the triangles $\triangle_1, \ldots, \triangle_k$, $\overline{CD}$ either does not divide it, or divides it into 2 triangles, or divides it into a triangle and a convex quadrilateral. For each such quadrilateral, let a diagonal be drawn which divides it into 2 triangles. Then each of the triangles $\triangle_1, \triangle_2, \ldots, \triangle_k$ is thus "divided" into 1, 2, or 3 smaller triangles. Since for each triangle $\triangle_j$, the line CD cannot intersect all 3 of its sides (not counting the end points), the division of $\triangle_j$ into other triangles must leave at least one side without new vertices. Thus, by the previous lemma, each triangle $\triangle_j$ has an area equal to the sum of the areas of the triangles into which it is divided. Therefore, if we refer to these new triangles as the "tiny" triangles, then the sum of the areas of the small triangles is equal to the sum of the areas of the tiny triangles. Now, we have $\triangle ABC$ divided into $\triangle ADC$ and $\triangle BDC$. Each tiny triangle lies entirely in $\triangle ADC$ or $\triangle BDC$, and each of these is composed of the tiny triangles in it. If we examine $\triangle ADC$, we see that $\overline{AD}$ contains no more than $m - 1$ (possibly fewer) of the vertices of the tiny triangles. $\overline{AC}$ contains at least m, and $\overline{CD}$ contains an unknown number. Also, no vertex of any of the tiny triangles lies inside $\triangle ADC$. Thus, we see that this decomposition of $\triangle ADC$ falls into a previous case of the lemma, that is, one in which the parameter of the induction is $0, 1, 2, \ldots, m - 2$, or $m - 1$. It follows that $\triangle ADC$ has for its area the sum of the areas of the tiny triangles which lie in it, and the same can be said for $\triangle BCD$, by the same reasoning. Thus, the sum of the areas of the tiny triangles is equal to the sum of the areas of $\triangle ADC$ and $\triangle BDC$. Since we have shown that the sum of the areas of the tiny triangles equals the sum of the areas of the small triangles, and since we know that the sum of the areas of $\triangle ADC$ and $\triangle BDC$ equals the area of $\triangle ABC$, we see that the lemma holds in this instance. Thus, the lemma is true for the value m of the parameter, and the induction step is complete. By the principle of induction, now, the lemma is proven. Q.E.D.

Now we can prove the theorem:

PROOF OF THEOREM 19: We let $\triangle$ be the large triangle, and $\triangle_1, \triangle_2, \ldots, \triangle_k$ the small ones. Let n (the parameter of the induction) be the number of vertices of the triangles $\triangle_1, \ldots, \triangle_k$ which lie *in the interior* of $\triangle$. If $n = 0$, then we know the theorem, by the previous lemma. Suppose we know that the theorem holds for $n = 0, 1, \ldots, m - 1$. Now we want to show that it holds for $n = m$. Let D_1 be any vertex of one of the small triangles which lies in the interior of $\triangle = \triangle ABC$. Let D be the point at which $\overline{CD_1}$ (extended) meets $\overline{AB}$. For each of the small triangles which is cut into a triangle and a quadrilateral by $\overline{CD}$, let a diagonal of the quadrilateral be drawn, and the 3 triangles thus formed designated as "tiny" triangles. If $\overline{CD}$ cuts a small triangle into 2 triangles, then designate each of them as a "tiny" triangle. Each small triangle which is *not* cut by $\overline{CD}$, we will call a "tiny" triangle also. As we say in the previous lemma, each small triangle has for its area the sum of the areas of the tiny triangles it contains. Thus, as before, the sum of the areas of the small triangles is equal to the sum of the areas of the tiny triangles.

Illustration 56

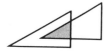

Now, let us examine the triangles $\triangle ADC$ and $\triangle BDC$. Each is made up of the tiny triangles which lie in it, and every vertex of a tiny triangle is also a vertex of a small triangle, except possibly for the vertices on $\overline{CD}$, which may or may not be. Thus, if P is a vertex of a tiny triangle and P lies inside $\triangle ADC$, then P is a vertex of a small triangle, P lies inside $\triangle ABC$, and $P \neq D_1$. Therefore, there are at most $m - 1$ such vertices inside $\triangle ADC$, and the same reasoning holds for $\triangle BDC$. By the hypothesis of the induction, then, we know that each of these two triangles has for its area the sum of the areas of the tiny triangles it contains. Since the sum of these two areas is the area of $\triangle ABC$, we have the sum of the areas of the small triangles being equal to the area of the large triangle.   Q.E.D.

Now we are ready for the major theorems in this section.

22 Theorem

Let A be a polygonal figure, and let A be triangulated in some way. Let a be the sum of the areas of the triangles. Then if A is triangulated in another way, the sum of the areas of the triangles in the second triangulation will also be a.

PROOF: Let $\triangle_1, \triangle_2, \ldots, \triangle_n$ be one triangulation, and let $\triangle'_1, \triangle'_2, \ldots, \triangle'_m$ be the other. Then if the triangles $\triangle_j$ and $\triangle'_k$ intersect so that some point lies inside both, then the intersection is a triangle, a quadrilateral, a pentagon, or a hexagon (see Illustration 56). In any case, if P and Q are two points lying in both $\triangle_j$ and $\triangle'_k$, then $\overline{PQ}$ lies in $\triangle_j$ and $\triangle'_k$, and so is in the intersection. Therefore, the intersection is convex. Thus, each side of each triangle can contribute at most one side to the boundary of the intersection. For each such figure, we perform a triangulation. Then we have a new triangulation of A, and it involves no more than $4 \cdot m \cdot n$ "little" triangles. Each triangle $\triangle_j$ is composed of the "little" triangles lying in it (no more than $4m$ of them) and each $\triangle'_k$ is composed of no more than $4n$ of them. Thus, we see that the sum of the areas of $\triangle_1, \triangle_2, \ldots, \triangle_n$ is the sum of the areas of the "little" triangles, and so is the sum of the areas of $\triangle'_1, \triangle'_2, \ldots, \triangle'_m$, which proves the theorem.   Q.E.D.

Illustration 57

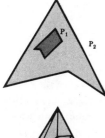

23 Definition

The number a, given in Theorem 22 will be taken as the area of the polygonal figure A.

At this point, we should be able to see that the area we have defined for polygonal figures satisfies the three criteria we set down for an

Illustration 58

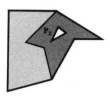

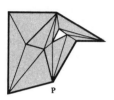

area function. It is clearly nonnegative (A1). If a polygonal figure P_1 is contained in a polygonal figure P_2, then we can triangulate P_1 and also the part of P_2 which is not in P_1, and thus produce a triangulation of P_2 which, when restricted to P_1, is a triangulation of P_1, as in Illustration 57. It then follows directly that the sum of the areas of the triangles contained in P_1 is no more than the sum of the areas of those in P_2, thus showing A2.

If the figures P_1 and P_2 combine to make up the figure P, as in Illustration 58, and have no overlap, then any triangulations of P_1 and P_2 will together give a triangulation of P, and thus we have Area(P) = Area(P_1) + Area(P_2), which is A3. A4 is clear from the definition. Thus, the well-known "area" for polygonal figures is indeed an area function in our sense.

Section 6 Area in General

Our next task is to extend this function to "figures" which are not polygons, or polygonal, in the sense used above: circles, ellipses, figures with infinitely many sides, the interiors of certain curves, such as those in Illustration 59.

To do this, we return to the use of rectangles. Just as triangles were better suited to polygons, so rectangles are better suited to this later work. If we have a circle, for instance, we are interested in

Illustration 59

knowing whether there are many ways to define its area or just one, consistent with what has gone before. The classical proof, if you have forgotten it, goes as follows:

24   Theorem

The area of the circle is equal to one-half the product of its perimeter by its radius.

PROOF: Let n be any integer which is at least 3, and let a regular n-gon be inscribed in the circle (call the circle K) and another around K. Illustration 60 shows the case for $n = 6$. Then the area of the circle, if there were one, would be some number between the areas of these two n-gons. Let us *estimate* the areas of the n-gons, using Illustration 61. The side of the inscribed n-gon is no longer than the arc which it cuts off (a straight line is the shortest distance between two points), and thus the side is no longer than p/n, where p is the perimeter of the circle. Therefore $a \leq p/2n$, and the triangle ABC has for its area $\frac{1}{2}a \cdot b \leq \frac{1}{2}(p/2n) \cdot r = pr/4n$. There are $2n$ such triangles, which tells us that the area of the inscribed polygon is no more than $\frac{1}{2}pr$. On the other hand, the perimeter of the outer polygon is at least as great as that of the circle (how would you prove this?), so that $d \geq p/2n$ and the area of $\triangle ABD \geq rp/4n$, in the same way. Thus, the area of the circumscribed polygon should be no less than $\frac{1}{2}rp$, in the same way.

We now note that the triangles $\triangle ABC$ and $\triangle ABD$ are similar, and that $a/d = b/r$. Furthermore, $b^2 = r^2 - a^2 \geq r^2 - (p/2n)^2$. Thus $b \geq r\sqrt{1 - (p/2nr)^2}$, and therefore $a \geq d\sqrt{1 - (p/2nr)^2}$. It follows that $a \cdot b \geq r \cdot d(1 - (p/2nr)^2)$, so that the area of the inner polygon (nab) exceeds the area of the outer polygon multiplied by $(1 - (p/2nr)^2)$. Thus,

$$\text{area of circle} \geq \text{area of inner polygon}$$
$$\geq (\text{area of outer polygon})\left(1 - \left(\frac{p}{2nr}\right)^2\right)$$
$$\geq \tfrac{1}{2}pr\left(1 - \left(\frac{p}{2nr}\right)^2\right).$$

Similarly,

$$\text{area of circle} \leq \text{area of outer polygon}$$
$$\leq \frac{(\text{area of inner polygon})}{(1 - (p/2nr)^2)}$$
$$\leq \frac{\tfrac{1}{2}pr}{(1 - (p/2nr)^2)}.$$

Illustration 60 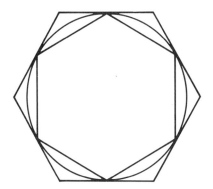 *Illustration 61*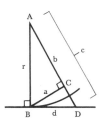

Thus, for every n, the area of the interior of the circle must lie between $\frac{1}{2}(pr)(1 - (p/2nr)^2)$ and $\frac{1}{2}pr/(1 - (p/2nr)^2)$. There is, of course, only one number which lies between those bounds for all values of n, namely, $\frac{1}{2}pr$.

Q.E.D.

Please note that this proof shows only that the area of the circle cannot be other than $\frac{1}{2}pr$. If another analysis should show that it could not be other than, for example, $3r^2$, it would then follow that there was no area for a circle possible. We shall now concern ourselves with this question:

Given a figure K, we all agree that if P_i is a polygon lying inside K, and P_0 is a polygon containing K, then Area$(P_i) \leq$ Area(K) $\leq$ Area(P_0), if Area(K) has meaning. For which figures K does this determine a unique possible value for Area(K)?

25 EXERCISES

1 Let P_0 be a regular polygon circumscribed about a circle K. Can you show that the length of P_0 is at least as large as the perimeter of K? What do we mean by the perimeter of a circle?† Is the problem any easier if P_0 is known to be a square?

† The most common definition of the perimeter of a circle is that it is the least number which is greater than or equal to the perimeter of every inscribed polygon. If we adopt this definition, then this exercise asks you to show that every circumscribed polygon has a longer perimeter than any inscribed polygon. Does this make the problem any easier? Make similar adjustments in the other exercises in this section.

2 Suppose P_0 in Problem 1 is not a regular polygon but still has every side tangent to the circle. Is the statement still true?
3 Same question, but with P_0 a convex polygon containing K.
4 Is the same statement true if P_0 is not convex?
*5 If we look at *interior* polygons, we can still ask questions like Problems 1–4. What are the questions? What are the answers?

Actually, we shall not be using all polygons yet. We shall first take up the somewhat easier case where the interior and exterior polygons (or polygonal areas) belong to $\mathcal{G}$. Thus, we consider the following situation:

Suppose A is any grouping or collection of points in the plane. Suppose we were now to include A in one of the figures in $\mathcal{G}$; call that figure B. Also, let C be a figure from $\mathcal{G}$ so chosen that every point inside C is a point of A (see Illustration 62). To do this, it may be necessary for C to be a single point, in which case, we will say that its area is 0. Now, let us look at the difference in area between B and C. If t is any real number with $t \geq \text{Area}(B) - \text{Area}(C)$, then we construct a class of figures, which we call $\mathcal{Q}_t$, which includes A. Let us be more formal with this definition, since it is a little tricky.

26 *Definition of $\mathcal{Q}_t$*

Let t be a real number greater than 0. A collection A of points of the plane is included in $\mathcal{Q}_t$ if we can find two figures B and C in $\mathcal{G}$, with A included inside B and every point inside C a member of A, and with the additional property that $\text{Area}(B) - \text{Area}(C) \leq t$. We see immediately that if $t_1 > t_2$ and some figure A is in $\mathcal{Q}_{t_2}$, then A is in $\mathcal{Q}_{t_1}$ also. Now the figures we are really interested in are those which are in *every one* of the collections $\mathcal{Q}_t$. As an example, let us look

Illustration 62

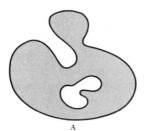
A

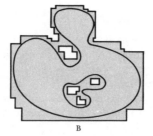
B

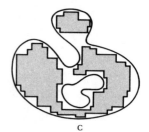
C

Section 6   Area in General

again at the isoceles right triangle. We put it between two "staircases," whose areas were $\frac{1}{2}\left(1 + \frac{1}{n}\right)$ and $\frac{1}{2}\left(1 - \frac{1}{n}\right)$, respectively. Thus, the triangle belonged to $Q_{1/n}$, by our definition, for every $n \geq 3$. Since, for every $t > 0$, we can find some n so large that $1/n < t$, we see that the triangle belongs to Q_t for all $t > 0$.

27   Definition
A figure belongs to the category Q if it belongs to Q_t for every $t > 0$.

If a figure belongs to Q, then we want to define its area. We look at the areas of all the figures from G which contain it and at all those figures from G which are included in it. Let B be a containing figure, and let C be contained. Then C is also contained in B, so that

$$\text{Area}(C) \leq \text{Area}(B).$$

Now, the possible areas for the outer figure include, with every number in it, all larger numbers as well, and the possible areas for the inner figures include, with each nonnegative number which is an example, all smaller nonnegative numbers (if any). Let us call these two classes of real numbers *upper numbers* and *lower numbers* for the figure A. As a matter of form, let us include all the negative numbers among the lower numbers. Then we now know:

(i) Every number greater than an upper number is an upper number.
(ii) Every number less than a lower number is a lower number.
(iii) We can never have a lower number greater than an upper number.

Furthermore, if A is in Q, then it is in every category Q_t, so that

(iv) For every $t > 0$, we can find an upper number b and a lower number c so that $b - c \leq t$.

It is a fundamental property of the real numbers (a property *not* shared by all ordered systems) that (i), (ii), (iii), and (iv) combine to ensure the existence of a *single* real number t_0 with the property that every number greater than t_0 is an upper number; every number less than t_0 is a lower number. We shall assume this property of the real numbers, and we shall accept t_0 as the area of A. Note that by the same sort of reasoning as that used in the case of the isoceles

Illustration 63

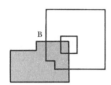

right triangle, t_0 is the only number which could *possibly* serve as the area. Our next task is to see that A1, A2, A3, and A4 apply to these figures and to this definition of the area.

28 Lemma

Let A and A' be two figures with no points in common, and assume that A is in Q_t, and A' is in $Q_{t'}$. Designate by D the figure made by putting A and A' together, that is, by thinking of them together as a single figure. Then D is in $Q_{t+t'}$.

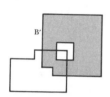

PROOF: Choose B to include A, and C included in A (we will use the symbols $B \supset A$ and $A \supset C$), so that B is in $\mathcal{G}$, C is in $\mathcal{G}$, and Area(B) − Area$(C) \leq t$. Choose C' and B' so that $B' \supset A'$, $A' \supset C'$, and Area(B') − Area$(C') \leq t'$. Now, C and C' can be taken together to constitute a figure F in $\mathcal{R}$. Since A does not overlap A', we see that C does not overlap C', and Area$(F) =$ Area$(C) +$ Area(C'). If we take B and B' together to constitute a figure E, B may overlap B'. In that case, we will consider any point which is in B or B' to be in E. We use the symbol $B \cup B'$ to designate this combination. Now, the area of $B \cup B'$ is no more than Area$(B) +$ Area(B'), a fact which we shall not prove here but which *does require proof*. Thus, Area$(E) \leq$ Area $(B) +$ Area(B'). Now, E belongs to $\mathcal{G}$, as does F, and we have $E \supset D$ and $D \supset F$. To show that D is in $Q_{t+t'}$, we note that

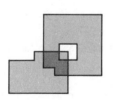

$$\text{Area}(E) - \text{Area}(F) \leq (\text{Area}(B) + \text{Area}(B')) - (\text{Area}(C) + \text{Area}(C'))$$
$$= (\text{Area}(B) - \text{Area}(C)) + (\text{Area}(B') - \text{Area}(C'))$$
$$\leq t + t'. \quad \text{Q.E.D.}$$

29 Corollary

If A, A', and D are as chosen above, and A and A' are both in Q, then D is also in Q.

PROOF: Prove it yourself.

30 EXERCISE

In Lemma 28, use what you already know to prove that Area$(E) \leq$ Area$(B) +$ Area(B').

Our next theorem in this context is a little harder:

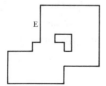

31 Theorem

If $E_1 \in Q_t$ and $E_2 \in Q_t$, then $E_1 \cap E_2$ (the collection of points which are in both) is in $Q_{t+t'}$.

Section 6   Area in General   191

PROOF: We first choose sets G_1, F_1, G_2, and $F_2 \in \mathcal{G}$, with $F_1 \subset E_1 \subset G_1$, $F_2 \subset E_2 \subset G_2$, $A(G_1) - A(F_1) \leq t$, $A(G_2) - A(F_2) \leq t'$, where we use now the symbol $A(G_1)$ instead of Area (G_1) and the symbols $A(F_1)$, $A(G_2)$, etc., in the same way. We shall use this new notation for the remainder of the chapter. Now, let us look at all the points which lie in both G_1 and G_2, that is, $G_1 \cap G_2$, and call it G_0. Let us similarly look at $F_1 \cap F_2$ and call it F_0. Every point of F_0 is in both F_1 and F_2 and is thus in both E_1 and E_2, so that $F_0 \subset E_1 \cap E_2$. Similarly, every point of $E_1 \cap E_2$ is in both E_1 and E_2 and so in both G_1 and G_2 and therefore in G_0 (see Illustration 64). Note that $E_1 \cap E_2$, F_0, and G_0 need not be in one piece (see Illustration 65). We now want to know something about $A(G_0) - A(F_0)$. This is a little tricky, since we do not know anything about either $A(G_0)$ or $A(F_0)$. However, let us designate by H_0 the collection of all the points in G_0 which are not in F_0 (and which we write as $G_0 \backslash F_0$). It is clear that $A(F_0) + A(H_0) = A(G_0)$, so that $A(G_0) - A(F_0) = A(H_0)$.

Illustration 64

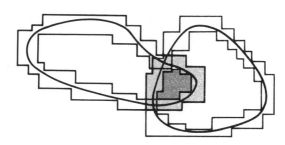

Illustration 65

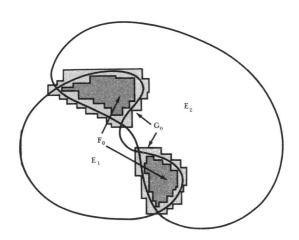

Illustration 66

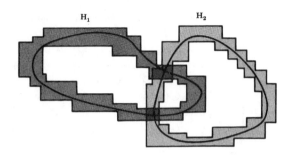

Illustration 67

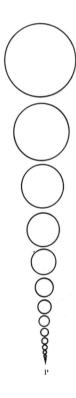

Suppose some point P is contained in H_0. Then P is in $G_0 = G_1 \cap G_2$, but not in $F_0 = F_1 \cap F_2$. Therefore, P is not in F_1, or P is not in F_2 or P is not in either F_1 or F_2. If P is not in F_1, then since P *is* in G_1, P is in $G_1 \setminus F_1$, which we shall call H_1. If P is not in F_2, then P is in $H_2 = G_2 \setminus F_2$. If P is in neither F_1 nor F_2, then P is in $H_1 \cap H_2$. Now, $G_1 = F_1 \cup H_1$, so that $A(H_1) = A(G_1) - A(F_1)$, and similarly, $A(H_2) = A(G_2) - A(F_2)$. Thus, $A(H_1) \leq t$ and $A(H_2) \leq t'$. The area of $H_1 \cup H_2$ is no larger than $t + t'$, though it might be smaller, due to the overlap, if any. Since every point of H_0 is in H_1 or H_2 or both, $A(H_0) \leq A(H_1 \cup H_2) \leq t + t'$. Thus, $A(G_0) - A(F_0) \leq t + t'$, and $E_1 \cap E_2 \in \mathcal{Q}_{t+t'}$. Q.E.D.

32 Corollary

If E_1 and E_2 belong to $\mathcal{Q}$, then $E_1 \cap E_2$ belongs to $\mathcal{Q}$.

We shall now look at another theorem of the same type, where the reasoning is very similar, but much trickier. Here, we want to show the same idea for $E_1 \setminus E_2$, the aggregate which is contained in E_1 but excluded from E_2. We should note, at this point, that our "figures" by this time might come in many pieces, possibly even infinitely many. As an example of a "figure" of infinitely many pieces that belongs to $\mathcal{Q}$, we might look at a sequence of circles like those in Illustration 67. The circles in Illustration 67 are converging to the point P and are getting smaller and smaller. In fact, any rectangle with P lying inside it also contains all but finitely many of the circles. Can you show that this "figure" belongs to $\mathcal{Q}$?

33 EXERCISE

Show that the figure in Illustration 67 belongs to $\mathcal{Q}$.

34 Theorem

If E_1 belongs to $\mathcal{Q}_t$, and E_2 belongs to $\mathcal{Q}_{t'}$, then $E_1 \backslash E_2$ belongs to $\mathcal{Q}_{t+t'}$.

PROOF: As before, choose G_1, F_1, G_2, and F_2 from $\mathcal{G}$, so that $F_1 \subset E_1 \subset G_1$, $F_2 \subset E_2 \subset G_2$, $A(G_1) - A(F_1) \leq t$, $A(G_2) - A(F_2) \leq t'$. Define $G_0 = G_1 \backslash F_2$ and $F_0 = F_1 \backslash G_2$. Then $F_0 \subset E_1$, $E_2 \subset G_0$, and $A(H_0) = A(G_0) - A(F_0)$, where $H_0 = G_0 \backslash F_0$. As before, we can show that $H_0 \subset H_1 \cup H_2$, where $H_1 = G_1 \backslash F_1$ and $H_2 = G_2 \backslash F_2$, so that $A(H_0) \leq A(H_1 \cup H_2) \leq A(H_1) + A(H_2) \leq t + t'$. Q.E.D.

Illustration 68

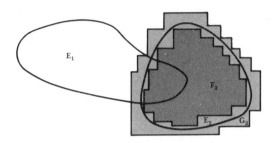

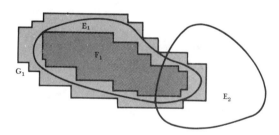

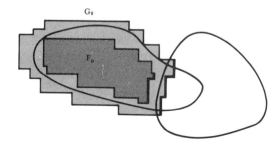

SECOND PROOF: We have another way to show this theorem, which is perhaps more instructive. We select a large rectangle R from $\mathcal{Q}$, large enough to include in its interior both G_1 and G_2, where G_1, G_2, F_1, and F_2 are defined as before. (Show that there is such a rectangle!) We would like to note that $R \backslash E_2 \in \mathcal{Q}_{t'}$. In fact, $R \backslash G_2 \in \mathcal{G}$, $R \backslash F_2 \in \mathcal{G}$, and $R \backslash G_2 \subset R \backslash E_2 \subset R \backslash F_2$. Furthermore, $A(R \backslash F_2) - A(R \backslash G_2) = [A(R) - A(F_2)] - [A(R) - A(G_2)] = A(G_2) - A(F_2) \leq t'$. We now observe the interesting fact that $E_1 \backslash E_2 = E_1 \cap (R \backslash E_2)$. Thus $E_1 \backslash E_2 \in \mathcal{Q}_{t+t'}$, by Theorem 31, since $E_1 \in \mathcal{Q}_t$, and $R \backslash E_2 \in \mathcal{Q}_{t'}$.   Q.E.D.

35   Corollary

If $E_1 \in \mathcal{Q}$ and $E_2 \in \mathcal{Q}$, then $E_1 \backslash E_2 \in \mathcal{Q}$.

There is a second corollary.

36   Corollary

If $E_1 \in \mathcal{Q}$, and $E_2 \in \mathcal{Q}$, then $E_1 \cup E_2 \in \mathcal{Q}$.

PROOF: Notice that we have already proved this under the hypothesis that E_1 and E_2 do not overlap. We shall now give a quick proof of this more general statement. Since E_1 and E_2 are in $\mathcal{Q}$, $E_1 \backslash E_2$ is also in $\mathcal{Q}$. Now, E_2 and $E_1 \backslash E_2$ do not overlap, so $E_1 \cup E_2 = (E_1 \backslash E_2) \cup E_2$ is also in $\mathcal{Q}$.   Q.E.D.

If we wanted closer estimates, we could show that:

37   Theorem

If $E_1 \in \mathcal{Q}_t$, and $E_2 \in \mathcal{Q}_{t'}$, then $E_1 \cup E_2 \in \mathcal{Q}_{t+t'}$.

PROOF: Prove it yourself, or show that $E_1 \cup E_2 \in \mathcal{Q}_{t+2t'}$, which is easier.

38   *Definition*

Now we see that we can define an area for every figure which belongs to $\mathcal{Q}$. If E belongs to $\mathcal{Q}$, then we can take as *upper numbers* for E all the numbers which are the areas of figures G which belong to $\mathcal{G}$ and contain E. We can take for *lower numbers* for E all the areas of figures of $\mathcal{G}$ contained in E, together with all negative numbers and 0. Every number greater than an upper number is an upper number, and every number less than a lower number is a lower number. Furthermore, it is possible to find an upper number and a lower number as close together as we wish, since $E \in \mathcal{Q}$. As we have noted,

Section 6   Area in General   195

there is exactly one real number x with the property that every number greater then x is an upper number and every number less than x is a lower number, and we set the area of E equal to x.

We shall now show that if the figures in $\mathcal{Q}$ are all assigned an area in this way, then the area satisfies A1 to A4.

39   Theorem

The condition A1 holds.

PROOF: For every $E \in \mathcal{Q}$, and every nonpositive number y, y is a lower number for the area of E, by definition. Thus, the area of E must be nonnegative.   Q.E.D.

40   Theorem

The condition A2 holds.

PROOF: Suppose this were false. Then we could find $E_1 \in \mathcal{Q}$, $E_2 \in \mathcal{Q}$, and $E_1 \subset E_2$ such that $A(E_1) > A(E_2)$. Let a real number y be chosen so that $A(E_1) > y > A(E_2)$. By the definition of $A(E_1)$, y is a lower number for $A(E_1)$, and therefore there is an $F_1 \in \mathcal{G}$ such that $A(F_1) = y$ and $F_1 \subset E_1$. Then $F_1 \subset E_2$ also, so that y is a lower number for E_2. But this contradicts the assertion that $y > A(E_2)$.   Q.E.D.

41   Theorem

The condition A3 holds.

PROOF: Suppose E_1 and E_2 belong to $\mathcal{Q}$, and have no points in common. Then $E_1 \cup E_2$ also belongs to $\mathcal{Q}$ (Corollary 29). Let $x_1 = A(E_1)$, $x_2 = A(E_2)$, $x = x_1 + x_2$, and $y = A(E_1 \cup E_2)$. The theorem states that x always equals y. If it were possible that $x \neq y$, let E_1 and E_2 be chosen so that $x \neq y$. Then $|x - y|$, the absolute distance between the numbers x and y, is not 0. Let $t = \frac{1}{4}|x - y|$. Then either $y = x + 4t$ or $y = x - 4t$. If we could succeed in showing that $x - 2t \leq y \leq x + 2t$, we would have a contradiction, and this contradiction would expose the falsity of the assumption $x \neq y$.

Since E_1 and E_2 belong to $\mathcal{Q}$, both belong to $\mathcal{Q}_t$. Let G_1, F_1, G_2, and F_2 be chosen from $\mathcal{G}$ so that $F_1 \subset E_1 \subset G_1$, $F_2 \subset E_2 \subset G_2$, $A(G_1) - A(F_1) \leq t$, and $A(G_2) - A(F_2) \leq t$, then

$$A(G_1) - t \leq A(F_1) \leq x_1 \leq A(G_1) \leq A(F_1) + t,$$

and $$A(G_2) - t \leq A(F_2) \leq x_2 \leq A(G_2) \leq A(F_2) + t.$$

Let $F = F_1 \cup F_2$, and $G = G_1 \cup G_2$. Then F and G belong to $\mathcal{G}$, and F_1 does not intersect F_2. We immediately see that $F \subset E_1 \cup E_2 \subset G$, so that $A(F) \leq y \leq A(G)$. Also, $A(F) = A(F_1) + A(F_2)$, while $A(G) \leq A(G_1) + A(G_2)$.

Combining $x_1 \leq A(F_1) + t$ and $x_2 \leq A(F_2) + t$, we get

$$x - 2t = (x_1 - t) + (x_2 - t) \leq A(F_1) + A(F_2) = A(F).$$

Combining $A(G_1) - t \leq x_1$ and $A(G_2) - t \leq x_2$, we get

$$A(G) \leq A(G_1) + A(G_2) \leq (x_1 + t) + (x_2 + t) = x + 2t.$$

Thus, $x - 2t \leq A(F) \leq y \leq A(G) \leq x + 2t$ or $x - 2t \leq y \leq x + 2t$, as promised.

This is our contradiction. The assumption $x \neq y$ is unmasked as false, and the theorem is proved. Q.E.D.

42 Theorem

The condition A4 holds.

PROOF: The assertion holds by definition for rectangles from $\mathcal{Q}$. For other rectangles, it follows from the material immediately following.

It should now be easy to show that:

(i) Every triangle belongs to $\mathcal{Q}$.
(ii) Every convex polygon belongs to $\mathcal{Q}$. (Use induction.)
(iii) Every polygon (convex or reflex) belongs to $\mathcal{Q}$.
(iv) Every polygonal figure belongs to $\mathcal{Q}$.
(v) Every circle belongs to $\mathcal{Q}$.

43 EXERCISES

1 Prove (i) to (v) above.
2 Is a line segment in $\mathcal{Q}$?
3 Is the boundary of a triangle in $\mathcal{Q}$?
*4 Is the boundary of an ellipse in $\mathcal{Q}$? How do you know?

It is interesting to note that we might have done the same thing for the classes of figures $\mathcal{Q}_t^*$ and $\mathcal{Q}^*$, where $\mathcal{Q}_t^*$ is defined to be similar to $\mathcal{Q}_t$, except that the figures chosen for G and F would be polygons instead of belonging to $\mathcal{G}$. Is the class $\mathcal{Q}_t^*$ the same as $\mathcal{Q}_t$? Is the class $\mathcal{Q}^*$ the same as $\mathcal{Q}$?

Illustration 69

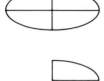

Now we shall go a bit deeper into the question of which figures are in Q. To begin with, we want to see that every ellipse is in Q. If we have proved above that Q and Q* are the same, then we can see that rotating any figure does not take it out of Q or put it into Q. So let us take our ellipse in "standard position," with the major axis horizontal, as in Illustration 69. We shall show that the upper right-hand quarter of the ellipse is in Q, and thus, the ellipse itself is. Divide the major half-axis into n equal pieces, as in Illustration 70, where n is any positive integer, and denote the $n+1$ points of division (including the end points) as $a_0, a_1, \ldots, a_n$. At each point a_i, let b_i be the length of the line perpendicular to the major axis joining a_i to a point on the ellipse. We can choose $G \in \mathcal{G}$ as the union of the rectangles whose base, for each i, is the line segment from a_{i-1} to a_i and whose height is $b_{i-1}, i = 1, 2, \ldots, n$. We choose $F \in \mathcal{G}$ as the union of the rectangles whose heights on those bases are b_i. Then $F \subset E \subset G$. We note that the area of the ith rectangle of G is $\frac{1}{n} \cdot \frac{M}{2} \cdot b_{i-1}$, where M is the length of the major axis of the ellipse. Thus, $A(G) = \frac{M}{2n}(b_0 + b_1 + b_2 + \cdots + b_{n-1})$. In a similar way, the area of the ith rectangle of F is $\frac{1}{n} \cdot \frac{M}{2} \cdot b_i$, so that $A(F) = \frac{M}{2n}(b_1 + b_2 + \cdots + b_{n-1})$.

Illustration 70

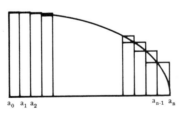

Illustration 71

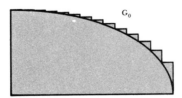
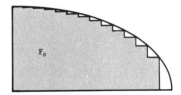

Illustration 72

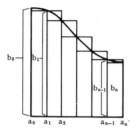 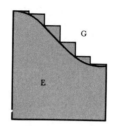 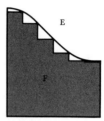

Illustration 73

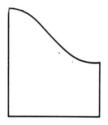 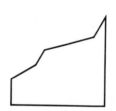

Illustration 74

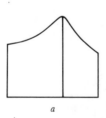 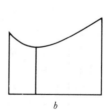 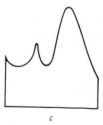

a   *b*   *c*

Illustration 75

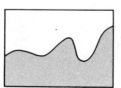

Thus, $A(G) - A(F) = \dfrac{M}{2n} \cdot b_0$. It follows that the quarter-ellipse is in $\mathcal{Q}_{\frac{M}{2n} \cdot b_0}$ for every value of $n \geq 1$. Since for every $t > 0$, $t > \dfrac{M}{2n} \cdot b_0$ for some $n \geq 1$, we see that the quarter ellipse is in $\mathcal{Q}_t$ for all t, and thus is in $\mathcal{Q}$. Thus, the whole ellipse is also in $\mathcal{Q}$.

Now, what property of the ellipse (or rather, of the quarter-ellipse) did we use? Only that $b_0 \geq b_1 \geq b_2 \geq \cdots \geq b_{n-1} \geq b_n \geq 0$. Suppose we choose another figure of the same kind, and where $b_n > 0$, as in Illustration 72. Then

$$A(G) = b_0(a_1 - a_0) + b_1(a_2 - a_1) + \cdots + b_{n-1}(a_n - a_{n-1})$$
$$= (b_0 + b_1 + b_2 + \cdots + b_{n-1}) \dfrac{a_n - a_0}{n}$$

and

$$A(F) = b_1(a_1 - a_0) + b_2(a_2 - a_1) + \cdots + b_n(a_n - a_{n-1})$$
$$= (b_1 + b_2 + \cdots + b_n) \dfrac{a_n - a_0}{n}.$$

Thus $A(G) - A(F) = (b_1 - b_n) \dfrac{a_n - a_0}{n}$, so that $E \in \mathcal{Q}_t$ for every $t > 0$, and $E \in \mathcal{Q}$.

Any region then which has a rectangular shape, except for the top, and that going down from left to right, belongs to $\mathcal{Q}$. It follows that the same must be true if it rises from left to right, or rises and then falls, or falls and then rises, or falls, rises, falls, rises, and falls, in that order. A line or curve which rises from left to right is called *isotonic*. One which falls from left to right is *antitonic*, and the two are the two cases of *monotonic* (not monotonous, please). The curve in Illustration 74(c) is monotonic in each of five pieces, and is called *piecewise monotonic*. We have shown:

44 Theorem

The part of a rectangle below a piecewise monotonic curve is in $\mathcal{Q}$.

45 Corollary

The part of a rectangle above a piecewise monotonic curve is also in $\mathcal{Q}$.

Illustration 76

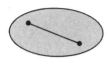

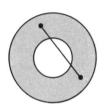

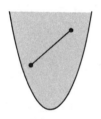

46 Definition

We have defined what a convex polygon is. Now, we shall define a *convex region*. A convex region is a collection of points in the plane such that if the end points of a line segment both lie in it, then the whole line segment lies in it also. The interior of a circle, an ellipse, or a convex polygon is a convex region. The area between two concentric circles is *not* a convex region. The interior of a parabola *is* convex but is not *bounded*. A figure is *bounded* if it can be put into a rectangle, though possibly a large one. Every figure in Q is bounded.

47 Theorem

Every bounded, convex figure is in Q.

PROOF: Let K be a convex figure, and let R be the smallest rectangle from G that we can put around K. (How do we know there is a smallest one?) Then the figure K touches the two sides of the rectangle, otherwise it is not the smallest. Let C_1 be the curve joining the sides of the rectangle over the top of K and C_2 the same sort of curve under the bottom (see Illustration 77). Let K_1 be the part of R lying under C_1, and K_2 the part lying over C_2. C_1 and C_2 are piecewise monotonic, and so K_1 and K_2 are in Q. Therefore, so is $K = K_1 \cap K_2$.   Q.E.D.

48 EXERCISE

Show $Q^* = Q$.

Illustration 77

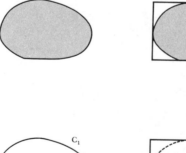

Section 7 Pathology

Illustration 78

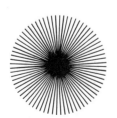

Perhaps it is time to look at a figure which is *not* in Q. Draw two perpendicular diameters in a circle, and then draw the diameters which bisect the right angles, and then the diameters which bisect those, and then the diameters which bisect those, etc., forever. This process will not take up all of the circle; for instance, the diameter making an angle of 60° to any of these will never be included. Now, if the original circle had radius 1, then we can put a figure from G around our "puff-ball" with area as close as we please to π, but we cannot put any figures from G *into* it with the sole exception of the empty set. In any case, the figure is not in Q_t for $t \leq \pi$, so it is not in Q. You might think that any figure which is the interior of a curve is in Q; that is not true in the sense in which mathematicians mean "curve," but it is true for curves which have finite length. The fact that a bounded curve could have infinite length is perhaps surprising, but it is, in any case true. Here is an example of a curve whose interior is not in Q.

49 Example

We start with the observation that for any integer n, the infinite series

$$\frac{1}{n} + \frac{1}{n^2} + \frac{1}{n^3} + \cdots = \frac{1}{n} \cdot \frac{1}{1 - (1/n)} = \frac{1}{n-1}.$$

Choose some n, as large as you like, but keep it fixed for the rest of this discussion. What we shall do is to put a curve C into the unit square so that every figure from G which lies in its interior has area at most $\frac{1}{n-1}$, while every figure from G which contains C has area at least $1 - \frac{1}{n-1}$. First, we place a cross in the square, as shown in Illustration 79. The cross is centered, and is made so thin that its total area is no more than $1/n$. We make a note that all of the cross is reserved for the *inside* of C. The ends of the cross are small line segments, and these will belong to C. C will go around the square from one of these ends to the other, though not as simply as the dotted line seems to indicate. For our next step, we cut each of the quarters of the square into three equal rectangles with two horizontal lines, and thicken these lines to form thin, thin rectangles. Each of these rectangles is less than $\frac{1}{2}$ in length, and there are 8 of them. If we take their widths to be $1/4n^2$, then their total area is less than $8 \cdot \frac{1}{2} \cdot \frac{1}{4n^2} = \frac{1}{n^2}$. The new rectangles are reserved for the outside and the inside of C, respectively, in such a way that C is committed to going through each of the remaining rectangles in a sort of "diagonal" fashion (see Illustration 79). The appropriate ends of the new rectangles are now fixed to be in C. Now, we divide each of the remaining 12 rectangles into 3 parts (see Illustration 80), using 2 thin rectangles for each division, or a total of 24 rectangles. Each is of length less than $\frac{1}{3} \cdot \frac{1}{2} = \frac{1}{6}$, and if we take their widths to be $\frac{1}{4n^3}$, then their total area

Illustration 79

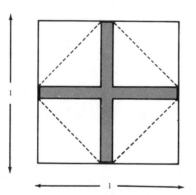
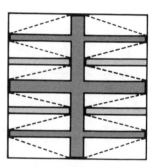

Illustration 80

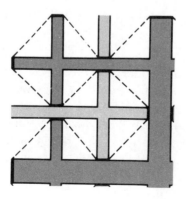
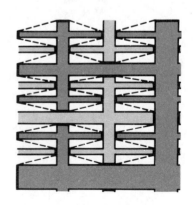

Section 7   Pathology

is less than $24 \cdot \frac{1}{6} \cdot \frac{1}{4n^3} = \frac{1}{n^3}$. We then assign each to the inside or outside of the curve-to-be, in the same way as before. We continue this process without end. At each stage, we cut each of the remaining rectangles into thirds, using very thin rectangles instead of lines. There are many of these rectangles (how many at the kth step?) and they all have the same length (what is it, at the kth step?). If we choose the width properly, we can make the sum of the areas of these rectangles add up to no more than $1/n^k$. At each stage, the curve C becomes better and better defined; in the limit, it is a simple closed curve, in the sense that mathematicians use the term.

In Illustration 81, we see the figure C_1, which consists of 4 line segments and four gray rectangles. The four line segments are part of C, and the rest of C is contained within the gray rectangles, some-

Illustration 81

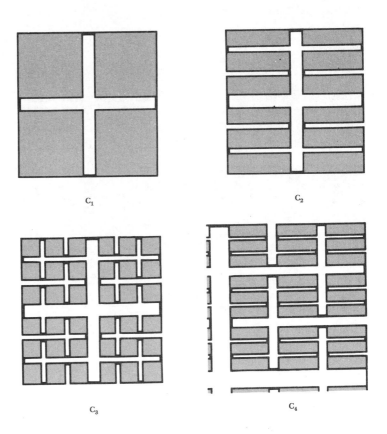

Illustration 82

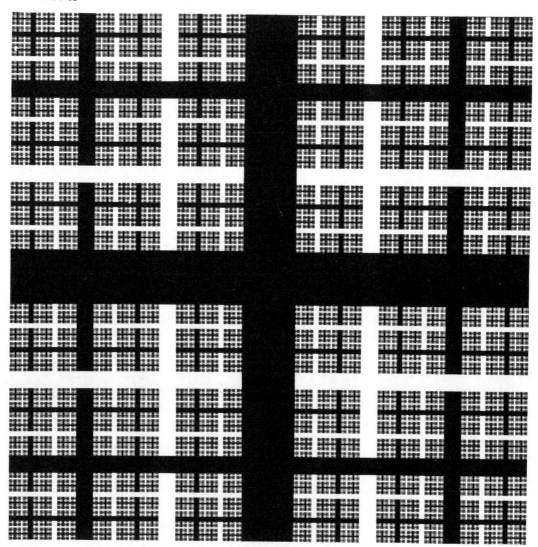

where. The figure C_2 gives us a better idea. It consists of 12 line segments and 12 gray rectangles, and again the curve C threads its way through somehow. In the same way C_3 is a refinement of what we know from C_2, and C_4, which shows only one quarter of the figure, is a refinement of our knowledge in C_3, and so we go, generating the figures we see here. Illustration 82 is a representation of what the interior of C looks like. Naturally, it is only approximate.

Each set $C_1, C_2, C_3, \ldots$, etc., is a refinement of the previous one, and the rectangles are getting smaller and smaller. In the meantime, we might notice the "inside" of $C_1, C_2, \ldots$, etc., which we shall call $F_1, F_2, \ldots$. We see that

$$A(F_1) = \frac{1}{n}, A(F_2) \leq \frac{1}{n} + \frac{1}{n^2}, A(F_3) < \frac{1}{n} + \frac{1}{n^2} + \frac{1}{n^3},$$

etc. The interior of the curve C will be all the F_1, F_2, F_3, etc., put together. NOW WE CLAIM that any figure F taken from $\mathcal{G}$ and lying inside C must actually be contained in one of the F_j. If, in fact, $F \subset F_m$, then

$$A(F) \leq A(F_m) \leq \frac{1}{n} + \frac{1}{n^2} + \cdots + \frac{1}{n^m} < \frac{1}{n-1}.$$

Thus, if $F \in \mathcal{G}$ lies inside C, then $A(F) < \frac{1}{n-1}$. Let G'_n denote that part of the square lying "outside" C_n. By the same reasoning, if $G' \in \mathcal{G}$ lies within the square but outside C, then G' is contained in some G'_m, and

$$A(G') \leq A(G'_n) < \frac{1}{n^2} + \frac{1}{n^3} + \cdots + \frac{1}{n^m} < \frac{1}{n(n-1)}.$$

If $G \in \mathcal{G}$, and the interior of C lies within G, then $S \backslash G$ lies outside C (S is the square) and $A(S \backslash G) < \frac{1}{n(n-1)}$. It follows that $A(G) > 1 - \frac{1}{n(n-1)}$. Thus, the interior of the curve C does not belong to $\mathcal{Q}$.

So our proof would now be complete, if we had shown, in fact, that every $F \in \mathcal{G}$ which lies inside C lies inside some F_m, and similarly for G. Since $F \in \mathcal{G}$, F is the finite union of rectangles in standard position. Let t be the smallest height or length of any of these rectangles. Then for every point P in F, there is a horizontal line segment through P of length at least t, and also a vertical such segment.

Suppose a point P' belongs to one of the rectangles added at the nth stage. How long a line, vertical or horizontal, can one pass through P' without leaving the interior of C? Along the direction of the long dimension of the rectangle, the distance would be twice the length of the rectangle, plus the width of the previous rectangle. In the transverse direction, any line which was longer than the width of the rectangle plus the sum of the widths of the two adjacent gray rectangles would contain a point from "outside." So it seems that we are committed to knowing the dimensions of the rectangles we add to F_m to make F_{m+1}, and also the dimensions of the "gray" rectangles at each step.

All the gray rectangles, at each step, have the same dimensions. Let us denote these horizontal and vertical dimensions, at the mth step by h_m and v_m, respectively. If m is even, then C_{m+1} is made from C_m by cutting the gray rectangles *vertically*, as in Illustrations 84(c) and (d), and $h_{m+1} < \frac{1}{3}h_m$, while $v_{m+1} = v_m$. If m is odd, then C_{m+1} is made from C_m by cutting the gray rectangles *horizontally*,

Before 1900 there were few research mathematicians in America. (The American Mathematical Society was not founded until 1888.) William Fogg Osgood (1865–1943) was among those few. Like many American mathematicians of this period he received his Ph.D. in Germany, then returned to Harvard where he remained until his retirement. In 1903 he invented the "curve of positive area," described in Example 50. His textbooks on calculus and function theory were standard for decades. (Courtesy of the Department of Mathematics, Harvard University.)

Illustration 83

Illustration 84

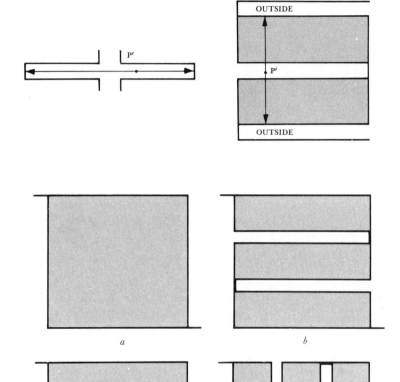

as in Illustrations 84(a) and (b), and $v_{m+1} < \frac{1}{3}v_m$, while $h_{m+1} = h_m$. Thus, for all m, $h_{m+2} < \frac{1}{3}h_m$, and $v_{m+2} < \frac{1}{3}v_m$. Let us now examine the rectangles added to F_m to make F_{m+1} (which are the same size as the rectangles added to G_m to make G_{m+1}). We shall call their horizontal dimensions $\bar{h}_m$ and their vertical dimensions $\bar{v}_m$. If m is odd, then $\bar{v}_m < \frac{1}{2}v_m$, and $\bar{h}_m = h_m$. If m is even, then $\bar{v}_m = v_m$ and $\bar{h}_m < \frac{1}{2}h_m$.

Let us review. We start with the cross, which is the original square less four gray squares, each h_1 by v_1. We remove from each such square two rectangles $\bar{h}_1$ by $\bar{v}_1$, thus leaving in each square three rectangles h_2 by v_2. We remove from each of these rectangles two rectangles $\bar{h}_2$ by $\bar{v}_2$, leaving three rectangles h_3 by v_3, and so on.

If m is odd, then we have these formulas:

$$h_{m+1} = h_m = \bar{h}_m,$$
$$v_{m+1} < \tfrac{1}{3}v_m, \qquad \bar{v}_m < \tfrac{1}{2}v_m.$$

If m is even, we have these formulas:

$$h_{m+1} < \tfrac{1}{3}h_m, \qquad \bar{h}_m < \tfrac{1}{2}h_m,$$
$$v_{m+1} = v_m = \bar{v}_m.$$

Assume now that P' is a point which belongs to F_{m+1}, but not to F_m.

If m is odd, then the longest horizontal line that can be drawn through P has length $2\bar{h}_m + \bar{h}_{m-1}$, and the longest vertical line is no more than $\bar{v}_m + 2v_{m+1}$. These two numbers come to no more than $2h_m + \tfrac{1}{2}h_{m-1}$ and $\tfrac{1}{2}v_n + 2v_{n+1}$, respectively.

If m is even, the same analysis will show that the longest horizontal and vertical lines through P' are of length no more than $\bar{h}_m + 2h_{m+1}$ and $2\bar{v}_m + \bar{v}_{m-1}$. These numbers are no more than $\tfrac{1}{2}h_m + 2h_{m+1}$ and $2v_m + \tfrac{1}{2}v_{m-1}$, respectively. We know that

$$h_1 = v_1 < \tfrac{1}{2},$$
$$h_2 = h_1, \qquad\qquad v_2 < \tfrac{1}{3}v_1,$$
$$h_{2m+1} < \tfrac{1}{2}(\tfrac{1}{3})^m \qquad v_{2m+1} < \tfrac{1}{2}(\tfrac{1}{3})^m,$$
$$h_{2m} < \tfrac{1}{2}(\tfrac{1}{3})^{m-1} \qquad v_{2m} < \tfrac{1}{2}(\tfrac{1}{3})^m.$$

Thus, in every case,

$$h_m < \tfrac{1}{2}(\tfrac{1}{3})^{\tfrac{m}{2}-1} \qquad v_m < \tfrac{1}{2}(\tfrac{1}{3})^{\tfrac{m}{2}-1},$$

whether m is odd or even. It follows that if P' is in F_{m+1}/F_m, then the longest horizontal or vertical line through P' is no more than the larger of these numbers:

$$2 \cdot \tfrac{1}{2} \cdot (\tfrac{1}{3})^{\tfrac{m}{2}-1} + \tfrac{1}{2} \cdot \tfrac{1}{2} \cdot (\tfrac{1}{3})^{\tfrac{m-1}{2}-1} = \tfrac{7}{12}(\tfrac{1}{3})^{\tfrac{m-3}{2}},$$
$$\tfrac{1}{2} \cdot \tfrac{1}{2}(\tfrac{1}{3})^{\tfrac{m}{2}-1} + 2 \cdot \tfrac{1}{2}(\tfrac{1}{3})^{\tfrac{m+1}{2}-1} = \tfrac{7}{12}(\tfrac{1}{3})^{\tfrac{m}{2}-1}.$$

We see at once that if m is large enough, say $m > M$, we have

$$\tfrac{7}{12}(\tfrac{1}{3})^{\tfrac{m-3}{2}} < t, \text{ that is, } \tfrac{49}{144} < 3^{m-3}t^2.$$

Thus, we know that for $m > M$, no point of F belongs to F_{m+1}/F_m. It follows that $F \subset F_M$. A similar argument holds for G.

Q.E.D

References

1. Carl B. Boyer, *The History of the Calculus and Its Conceptual Development*, Dover Publications Inc., New York, 1959.
2. R. C. Buck, *Advanced Calculus* (2d ed.), McGraw-Hill Book Company, New York, 1965, pp. 97–99.
3. C. Carathéodory, *Vorlesungen Über Reele Funktionen* (2d ed.), B. G. Teubner, Leipzig and Berlin, 1927.
4. C. Jordan, *Cours d'analyse* (2d ed.), Gauthier-Villars, Paris, 1893.
5. Ervin O. A. Kreyszig, *Differential Geometry*, University of Toronto Press, Toronto, 1950.
6. H. Lebesgue, *Intégrale, longueur, aire*, Thesis, Paris, 1902; also appears in *Annali Mat. Pura e Appl.*, (3) **7** (1902), pp. 231–359.
7. ———, *Leçons sur l'intégration* (2d ed.), Gauthier-Villars, Paris, 1928.
8. J. von Neumann, *Functional Operators*, Measure and Integration, Vol. 1, Annals of Math. Studies 21, Princeton University Press, Princeton, 1950.
9. W. F. Osgood, "A Jordan curve of positive area" [sic], *Trans. Amer. Math. Soc.*, **4** (1903), pp. 107–112.

EXCURSIONS INTO MATHEMATICS

CHAPTER 4 **SOME EXOTIC GEOMETRIES**

BY DONALD W. CROWE

A page from Isaac Barrow's (1660) 'pocket edition' of Euclid's ELEMENTS. Barrow was Newton's teacher at Cambridge and represented the old 'geometrical' school of calculus, in contrast to Newton's more algebraic approach. Having recognized Newton as his superior, Barrow gave up his professorship in 1669 in favor of Newton.

Notice that Euclid's proof of Proposition 1 is already inadequate, for none of his axioms guarantees that the two circles D and F actually have a point C in common. (Courtesy of Zeb Delaire.)

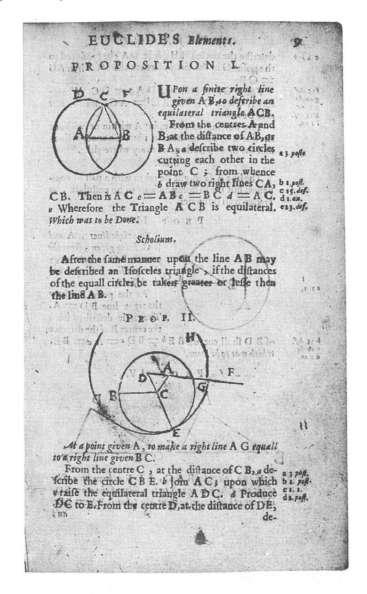

Section 1 Historical Background

A revolution in mathematics began about 1900, shortly before the theories of relativity and quantum mechanics revolutionized physics. Like the revolution in physics, the revolution in mathematics was a new way of looking at the subject. In mathematics, this new way was the *axiomatic method*. But again like quantum theory in physics, the roots of the revolution go back to the Greeks. In fact, the pattern of the axiomatic method was established by Euclid (about 300 B.C.) in the *Elements of Euclid*.†

For our purpose, the accomplishment of Euclid can be described as follows: There was, in classical Greece, a substantial body of geometric knowledge, partly inherited from the Egyptians and Babylonians, but mostly the work of Greek mathematicians and philosophers themselves—notably Thales, Pythagoras, Democritus, Theaetetus, and Eudoxus. Euclid incorporated this work into a treatise, the *Elements*. The quality of this treatise, which was by no means designed to be the high school textbook which it later became, especially in England, is suggested by G. H. Hardy in his fascinating little book, *A Mathematician's Apology* (p. 21): "Oriental mathematics may be an interesting curiosity, but Greek mathematics is the real thing. . . . The Greeks, as Littlewood said to me once, are not clever schoolboys or 'scholarship candidates' but 'fellows of another college.' So Greek mathematics is 'permanent,' more permanent even than Greek literature. Archimedes will be remembered when Aeschylus is forgotten, because languages die and mathematical ideas do not."

However, for modern mathematics, it is Euclid's organizing principle which is of particular importance, not the actual geometric content of the *Elements*. This principle is to take certain fundamental geometric facts, called *postulates*, as "rules of the game." These are to be accepted without proof, and all other geometric facts are to be deduced from them (and certain general "self-evident truths," called *common notions*) by means of the rules of logic, also accepted without

---

† The excellent English edition of T. L. Heath [4] is available as an inexpensive paperback.

proof. That is, once the basic axioms† are agreed upon as an accurate (partial) description of the actual physical world, other geometric properties are deduced from them *without further reference to the outside world*. For the purposes of modern mathematics, the particular axioms which Euclid used are not really important, but for definiteness and for historical perspective, we list them here, rephrased from Heath [4, p. 155]. The reader will note that several terms, such as "point," "line," "circle," "right angle," and "congruence," have not been defined. Although Euclid's "definitions" are now recognized to be inadequate, the reader will find them and the axioms (i.e., postulates and common notions), together with a detailed discussion of their significance, in [4, pp. 153–155]. For easy reference, we have also stated them in the appendix to this chapter.

EUCLIDS' AXIOMS (POSTULATES)

1 A straight line segment can be drawn joining any two points.

2 Any straight line segment can be extended continuously in a straight line.

3 Given any straight line segment, a circle can be drawn having the segment as radius and one end point as center.

4 All right angles are congruent.

5 If a straight line meets two other straight lines so that the sum of the interior angles on the same side is less than two right angles, then the two straight lines meet on that side on which the angles are less than two right angles.

Of the long list of definitions by which Euclid preceded his axioms, many sound very awkward today. Euclid apparently did not fully realize the obvious fact that *it is not possible to define everything*. That is, some terms must be taken as undefined.

This is easier to understand by an actual example. Under "dead" in *Webster's Seventh New Collegiate Dictionary*, the first definition is "deprived of *life*." So we look up "life," to find "the qual-

---

† Current mathematical usage refers to the fundamental assumptions of any mathematical theory as *axioms*. In this sense, Euclid's postulates and common notions are axioms.

ity that distinguishes a vital and functional being from a *dead body* . . . ,"

¹dead \'ded\ *adj* [ME *deed*, fr. OE *dēad*; akin to ON *dauthr* dead, *deyja* to die — more at DIE] **1** : deprived of life : having died

¹life \'līf\ *n, pl* lives \'līvz\ [ME *lif*, fr. OE *līf*; akin to OE *libban* to live — more at LIVE] **1 a** : the quality that distinguishes a vital and functional being from a dead body or purely chemical matter

and we have already gone full circle back to the original word we were trying to define. However, dictionary makers are becoming more sophisticated, and it is no longer possible to pick a word completely at random and expect this to happen. For example, the same dictionary defines "point" in essentially the same way we shall, as "an undefined geometric element . . . ."

¹point \'point\ *n* [ME, partly fr. OF, puncture, small spot, point in time or space, fr. L *punctum*, fr. neut. of *punctus*, pp. of *pungere* to prick; partly fr. OF *pointe* sharp end, fr. (assumed) VL *puncta*, fr. L, fem. of *punctus*, pp. — more at PUNGENT] **4 a (1)** : an undefined geometric element of which it is postulated that at least two exist and that two suffice to determine a line

Two particularly important objections have been raised against Euclid's *Elements*. They are:

6 *First Objection.* The fifth axiom, the "parallel postulate," is much more complicated in its statement than the others. It looks as if it should be proved as a theorem, not assumed as one of the unproved "rules of the game."

This Axiom 5 is called the *parallel postulate* because it is essentially equivalent to the simpler statement:

(**E**) Given any line l and any point P not on l, then there is exactly one line through P parallel to (i.e., not meeting) l.

In fact, in what follows, a reference to Axiom 5 is to be understood as a reference to this modified version (E). We can legitimately refer to (E) as the *Euclidean axiom*, since it is the characterizing property of *Euclidean geometry*.

7 *Second Objection.* The axioms of Euclid are incomplete. That is, it is in fact not possible to prove the theorems of Euclid on the basis of the few axioms he laid down. This is already illustrated in the very first proposition of Euclid, which says:

Proposition 1

An equilateral triangle can be constructed with any given segment AB as one side.

The proof proceeds as follows: Construct a circle with center A and radius AB. Then construct a circle with center B and radius AB. (See Illustration 1.) Let C be a point which the two circles have in common. Then ABC is the required equilateral triangle.   Q.E.D.

This is, of course, an obvious and elegant way to construct an equilateral triangle. But logically speaking, it is unsound, for there is no axiom of Euclid which even remotely suggests that the two circles do indeed have a point C in common. This, along with other inadequacies of Euclid's axioms, is discussed again in Section 6. (Problem 3 of Exercises 71 shows that sometimes Euclid's construction can lead to a very peculiar "equilateral triangle" indeed.)

From a purely mathematical point of view, only the second of these objections is relevant. With reference to the First Objection (6), it is, of course, more aesthetic to take simple statements as axioms. But there is no purely mathematical reason not to take very complicated statements as axioms. In fact, the history of a particular mathematical theory is very often that its final axiomatization utilizes an elaborate statement whose truth was only discovered after the subject had been developed for many years. Nevertheless, this First Objection had an enormous influence on the historical development of mathematics and physics, and indeed on philosophy. For it was the attempt to remove this objection that led to the discovery of non-Euclidean geometries. This paved the way for a rejection of the Kantian notion that the nature of the human mind was such that Euclidean space was the only space conceivable. This, in turn, opened the way for an acceptance of the theory of relativity, with its unorthodox geometric framework. In Sections 2 to 5, we shall examine in some detail the mathematical and historical consequences of the questions raised by this First Objection.

8   EXERCISES

1 Read the appendix to this chapter.
2 Read parts of Volume I of Heath's edition of *Euclid* [4], especially Propositions 1, 2, 3 (pp. 241–247), the congruence

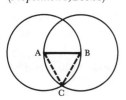

Illustration 1

Euclid's construction of an equilateral triangle (Proposition 1, Book I)

theorems (Propositions 4, 8, 26), the triangle inequality (Proposition 20), and the notes on the parallel postulate (pp. 202–220). The extensive historical notes are also of interest, as are the notes on the definitions (pp. 153–194).

Some early writers† attempted to prove Axiom 5 as a theorem "by contradiction." That is, they tried to show that the assumption of the falsity of Axiom 5 would lead to an obvious absurdity. Two possible ways of assuming the falsity of Axiom 5 have turned out to be of special importance. They are the assumptions:

(**S**) Given any line l and any point P not on l, then every line through P meets l.

(**H**) ‡ Given any line l and any point P not on l, then at least two different lines through P do not meet l.

Section 2 Spherical Geometry

In this section we consider briefly the assumption (S). Does it sound absurd? Of course it does! But in fact each of us is well acquainted with a geometry in which (S) holds, for *this is, in fact, the geometry of the very world on which we live*. For, suppose that we regard the shortest path between two points on the surface of the earth (which we consider for this purpose as a perfect sphere) as part of a *line*. Thus, for example, if a railroad goes "straight" from Chicago to Kansas City, we usually agree to call its track part of a (straight) line. Of course, from the point of view of space, it is not a "line" at all but an arc of a huge circle (a so-called *great circle*) drawn on the surface of the earth and having the earth's center as its center.

This is still more easily visualized if we imagine the "straight-line" route taken by an airplane flying from New York to London. (See Illustration 2(a).) If this route is traced on the surface of a globe, it is again an arc of a great circle. Nevertheless, it is convenient for many purposes to call these great-circle arcs (parts of) "lines." The geometry of the points and these "lines" on a sphere is called "spherical geometry."

---

† Notably Saccheri. See Section 4.
‡ H stands for *hyperbolic*, the mathematical term for this geometry.

Illustration 2

It is easy to see that lines of this spherical geometry do, in fact, represent case (S). For the "lines" are great circles, that is, paths where planes† through the center of the sphere meet its surface, as

a

A straight line on the sphere is an arc of a great circle.

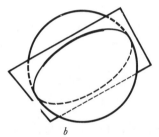

b

A great circle on a sphere is one whose plane contains the center of the sphere.

Two great circles have exactly two (antipodal) points in common.

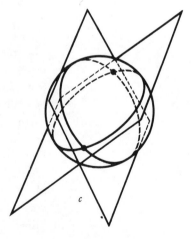

c

---

† By an unfortunate accident, the English word for a flat surface ("plane") is the same as the word for a machine which flies through the air. We shall try to avoid confusion by always referring to the latter as an "airplane."

in Illustration 2(b). But any two such planes have two (antipodal) points in common with the surface of the sphere—namely, the two points where their line of intersection meets the surface of the sphere. (See Illustration 2(c).) Hence any two (distinct) great circles have exactly two points in common. That is, *no two great circles are parallel*. More vividly, every airplane which goes through London (point L in Illustration 3) and travels continually in a "straight line" must eventually cross the equator—as well as the 90 degree meridian or any other great circle whatsoever! This means that the geometry of the earth's surface is an example of a "non-Euclidean geometry," which has, of course, been known and understood for hundreds of years.

Illustration 3

Any straight-line route from London eventually crosses the equator, ℓ.

A fundamental theorem of Euclidean geometry is:

9E Theorem

The sum of the angles of a triangle is two right angles.

This theorem is *not* true in spherical geometry, for we can readily find a spherical triangle whose angle sum is *three* right angles. For example, consider the triangle of Illustration 4, bounded by the equator, the Greenwich (0°) meridian, and the 90° meridian (through New Orleans), which meets the equator at the Galapagos Islands. That is, denoting the North Pole by N, the Galapagos Islands by G, and a point on the equator straight south of Greenwich by A, each of the angles at N, G, A is a right angle, and the sum of the angles of triangle NGA is three right angles.

The theorem (corresponding to Theorem 9E) which *is* true for spherical geometry is:

Illustration 4   *The angle sum of this spherical triangle is three right angles.*

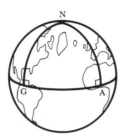

9S Theorem

The sum of the angles of a triangle is greater than two right angles.

The reader should notice that, in contrast to Euclidean geometry, not all triangles have the same angle sum. (See Problem 1 of Exercises 10.) In particular, if the triangle NGA is bisected by the 45° meridian from the North Pole N to the equator, each of the two new triangles has angle sum 225°, whereas the angle sum in NGA itself is 270°.

However, many other Euclidean theorems *are* true in spherical geometry. The reader should verify intuitively that Euclid's Propositions 12, 15, 4, and 24 hold in spherical geometry. (These are stated in the appendix to this chapter, and also as Theorems 12 to 15 in the next section.) However, Euclid's Proposition 16, "*An exterior angle in any triangle is greater than either opposite interior angle,*" does *not* always hold in spherical geometry. This illustrates the fact that the theorems of spherical geometry follow the Euclidean pattern closely enough to be a useful guide to what theorems to expect under case (H). In fact, this was our primary motivation for introducing these few remarks about spherical geometry. Spherical geometry as such will not be of any other importance to us in what follows.

10 EXERCISES

1 Theorem 9S says that the angle sum in a triangle is always greater than two right angles. Can it be close to two right angles? In particular, is there a spherical triangle whose angle sum is only $\pi/10^{10}$ greater than two right angles?

2 (a) Find a spherical triangle (that is, a triangle on the surface of the sphere having arcs of great circles as sides) for which Euclid's Proposition 16 holds.

(b) Find a spherical triangle for which Proposition 16 does *not* hold.

3 Look up the proof of Euclid's Proposition 16 in [4] or elsewhere. How does Euclid's proof of this theorem fail in spherical geometry?

Section 3 Absolute Geometry

Euclid himself was certainly aware of the special nature of his parallel postulate, and he arranged his *Elements* so as to postpone its use as long as possible. In fact, the proofs of the first 28 propositions† of the *Elements* make no use of the parallel postulate. "When we consider the countless successive attempts made through more than twenty centuries to prove the Postulate, many of them by geometers of ability, we cannot but admire the genius of the man who concluded that such a hypothesis, which he found necessary to the validity of his whole system of geometry, was really indemonstrable." [4, p. 202] For just these reasons, H. S. M. Coxeter has referred to Euclid himself as "the first non-Euclidean geometer."

However, although Euclid proved his first 28 propositions without using the parallel postulate explicitly, he *did* intend to rule out the spherical case. This is apparent from the fact that among the first 28 propositions is:

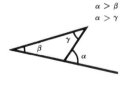

Illustration 5

11 Theorem (Proposition 16)

An exterior angle in any triangle is greater than either opposite interior angle. (See Illustration 5.)

We have seen from Section 2 that this cannot be a theorem of spherical geometry, since an example was found (Problem 2(b) of Exercises 10) for which it does not hold.

† These propositions are all stated in the appendix to this chapter.

The geometry which assumes only that (S) does not hold but which does not specify which of (E) or (H) holds has come to be called *absolute geometry*. Its fundamental assumption about parallels can be stated as:

(**EH**) Given any line l and any point P not on l, then at least one line through P does not meet l.

Illustration 6

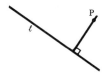

Thus, among the theorems of absolute geometry are all of the first 28 propositions of Euclid. For reference, we state some of the most important of these.

12 Theorem (Proposition 12)

From a point not on a given line it is possible to draw a line perpendicular to the given line (see Illustration 6).

13 Theorem (Proposition 15)

If two lines intersect, their vertical angles are congruent (see Illustration 7).

14 Theorem (Proposition 4)

If two sides and the included angle of one triangle are congruent, respectively, to two sides and the included angle of another, then the other angles and sides of the two triangles are congruent, respectively.

Illustration 7

15 Theorem (Proposition 24)

If two sides of one triangle are congruent, respectively, to two sides of another, and the included angle in the first triangle is greater than the included angle in the second, then the third side of the first triangle is greater than the third side of the second.

16 Theorem (Proposition 26)

If two angles and a side of one triangle are congruent, respectively, to two angles and a side of another, then the other sides and angle of the two triangles are congruent, respectively.

Section 3   Absolute Geometry   223

17   Theorem (Proposition 20)
In any triangle, the sum of the lengths of two sides is greater than the length of the third.† (See Illustration 8.)

We saw in Section 2 that the theorem of spherical geometry corresponding to the Euclidean angle-sum theorem 9E is:

9S   Theorem
The sum of the angles of a triangle is greater than two right angles.

What is the corresponding theorem in absolute geometry? Since absolute geometry is, in a sense, the opposite of spherical geometry, we might expect the corresponding theorem to be the opposite of 9S, namely,

18EH   Theorem
The sum of the angles of a triangle is *not* greater than two right angles.

This is indeed true, and there are two particularly elegant proofs of this theorem discovered by Legendre while he was trying to prove Axiom 5 as a theorem. We reproduce one of these proofs from [4, p. 216].

PROOF OF THEOREM 18EH: Suppose there is some triangle, say ABC, whose angle sum $\alpha + \beta + \gamma$ is greater than two right angles (see Illustration 9). Now extend one side say AC, past C. Call this line l. At C, draw a line making angle α with l. On this line, mark point B_1 so that $CB_1 = AB$. At

Illustration 8

$b + c > a$

† "Any ass knows this," said the early Epicurean philosophers. For just put a haystack at one corner of a triangle and an ass at another. The ass will certainly not go along *two* sides of the triangle to get to his hay! [4, p. 287]

B_1, draw a line meeting l at D, making angle β with CB_1, as shown in Illustration 10. Then triangle ABC is congruent to triangle CB_1D (Theorem 16). Now note that $\sphericalangle BCB_1 \ (= \beta_1) < \sphericalangle ABC \ (= \beta)$ (Why?), and hence by Theorem 15, segment BB_1 is shorter than segment AC. Let us say that $BB_1 + \Delta = AC$, where Δ is some segment.

Illustration 9

Illustration 10

Illustration 11

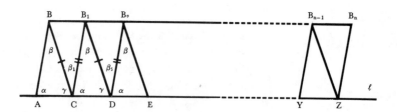

Now repeat this process at D, drawing DB_2 at angle α with l and making $DB_2 = AB$. At B_2, draw BE, making angle β with DB_2, so that triangle CB_1D is congruent to DB_2E, as before (see Illustration 11). Likewise, note that $\sphericalangle \beta_1 = \sphericalangle \beta_2$ (since each is $180° - (\alpha + \gamma)$). Hence $BB_1 = B_1B_2$. Therefore, $B_1B_2 + \Delta = CD$.

Now our goal is to show that by repeating this process often enough, we shall have a path $ABB_1B_2 \cdots B_nZ$, as in Illustration 11, from A to a point Z on l which is *shorter* than the straight-line path $ACDE \cdots YZ$ from A to Z. But this is impossible, since it follows from the "triangle inequality" (Theorem 17) that a polygonal path between two points is not shorter than the straight-line path between them (see Problem 2 of Exercises 19). This implies that the original "possibility" (that triangle ABC has angle sum greater than 180°) is in fact *impossible*, and hence our theorem will be proved.

It is clear that by taking enough of the "short segments" BB_1, $B_1B_2, \ldots, B_{n-1}B_n$, we shall eventually get a path shorter than the sum of the corresponding "long segments" $AC, CD, \ldots$. Exactly how many do we need? Enough to compensate for the extra pieces AB, B_nZ ($= AB$). That is, we need to take n large enough so that $n\Delta > 2AB$. But this certainly can be done.† This completes the proof that the sum of the angles in any triangle is $\leq 180°$.

19 EXERCISES

1 We have seen that Theorem 18EH is not true in spherical geometry. How does Legendre's proof fail in spherical geometry?

2 Use the triangle inequality (Theorem 17) to prove that a polygonal path between two points is not shorter than the straight-line path between them. [Hint: Use mathematical induction on the number of edges in the polygonal path.]

Assumption (EH) only enables us to prove that the sum of the angles of a triangle is not greater than two right angles. In a sense, this is because it may be the case, as in spherical geometry, that not all triangles have the same angle sum. We state this explicitly as:

20 Theorem

If we assume, in addition to the other axioms of absolute geometry, that all triangles have the same angle sum, say Σ, then that sum is two right angles.

PROOF: Let us abbreviate the sum of two right angles as $2R$. Now consider any triangle ABC with angle sum $\alpha + \beta + \gamma = \Sigma$. Let L, M, N be points on the sides of ABC (see Illustration 12). The segments NM, ML, LN divide ABC into four triangles, each having angle sum Σ. Thus the total

---

† In fact, another axiom is needed here. It is usually called the *Archimedean Axiom*. It states: If RS is an arbitrary segment and Δ is any other segment, then there is an integer n such that if n copies of Δ are laid end to end along a straight line, the resulting segment is longer than RS.

angle sum in these four triangles is, on the one hand, 4Σ and, on the other hand, $\alpha + \beta + \gamma +$ the angle sum at L, M, N. But the angle sum at each of L, M, N is $2R$ (that is, one "straight angle"). Hence

$$4\Sigma = \alpha + \beta + \gamma + 3 \cdot 2R,$$
so that $4\Sigma = \Sigma + 6R$ and $3\Sigma = 6R$.

Hence $\Sigma = 2R$, as required.

Illustration 12

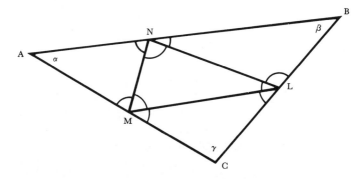

In the next section we shall see what happens if we impose the further assumption (H) on our absolute geometry.

Section 4   More History—Saccheri, Bolyai, and Lobachevsky

Many theorems of *absolute geometry* were proved by the Italian Gerolamo Saccheri (1667–1733), although this was not his goal at all. His goal was the same as that of others for centuries before him: to prove Euclid's parallel postulate as a theorem. He hoped to do this by assuming the falsity of the parallel postulate and deriving, on the basis of this assumption, a contradiction.

Of the two ways already discussed in Section 2 of assuming the falsity of the parallel postulate, Saccheri chose to assume postulate

(H), not (S).† In his book *Euclides ab omni naevo vindicatus* ("Euclid Freed of Every Flaw"), he proves a great many theorems on the basis of assumption (H), always hoping to reach a contradiction. In the end he becomes discouraged and asks the reader to agree that one of his theorems is, to quote Bonola [2, p. 43], "contrary to the nature of the straight line" and hence false. Hence (H) itself must be false. In fact, however, his "false theorem" is not false at all but, like Saccheri's other theorems, simply a theorem in a new kind of geometry, the *hyperbolic geometry* based on assumption (H). But because Saccheri was blinded by his devotion to the "truth" of Euclidean geometry, he missed being the discoverer of this new geometry.

The credit for recognizing that the geometry based on assumption (H) was an equally valid companion geometry to Euclid's is shared by the Hungarian Johann Bolyai and the Russian Nicolai Lobachevsky.

Bolyai János (in German: Johann Bolyai) (1802–1860) shares the credit with Lobachevsky for being the discoverer of non-Euclidean geometry. The son of a prominent Hungarian mathematician, János himself was an army officer by profession. Three streets in Budapest are named after the Bolyai family. (Permission of the Granger collection.)

† In fact, he assumes another condition equivalent to (H), not (H) itself.

Johann Bolyai (1802–1860) was the son of the Hungarian mathematician Wolfgang Bolyai, who had been a close friend in his student days of Gauss, the German mathematician universally acknowledged to be the greatest mathematician of his time. About the elder Bolyai himself, his (and Johann's) biographer, P. Stäckel, says that "the history of mathematical research in Hungary begins with Wolfgang Bolyai." The significance of this claim is considerable, since by 1960 Hungary was reported to have more mathematicians per capita than any other country in the world. Five mathematical research journals are published in major European languages in Hungary, besides others in the Hungarian language.

Johann learned mathematics at an early age from his father and became especially interested in trying to prove Euclid's parallel postulate while a student in Vienna. But soon after that, in 1823, Johann realized that he could develop a whole new geometry—as he put it, "a whole new universe from nothing"—on the basis of assumption (H). His excitement over this discovery leaps from the page of the letter he wrote announcing it to his father. Then, in 1832, Wolfgang included Johann's results as an appendix (see [1] or [2]) to his own large geometric treatise.

The elder Bolyai sent a copy of his treatise, with its appendix, to his old friend Gauss and waited for Gauss' reply, which came promptly. But the reply was a shock from which Johann seemed never to recover. For Gauss replied in the following words [2, p. 100]: "If I commenced by saying that I am unable to praise this work (by Johann), you would certainly be surprised for a moment. But I cannot say otherwise. To praise it would be to praise myself. Indeed the whole contents of the work, the path taken by your son, the results to which he is led, coincide almost completely with my meditations, which have occupied my mind partly for the last thirty or thirty-five years." He goes on to say that he had not intended to publish anything in his own lifetime since few people understood these ideas. The implication is that he did not want to bring scorn to his own famous name by supporting an opinion (namely, that Euclid's parallel postulate was unprovable) that was sure to be unpopular.

Nicolai Lobachevsky (1793–1856) spent his entire academic life at the University of Kazan in Russia—first as student, then as Assistant, and finally as Professor. He apparently first began thinking about a possible new geometry in 1823, although he had been interested in the possibility of proving Euclid's parallel postulate several years earlier. In 1826 he presented a formal lecture (in French) to the university explaining the principles of the new geometry. He devoted the rest of his life to this work, publishing papers on it in 1829 and intermittently thereafter until the years before his death in 1856. The booklet [6], from which we quote later, appeared in 1840 in German.

When the younger Bolyai had first informed his father of his new discoveries, Wolfgang had urged him to publish them soon. He made the remarkably prophetic observation that "many things have an epoch, in which they are found at the same time in several places, just as the violets appear on every side in the spring." [1, p. 99] For, by one of the remarkable coincidences of science, comparable to the nearly simultaneous invention of differential calculus by Newton and Leibniz, Bolyai and Lobachevsky made their revolutionary discoveries of the same new geometry within a year or two of each other. At the time, neither of them knew of the other's existence. In fact, apparently Lobachevsky never, throughout his entire life, heard of Bolyai or his discoveries. However, in 1848 Bolyai learned of Lobachevsky's work through a copy of [6] which had been called to his father's attention (indirectly) by Gauss. Johann's old resentment of Gauss flared up anew at this. He could see no other explanation for the similarity of Lobachevsky's results with his own than that Gauss had not been satisfied to destroy the value of Johann's own work by his earlier remarks but had passed the results on to Lobachevsky, who had presented it as his own. As mentioned already, however, in spite of Johann's convictions on this point, there is no evidence whatsoever that Lobachevsky ever heard of Bolyai or his work.

As a further quirk of the history of science, we can compare Bolyai's excitement ("I have discovered such magnificent things that I am myself astounded at them!") to Kepler's excitement when he realized, almost overnight, that the position of the planets could be described by relations between the five regular polyhedra. (Recall Illustration 2, Chapter 1.) The only difference was that Bolyai *had* discovered a whole new world, while Kepler's insight was completely false!

Section 5 Hyperbolic Geometry

A good introduction to the hyperbolic geometry which Bolyai and Lobachevsky developed on the basis of their denial of the parallel postulate, that is, on the basis of assuming axiom (H),† is given in Lobachevsky's own booklet [6] of 1840. Following a reminder to the reader of the theorems which are known from Euclid, independent of the parallel postulate, he says:

All straight lines which in a plane go out from a point can, with reference to a given straight line in the same plane, be divided into two classes—into *cutting* and *not-cutting*.

The *boundary lines* of the one and other class of those lines will be called *parallel to the given line*.

Nicolai Ivanovich Lobachevsky (1793-1856) was the co-discoverer, with J. Bolyai, of non-Euclidean geometry. He spent his entire academic life at the University of Kazan, having entered as a student two years after it opened. In 1819, the university fell on hard times, and Lobachevsky alone gave all the lectures in pure mathematics, astronomy, and theoretical physics. In 1827 he was elected Rector, a post he held for 19 years. In spite of these heavy duties, he developed his new geometry in such detail that it is often called "Lobachevsky geometry." (Permission of the Granger Collection.)

† In fact, Lobachevsky is only assuming the weaker axiom (EH) in the beginning of this excerpt.

Section 5 *Hyperbolic Geometry* 231

From the point A,† let fall upon the line BC the perpendicular AD to which again draw the perpendicular AE.

In the right angle EAD, either all straight lines which go out from the point A will meet the line DC, as for example AF, or some of them, like the perpendicular AE, will not meet the line DC. In the uncertainty whether the perpendicular AE is the only line which does not meet DC, we will assume it may be possible that there are still other lines, for example AG, which do not cut DC, how far soever they may be prolonged. In passing over from the cutting lines, as AF, to the not-cutting lines, as AG, we must come upon a line AH, parallel to DC, a boundary line, upon one side of which all lines AG are such as do not meet the line DC, while upon the other side every straight line AF cuts the line DC.

The angle HAD between the parallel HA and the perpendicular AD is called the parallel angle (angle of parallelism), which we will here designate by $\Pi(p)$, for $AD = p$.

Illustration 13

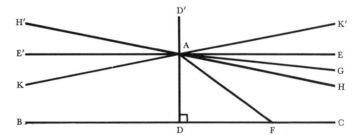

If $\Pi(p)$ is a right angle, so will the prolongation AE' of the perpendicular AE likewise be parallel to the prolongation DB of the line DC, in addition to which we remark that in regard to the four right angles, which are made at the point A by the perpendiculars AE and AD, and their prolongations AE' and AD', every straight line which goes out from the point A, either itself or at least its prolongation, lies in one of the two right angles which are turned toward BC, so that except the parallel EE', all others, if they are sufficiently produced both ways, must intersect the line BC.

If $\Pi(p) < \frac{1}{2}\pi$, then upon the other side of AD, making the same angle $KAD = \Pi(p)$, will lie also a line AK, parallel to the prolongation DB of the line DC, so that under this assumption we must also make a distinction of *sides in parallelism*.

All remaining lines or their prolongations within the two right angles turned toward BC pertain to those that intersect, if they lie within the angle $HAK = 2\Pi(p)$ between the parallels; they pertain on the other hand to the nonintersecting AG if they lie upon the other sides of the parallels AH and AK, in the opening of the two angles $EAH = \frac{1}{2}\pi - \Pi(p)$, $E'AK = \frac{1}{2}\pi - $

† See Illustration 13.

$\Pi(p)$, between the parallels and EE', the perpendicular to AD. Upon the other side of the perpendicular EE' will, in like manner, the prolongations AH' and AK' of the parallels AH and AK likewise be parallel to BC; the remaining lines pertain, if in the angle $K'AH'$, to the intersecting, but if in the angles $K'AE$, $H'AE'$, to the nonintersecting.

In accordance with this, for the assumption $\Pi(p) = \frac{1}{2}\pi$, the lines can be only intersecting or parallel; but if we assume that $\Pi(p) < \frac{1}{2}\pi$, then we must allow two parallels, one on the one and one on the other side; in addition, we must distinguish the remaining lines into nonintersecting and intersecting.

For both assumptions, it serves as the mark of parallelism that the line becomes intersecting for the smallest deviation toward the side where lies the parallel, so that if AH is parallel to DC, every line AF cuts DC, how small soever the angle HAF may be.

Aside from the particularly archaic translation, these facts are essentially as they would be presented in an elementary course on the subject today. For convenience, we shall rephrase the essential results as theorems and modify the terminology.

The attentive reader will note that several details in the following theorems are not treated very carefully. For example, Lobachevsky does not explain how to distinguish "right" from "left," and we make no effort to do this either. We rely on the reader's intuition, just as Lobachevsky did. The particular difficulties involved in distinguishing between "right" and "left" could be clarified somewhat by considering *rays* (or *half-lines*) instead of entire lines, but because these theorems all depend so much on intuition anyway, we have not done this. (In Illustrations 15 to 23, the arrows are intended to suggest the advantage of restricting our attention to a particular ray, as opposed to the whole line of which it is a part.)

Because of our reliance on intuition, several of the following theorems are provided with "reasons," not "proofs." We take up this problem again in Section 6. In the meantime, the reader should be satisfied if he has a good intuitive understanding of how Theorems 22 to 33 follow from the assumption (H) and to what extent they are *not* true in Euclidean geometry.

21 Theorem

Given a point A not on a line l ($= BC$), there are infinitely many lines through A which do not meet l. (In Illustration 13, AH, AG, AE' are examples of such lines.)

22 Theorem

At any point A not on a line l, there are two distinguished lines (AH and AK) among the infinitely many lines which do not meet l. Line AH on the right side, which separates the lines (such as AF) which meet l from those (such as AG) which do not, is called the *right parallel to l at A*. The corresponding line AK, on the other side of AD, is called the *left parallel to l at A*. The other lines at A which do not meet l (for example, AG) are called *hyperparallels* to l.

We also list the following facts about parallels here, though they are not included in our quotation from Lobachevsky.

Illustration 14

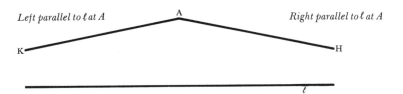

Illustration 15

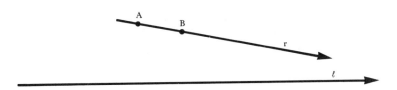

23 Theorem

(a) If a line r is the right (or left) parallel to line l at a point A not on l and if B is another point on r, then r is also the right parallel to l at B. (That is, parallelism does not really depend on the particular point used to define it; see Illustration 15.)

(b) If line r is a right (or left) parallel to line l, then l is a left (or right) parallel to r.

(c) If line *r* is a right (or left) parallel to line *s* and line *s* is a right (or left) parallel to line *q*, then line *r* is a right (or left) parallel to line *q*. (See Illustration 16.)

The proofs of these statements are somewhat technical and at the present informal stage of our treatment would not be much more convincing than the mere statement of the theorems. For this reason, we omit the proofs. The interested reader may consult Lobachevsky [6, Propositions 17, 18, 25] for (nearly complete) proofs.

Illustration 16

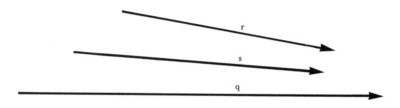

Illustration 17

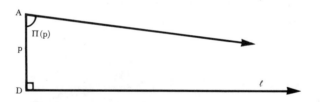

24 Definition

For any given segment *p* (= *AD*), an *angle of parallelism*, written $\Pi(p)$, is defined as follows: Erect a perpendicular *l* at *D* to *AD*. At *A* draw a (right or left) parallel to *l*. The angle made between this parallel and *AD* is the angle of parallelism of *AD*. (See Illustration 17. The arrows in the illustrations of this section are intended to indicate that the corresponding lines are right or left parallels to each other.)

25 Theorem

The angle of parallelism of any segment is always an acute angle.

Moreover, two segments of equal length have equal angles of parallelism.

We now proceed to some simple but striking consequences of these facts. We define an *asymptotic triangle* to be the figure formed by a segment AB and two lines at its ends, each the right (or left) parallel of the other (see Illustration 18). If we think of an angle zero at E (for "end"), we see that Theorem 25 says: The angle sum in a right-angled asymptotic triangle is less than two right angles.

Illustration 18

An asymptotic triangle

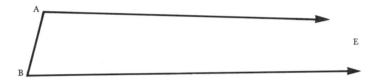

Illustration 19

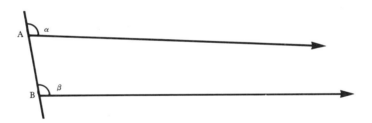

It should be clear what is meant by an "exterior angle" of an asymptotic triangle. Using this terminology, we have:

26 Theorem

An exterior angle of an arbitrary asymptotic triangle is greater than the opposite interior angle. (That is, $\alpha > \beta$ in Illustration 19.)

REASON: There are only three possibilities: (a) $\alpha < \beta$, (b) $\alpha = \beta$, and (c) $\alpha > \beta$. By ruling out the first two, we shall have the required result.

(a) If $\alpha < \beta$, we can draw a line in the interior of angle β, making an angle α with AB, as in Illustration 20. This line necessarily meets line AE, say at C. (Why?) But then in triangle ABC, we have an exterior angle, α, which is *equal* to an opposite interior angle, which is impossible by Theorem 11.

(b) If $\alpha = \beta$ and β is a right angle, we have an immediate contradiction to Theorem 25. (Why?) So we need only consider the case where β is

Illustration 20

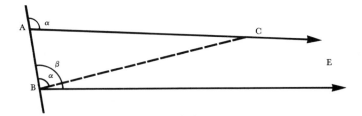

Illustration 21

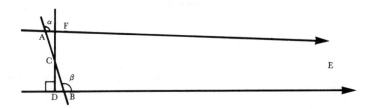

not a right angle. Let C be the midpoint of AB. (Do segments have midpoints? What axiom or theorem of Euclid guarantees this?) Drop a perpendicular from C to a point D on BE, and extend it in the other direction to meet AE at F, as in Illustration 21. (Why does it meet AE? If it does not, it is a parallel to AE—since it can be proved that any line between the two parallels AE, BE is parallel to each of them. But this is impossible since there would then be two distinct parallels (DC and DB) in the same direction to AE from the same point D.) Then triangles DBC and FAC are congruent (Why?). Hence F is a right angle also. That is, we have found a segment, DF, whose angle of parallelism is a right angle, F. This contradicts Theorem 25 again. Hence we know that only case (c), $\alpha > \beta$, can hold.

Section 5 *Hyperbolic Geometry*   237

27 EXERCISES

1 According to Theorem 26, an exterior angle of an asymptotic triangle is greater than the opposite interior angle. What is an asymptotic triangle in Euclidean geometry? Is Theorem 26 true in Euclidean geometry?

2 What is the sum of the angles of an asymptotic triangle in Euclidean geometry?

Illustration 22   *A Saccheri quadrilateral*

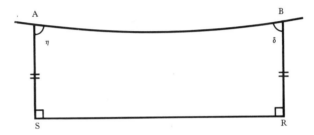

Illustration 23

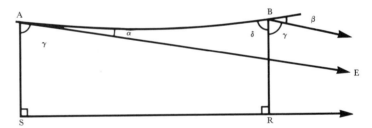

In Saccheri's work, a fundamental figure was a certain kind of quadrilateral, now usually referred to as a *Saccheri quadrilateral*. This is a quadrangle $ABRS$ in which $AS \perp SR$, $BR \perp RS$, and $AS = BR$. (See Illustration 22.)

28 Theorem

In a Saccheri quadrilateral, the angles η and δ, at A and B, respectively, are equal and acute.

REASON: By using Theorem 14 (Euclid's Proposition 4), it can be shown that $\eta = \delta$. So we have only to show that η (or δ) is acute. Draw right parallels to line SR from A and B, as in Illustration 23. Since $AS = BR$, these parallels make equal acute angles, say γ, with AS and BR. Let the rest of the interior angle at A be α and the rest of the exterior angle at B be β. Then EBA is an asymptotic triangle (Why?) with exterior angle β and opposite interior angle α. Hence $\beta > \alpha$, so that $\beta + \gamma > \alpha + \gamma = \delta$. Hence $180° = (\beta + \gamma) + \delta > 2\delta$, so that δ is acute, as required.

29   EXERCISES

1 What is the usual name for a Saccheri quadrilateral in Euclidean geometry?
2 How do we know there is such a figure as a Saccheri quadrilateral?
3 Prove in detail the first sentence of the Reason for Theorem 28. [Hint: Draw AR and BS. Then triangles ARS and BSR are congruent. This shows that *part* of η is equal to part of δ. To show that the other parts are equal, consider triangle ABS and BAR.]

An immediate consequence of this relatively innocent looking theorem is the more startling.

30   Theorem

The sum of the angles in any triangle is *less* than two right angles.

REASON: We show, in fact, that the sum of the angles in any triangle is the same as the sum of the two acute angles in some Saccheri quadrilateral. Let ABC be the triangle, with L and M the midpoints of the sides AC and BC, respectively, as in Illustration 24. Drop perpendiculars from A, B, and C to the line LM, meeting it at S, R, T, respectively, to get the figure of Illustration 25. This figure contains the entire proof, for triangle LTC is congruent to LSA (by Theorem 16), and hence $\alpha_1 = \alpha_2$; likewise, MTC is congruent to MRB, and hence $\beta_1 = \beta_2$. So the sum of the angles of triangle ABC is the sum of the acute angles A and B of the Saccheri quadrilateral $ABRS$, and hence is less than two right angles, as required.

An easy corollary to Theorem 30 is

31   Theorem

The sum of the angles in an arbitrary quadrilateral is less than two right angles.

Section 5 Hyperbolic Geometry   239

Illustration 24

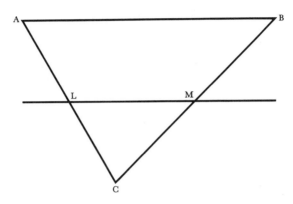

Illustration 25

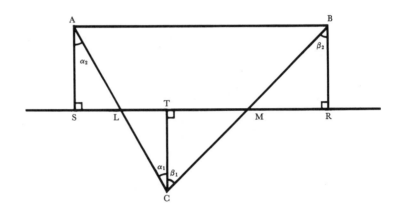

Theorem 31 has an even more remarkable consequence, namely:

Illustration 26

32 Theorem

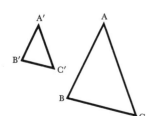

Any two *similar* triangles are necessarily *congruent*. That is, if three angles of one triangle are equal, respectively, to three angles of another, then the corresponding sides of the triangles are also equal.

REASON: Consider the similar triangles ABC and $A'B'C'$ of Illustration 26. Suppose, if possible, that they are *not* congruent, say $A'B' < AB$. Now copy $A'B'C'$ onto ABC; that is, lay off $AB'' = A'B'$ on AB and $AC'' = A'C'$ on AC.

The two possible cases $A'C' < AC$ and $A'C' > AC$ are shown in Illustration 27. (Why is $A'C' = AC$ impossible?) In case $A'C'(= AC'') < AC$,

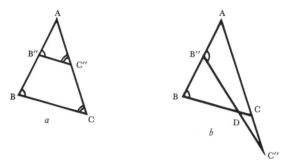

Illustration 27

the angles at B and B'' are equal and the angles at C and C'' are equal. That is, the sum of the angles in the quadrilateral $BB''C''C$ is four right angles, which contradicts Theorem 31. On the other hand, in case $A'C'(=AC'') > AC$ [Illustration 27(b)], the angles at B and B'' are equal. This contradicts the exterior-angle theorem (Theorem 11).

33 EXERCISES

1 Prove Theorem 31. [Hint: Divide the quadrangle into two triangles.]
2 Why is $AC = A'C'$ impossible, in the proof of Theorem 32?

It is natural to ask, when first encountering hyperbolic geometry, whether such a geometry has any physical significance (as spherical geometry has a natural application on the surface of the earth) or is just a mathematical and philosophical curiosity. It is part of mathematical folklore that Gauss himself attempted to determine whether the geometry of the universe was Euclidean or non-Euclidean by measuring the angles of a large triangle whose vertices were three mountain peaks. Depending on whether the angle sum of this triangle was less than, equal to, or greater than 180° its geometry would be hyperbolic, Euclidean, or spherical.

It is doubtful that Gauss actually made such an experiment, but even if he did, it would have been inconclusive. For enough is now known about the universe to show that even if its geometry is hyperbolic, the angle sum in such a small triangle would not differ from 180° by more than the error of measurement using the best instruments available today.

For anyone determined to believe that the universe is actually Euclidean, it is a discouraging fact that it is logically impossible to prove by measurement that a triangle is Euclidean, even though it is logically possible to prove by measurement that a triangle is hyperbolic or spherical. This is because any measurement involves some margin of error, even though it may be very small. This makes it impossible to prove that a triangle is Euclidean, since to be Euclidean requires that its angle sum be *exactly* 180°, while a whole range of values would prove it hyperbolic or spherical. For example, if the angle sum were measured to be $180.000° \pm 0.002°$, then the experiment would be inconclusive and any one of the three geometries might apply. On the other hand, if measurement gives $179.997° \pm 0.002°$, it can be concluded that the triangle is hyperbolic.

An actual physical application of hyperbolic geometry occurs in the theory of binocular vision as developed by R. K. Luneburg and A. A. Blank. In a classical experiment by W. Blumenfeld (1913), a subject is seated in a dark room facing two rows of lights extending directly out in front of him. With his head fixed (but his eyes freely moving, to get the benefit of binocular vision), he instructs an assistant to move the lights until they are in two exactly parallel rows.

After the experiment, these two supposedly parallel rows are examined. It can be assumed, on other evidence, that the subject's "perceived space" will have one of the classical geometries—hyperbolic, Euclidean, or spherical. Thus, it is only necessary to compare the perceived parallels with the way in which parallels would be perceived in each of these three geometries. The result, for most subjects, is that the perceived parallels are in closer agreement with hyperbolic geometry than with either Euclidean or spherical geometry.

A later, independent experiment, taking advantage of the different properties of the line joining the midpoints of two sides of a triangle in the three geometries, confirms that for most subjects the perceived midline is hyperbolic rather than Euclidean or spherical. More details of these experiments can be found in a series of articles by A. A. Blank in the *Journal of the Optical Society of America* from 1953 to 1961. (Especially **51** (1961), pp. 335–339, and **43** (1953), pp. 717–727.)

Section 6 New Beginnings

We have already called attention to the fact that we have given "reasons" for these theorems of hyperbolic geometry, not proofs. It is hoped that the questions scattered throughout the text will have served to call the attention of even the not-so-attentive reader to certain gaps in the "reasons" which keep us from being able to call them proofs.

For example, in the quotation from Lobachevsky, the lines through a given point A fall into two classes: those which meet l to the right of point D and those which do not. Then it is asserted, entirely without proof, that "in passing over from the cutting lines, as AF, to the not-cutting lines, as AG, we must come upon a line AH, parallel to DC, a boundary line, ..." How do we know there is a "boundary line" there? Perhaps at this very place there is a blank space of some sort. In exactly this way, the search for a finite decimal number whose square is 2 separates the finite decimals into two classes—those, such as 1, 1.4, 1.41, 1.414, whose square is less than 2 and those, such as 2, 1.5, 1.42, 1.415, whose square is greater than 2. But no matter how hard you look, you will not find a finite decimal exactly in the blank space between these two kinds of finite decimal. How do we know there are not blank spaces among the lines at A, in exactly the same way? The answer is that we don't, unless we assume it as an axiom. The missing axiom is an *axiom of completeness*.

Another omission was made in the Legendre proof that the sum of the angles of a triangle is not greater than two right angles. We assembled a number of triangles side by side. Then we took "n large enough so that $n\Delta > 2AB$." As already remarked there, it is not obvious that such an n can always be chosen. The missing axiom here is the so-called *Archimedean axiom of continuity*.

Referring again to the quotation from Lobachevsky, we might ask how we know that the two types of lines at A are not jumbled up. Perhaps there are some lines that meet l, then some which do not, then again some which meet, and so on. We need, in fact, to know that between two lines which meet l, there can be no line which does not meet l. Referring to Illustration 28, we need to know that any line through A, such as l', lying in the angle DAF necessarily meets l, in fact necessarily meets the segment DF. For this we could assume

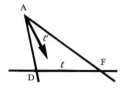

Illustration 28

some such axiom as: "If a line enters a triangle, it must also leave it." This is known as *Pasch's axiom*.

As a final example, we point out that the whole idea of *order* has been assumed without explicit mention. For example, we talked about the *right* and *left parallels* to a line l. How do we know there is a right and left? (Referring to spherical geometry, which part of the equator is the *right side* of the equator and which is the *left side?* Where does one leave off and the other begin?)

Another illustration of this is in our discussion of Euclid's defective Proposition 1, the construction of an equilateral triangle from a given side. In order to repair this defect, we needed some such axiom (or theorem) as: "If one circle contains a point on the inside of another circle and also a point on the outside of that circle, then the two circles have two points in common." But the mere statement of this axiom (or theorem) involves notions of "inside" and "outside" which do not appear explicitly in Euclid. We need some *axioms of order or betweenness*.

The task now begins to look difficult, and it is (or *was*). It took the entire last 75 years of the nineteenth century to make a satisfactory formulation of the underlying assumptions of Euclidean geometry (and, as a by-product, of hyperbolic geometry). Important observations were made by Pasch and Dedekind in Germany and Peano in Italy. The synthesis of this work was finally given by Hilbert in *Foundations of Geometry* [5], which of course rests strongly on the work of its predecessors, in the same way that Euclid's synthesis in the *Elements* did. Hilbert gives five groups of axioms. We have already seen the motivation for most of them, and we list them here for the record. The reader will see that Euclid's missing assumptions about completeness, continuity, and order, as well as Pasch's triangle axiom (mentioned in the preceding paragraphs), have all been included. Some of Euclid's "theorems" about congruence of segments and angles are also included as axioms by Hilbert.

HILBERT'S AXIOMS

The undefined terms are "point," "line," "lies on" (or "is incident with," or "is on," etc.), "congruence," and "point Y is between point X and point Z."

Hilbert's complete list of axioms also includes axioms about *planes*. But since our entire discussion is restricted to plane geometry,

David Hilbert (1862–1943) began his brilliant mathematical career in algebra, where his abstract solution to an old problem in the theory of invariants was described as "not mathematics, but theology." By 1900 his genius was demonstrated in his now-famous lecture on "23 fundamental problems of mathematics." This lecture influenced mathematical research for decades. His work on the foundations of geometry, described in Chapter 4, is the culmination of 2,500 years of mathematical investigation. His contributions to analysis ("Hilbert spaces") and mathematical logic are equally spectacular. Hilbert's presence at Göttingen (1895–1943) made that university the Mecca of the mathematical world. (Courtesy of the Smith Collection, Columbia University.)

we have omitted these axioms. Hilbert's book has been continually improved from the first edition (English translation in 1902 [5]) through the tenth (1968). We have incorporated some of these improvements.

34 Group I: Axioms of Incidence

I(a) Two distinct points A and B are incident with one and only one line l. We write $AB = l$ or $BA = l$.

I(b) Each line is incident with at least two points. There are at least three points, not all incident with the same line.

35 Group II: Axioms of Order

II(a) If point B is between points A and C, then A, B, and C are three distinct points on a line and B is also between C and A.

II(b) If A and C are two points on a line, then there is at least one point B between A and C.

II(c) Of any three points on a line, there is always one and only one which is between the other two.

II(d) (Pasch's Axiom). Let A, B, C be three points not on the same line, and let l be a line not passing through any of the points A, B, C. Then, if line l passes through a point of the segment AB, it will also pass through either a point of the segment BC or a point of the segment AC.

36 EXERCISES

1. "Segment" [as used in II(d) above] has not yet been defined. Give a reasonable definition, using Hilbert's terminology. (The end points of the segment should be thought of as part of the segment.)
2. If A and B are two different points, define the *ray* (or *half-line*) through B and having A as its end point. (The point A should be thought of as part of the ray.)
3. Hilbert proves, on the basis of these two groups of axioms alone, that any line l divides the remaining points of the plane into two parts, called *open half-planes*. Explain how you could say that "A and B do not lie in the same open half-plane" in Hilbert's terminology.

Some definitions are needed before we can state the next group of axioms.

37 Definition

1. Any point A on line l determines two rays, which have only A in common. Two points (different from A) in the same ray are said to *be on the same side of A on line l*.
2. Similarly, two points in the same one of the two open half-planes into which a line l divides a plane are said to *be on the same side of l in the plane*.

38 Definition

1. Two rays, h and k, having a common end point A but not lying in the same line, are said to form an angle (h, k), also written $\angle(h, k)$ or $\angle(k, h)$.
2. A point A is said to be an *interior point* of angle (h, k) if every line through A meets at least one of the rays h or k.

39 Group III: Axioms of Congruence

III(a) If A, B are two points on a line l and if A' is a point on the same or another line l', then on a given side of A' on the line l', we can always find one and only one point B' so that the segment AB (or BA) is congruent to the segment $A'B'$. We indicate this relation by writing $AB \equiv A'B'$.

III(b) If $AB \equiv A'B'$ and $AB \equiv A''B''$, then $A'B' \equiv A''B''$.

III(c) Let AB and BC be two segments of a straight line l which have no points in common aside from the point B, and furthermore, let $A'B'$ and $B'C'$ be two segments of the same or another straight line l' having, likewise, no point other than B' in common. Then, if $AB \equiv A'B'$ and $BC \equiv B'C'$, we have $AC \equiv A'C'$.

III(d) Let an angle (h, k) be given in the plane, and let a straight line l' be given in the same plane. Suppose also that, in the plane, a definite side of the straight line l' be assigned. Denote by h' a half-ray of the straight line l' emanating from a point O' of this line. Then in the plane there is one and only one ray k' such that the angle (h, k), or (k, h), is congruent to the angle (h', k') and at the same time all interior points of the angle (h', k') lie upon the given side of l'. We express this relation by means of the notation

$$\angle(h, k) \equiv \angle(h', k').$$

Every angle is congruent to itself; that is,

$$\angle(h, k) \equiv \angle(h, k)$$
or $\quad\angle(h, k) \equiv \angle(k, h).$

We say, briefly, that every angle in a given plane can be *laid off* upon a given side of a given ray in one and only one way.

III(e) If, in the two triangles ABC and $A'B'C'$, the congruences

$$AB \equiv A'B' \qquad AC \equiv A'C' \qquad \angle BAC \equiv \angle B'A'C'$$

hold, then the congruence

$$\angle ABC \equiv \angle A'B'C'$$

also holds.

40 Group IV: Euclid's Parallel Postulate

IV Through any point A, not incident with a line l, one and only one line can be drawn which does not intersect line l. This line is called the *parallel* to l through the given point A. (In case A is on l, we also say that *l is parallel to l at A*; that is, every line is parallel to itself.)

41 Group V: Axiom of Continuity (Axiom of Archimedes)

V   Let A_1 be any point upon a straight line between the arbitrarily chosen points A and B. Take the points $A_2, A_3, A_4, \ldots$ so that A_1 lies between A and A_2, A_2 between A_1 and A_3, A_3 between A_2 and A_4, etc. Moreover, let the segments

$$AA_1, A_1A_2, A_2A_3, A_3A_4, \ldots$$

be congruent to one another. Then, in this sequence of points, there always exists a certain point A_n such that B lies between A and A_n.

There is one further axiom, the axiom of completeness, which can be used to guarantee that the plane has no "empty spaces" in it.

Axiom of Completeness

To a system of points and straight lines satisfying Axioms I to V, it is impossible to add other elements in such a manner that the system thus generalized shall form a new geometry obeying all of Axioms I to V. In other words, the elements of the geometry form a system which is not susceptible of extension, if we regard the five groups of axioms as valid.

In view of the criticisms that have been made of Euclid's and later work, it might be thought that it would still be a difficult task to develop all of plane geometry from these axioms. It is! Using these or similar axioms, the job has been done by A. G. Forder in *Foundations of Euclidean Geometry* (1927; Dover reprint, 1958) and by K. Borsuk and W. Szmielew in *Foundations of Geometry* (1960). We do not propose to follow these developments in detail. The reader who has looked at either of these books is likely to be especially grateful for this decision.

This concludes our discussion of the "classical" geometries. In the following sections (7 to 11), we direct our attention to a "rudimentary" geometry, which satisfies only a few of Hilbert's axioms. Our only reference to classical Euclidean geometry will be as a guide to suggest what theorems we can try to prove in the rudimentary geometry.

Section 7 Analytic Geometry—A Reminder

In the following sections we shall discuss in detail a miniature geometry which has the incidence and parallelism properties given by Groups I and IV of Hilbert's axioms, but not many of the other properties. We shall rely on the reader's having some knowledge of the facts of ordinary analytic geometry. The present section is devoted to a summary of these facts.

Two intersecting *coordinate axes*, the *x* axis and the *y* axis, are singled out in the (Euclidean) plane. (Although these axes are often chosen to be perpendicular, this is not essential for our purpose.) The points on each coordinate axis are labeled by ordinary real numbers in their natural order, the point of intersection being 0 on both scales. (See Illustration 29.) Using this *coordinate system*, any point P is assigned a pair of *coordinates* (x, y) as follows:

Illustration 29

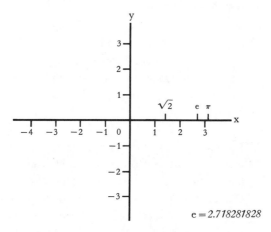

$e = 2.718281828$

(a) If P is on the x axis, then P has coordinates $(x, 0)$, where x is the number assigned on the x axis to P.

And similarly:

(b) If P is on the y axis, then P has coordinates $(0, y)$, where y is the number assigned on the y axis to P.

In particular, the point of intersection of the two axes has coordinates $(0, 0)$. It is called the *origin* and is usually designated as O. Finally:

(c) If P is not on either coordinate axis, we draw the line through P parallel to the y axis and the line through P parallel to the x axis. Suppose the points where these two lines meet the x and y axes, respectively, are $(x, 0)$ and $(0, y)$. Then we assign the coordinates (x, y) to the point P.

The student will recall that in terms of these coordinates, every line has an equation of the form

Illustration 30

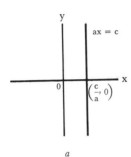

42 $$ax + by = c \quad (a, b \text{ not both } 0)$$

That is, the points of a line are exactly those points (x, y) satisfying Equation 42 for some fixed numbers a, b, and c. Conversely, every equation of form 42 represents a line. In particular, if $b = 0$, the line is $ax = c$ and is parallel to the y axis (or coincides with the y axis if $c = 0$), as in Illustration 30(a); if $a = 0$, the line is $by = c$ and is parallel to the x axis (or coincides with the x axis if $c = 0$), as in Illustration 30(b).

If $c \neq 0$ in the equation of a line $ax + by = c$, that is, if the line does not contain the origin $(0, 0)$, then we can divide both sides of Equation 42 by c and get $\frac{a}{c}x + \frac{b}{c}y = 1$, that is, $\frac{x}{c/a} + \frac{y}{c/b} = 1$. This can be written as

42* $$\frac{x}{a'} + \frac{y}{b'} = 1 \quad \text{where } a' = \frac{c}{a} \text{ and } b' = \frac{c}{b}.$$

Equation 42* is called the *intercept form* of the equation of the line $ax + by = c$ since it shows at a glance that the line 42* has x intercept $(a', 0)$ and y intercept $(0, b')$. That is, the line crosses the x axis at $(a', 0)$ and the y axis at $(0, b')$. Since two points determine a

Illustration 31

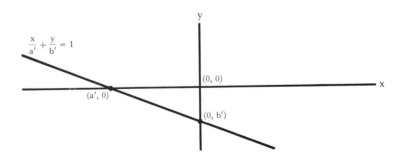

line, these intercepts determine the line completely, as shown in Illustration 31.

Every line which is not parallel to the *y* axis has a *slope*, denoted by the letter *m*. This is the measure of the "vertical movement" of the line compared with the "horizontal movement." Since for a line this is a constant, independent of where along the line we make the measurement, we can make the following formal definition:

43 *Definition*

If $P_1: (x_1, y_1)$ and $P_2: (x_2, y_2)$ are two distinct points on a line *l* which is not parallel to the *y* axis (that is, $x_1 \neq x_2$), the *slope m* of *l* is defined to be

$$m = \frac{y_1 - y_2}{x_1 - x_2}.$$

(It can be verified that *m* is independent of the choice of P_1 and P_2.)

Illustration 32

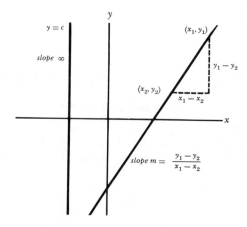

If l is a line parallel to the y axis, we say l has *infinite slope*, or has *slope infinity*, ∞. (See Illustration 32.)

If $b \neq 0$ (that is, if the line $ax + by = c$ is not parallel to the y axis), then Equation 42 can be solved for y to give

42** $$y = mx + r \qquad \left(m = \frac{-a}{b}, r = \frac{c}{b}\right).$$

Equation 42** is called the *slope-intercept form* of the equation of the line $ax + by = c$ ($b \neq 0$) since it shows at a glance that the line 42** has slope m and y intercept $(0, r)$. (See Problem 1 of Exercises 46.)

Note that parallel lines are those having the same slope m. Equation 42** is particularly useful when we are interested in all lines parallel to some fixed line. We simply write the equation of the given line in the form 42** and then obtain all lines parallel to it as $y = mx + r$, where r takes on all possible values (see Illustration 33).

Illustration 33

If m *is fixed and* b *varies, the lines* y = mx + b *are all parallel and meet the* y *axis at* $(0, b)$.

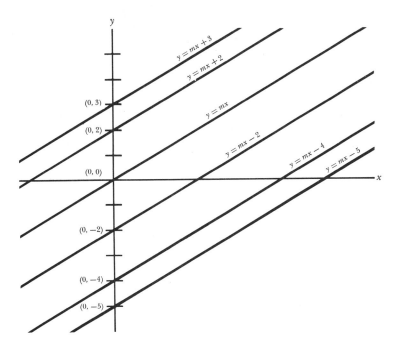

It is known that the equation $x^2 + y^2 = r^2$ represents a circle with center at $(0, 0)$ and radius $r \geq 0$, as in Illustration 34(a). (If $r = 0$, we speak of a *point-circle*, or *degenerate circle*.) More generally, the equation

44(a) $$(x - u)^2 + (y - v)^2 = r^2$$

represents a circle with center (u, v) and radius r. (See Illustration 34(b).) This circle equation

44(b) $$x^2 + y^2 - 2ux - 2vy + (u^2 + v^2 - r^2) = 0$$

is a special case of the *general quadratic equation*

45 $$ax^2 + bxy + cy^2 + dx + ey + f = 0.$$

It is known that a quadratic equation of this form always represents one of the three conic sections: an *ellipse* (of which a circle is a special case), a *hyperbola* or *parabola*, or a *degenerate form* of one of these, which may be two straight lines, a single straight line, a single point, or no points at all. These are called *conic sections* or, more briefly, *conics* because each of them (including the degenerate cases, except for the last one) is the set of points which a suitable plane in Euclidean space has in common with an ordinary (circular) *cone*. Illustration 35 shows these various possibilities.

Illustration 34   *Circle of radius* r, *center at* (u, v).

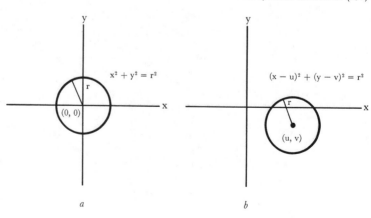

Section 7 Analytic Geometry—A Reminder

Illustration 35

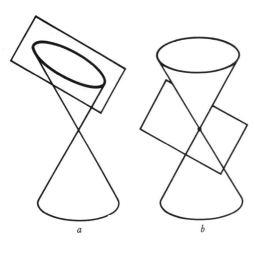

a — Ellipse

b — Single point (degenerate ellipse)

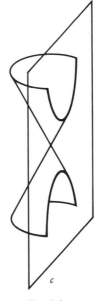

c — Hyperbola

d — Two intersecting straight lines (degenerate hyperbola)

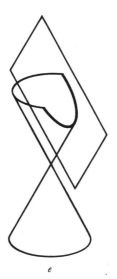

e — Parabola

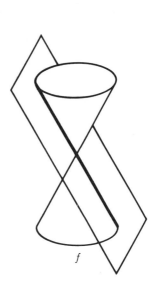

f — Single line (degenerate parabola)

46 EXERCISES

1. (a) Prove that $y = mx + r$ crosses the y axis at $(0, r)$.
   (b) Prove that the point $(r, mr + r)$ lies on the line $y = mx + r$.
   (c) Deduce that the line $y = mx + r$ has slope m. (Use Definition 43 with $P_1 = (r, mr + r)$ and $P_2 = (0, r)$.)

2. Find a quadratic equation (of the form of Equation 45) which represents
   (a) Two straight lines
   (b) A single straight line
   (c) A single point
   (d) No points at all

3. Sketch the "curves" represented by
   (a) $3x + 2y = 0$
   (b) $3x + 2y = 1$
   (c) $y = \frac{4}{3}x + 2$, $y = \frac{4}{3}x + 3$
   (d) $x^2 + y^2 = 6$
   (e) $(x - 3)^2 + (y + 2)^2 = 4$
   (f) $x^2 + y^2 - 6x + 4y + 9 = 0$ [Hint: "Complete the square."]

The subject which studies these matters was invented by Descartes and is called analytic geometry. In many respects it is really just algebra. (It might best be called *algebraic geometry* if this name were not currently reserved for a more esoteric subject which grew out of it.) The fact that geometry can be studied by means of algebra suggests that if we had a different kind of algebra from the usual one, we would have a different kind of geometry. We propose to follow that hint in the next few sections.

Section 8 Finite Arithmetics

Nearly everyone by now is familiar with a certain kind of finite arithmetic which has become so popular that it is often taught in grade schools. At that level the name of *clock arithmetic* is sometimes given to this arithmetic (or algebra) because the way hours are added is the way numbers are added in the finite arithmetic. But even more, multiplication is also performed in essentially the same way.

We illustrate this by the complete tables of addition and multiplication for the *integers modulo* 3, abbreviated I_3—that is, for the hours of a day which is only three hours long (Illustration 36).

Table 1. Addition table for I_3

+	0	1	2
0	0	1	2
1	1	2	0
2	2	0	1

Table 2. Multiplication table for I_3

·	0	1	2
0	0	0	0
1	0	1	2
2	0	2	1

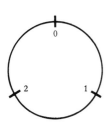

Illustration 36

Except for the entries in the boxes, Tables 1 and 2 are just like the tables for ordinary addition and multiplication. The entries in the boxes have to be different from the ordinary ones if we want our arithmetic to be a closed system, that is, if we want the sum or product of any two of the numbers 0, 1, 2 also to be one of these numbers. Thus $1 + 2$ cannot be 3, since 3 is not one of the original numbers 0, 1, 2. We take care of these sums or products which lead "outside the system" in a simple way: Whenever a computation by ordinary rules leads to a number which is not in the system, just subtract as many 3s as are needed to bring it back to the system. Thus $1 + 2 = 0$, since if we subtract a 3 from the result of computing $1 + 2$ in the ordinary way, we get 0.

The student should now verify that a similar rule, "Subtract 5 whenever a number other than 0, 1, 2, 3, 4 appears," gives the following Tables 3 and 4 of addition and multiplication for the *integers modulo* 5, I_5.

Table 3. Addition table for I_5

+	0	1	2	3	4
0	0	1	2	3	4
1	1	2	3	4	0
2	2	3	4	0	1
3	3	4	0	1	2
4	4	0	1	2	3

Table 4. Multiplication table for I_5

·	0	1	2	3	4
0	0	0	0	0	0
1	0	1	2	3	4
2	0	2	4	1	3
3	0	3	1	4	2
4	0	4	3	2	1

47 EXERCISES

1. Construct the addition and multiplication tables for the integers modulo n, when $n = 2, 4, 6, 7$.
2. In I_5, what is $3 - 2$? $2 - 3$? $0 - 4$? $2 - 4$? $1 - 3$?
3. Why does it make sense to say that in I_5, $-1 = 4$, $-2 = 3$, $-3 = 2$, and $-4 = 1$? [Hint: Recall that the *negative of a number* in ordinary arithmetic *is that number which must be added to it to give* 0.]

The addition tables (Tables 1 and 3) we have constructed for I_3 and I_5 have an important property which they share with the addition tables for I_n, for each value of n. That is that every element of I_n appears exactly once in every row and every column of the addition table for I_n. (Square arrays having this property are called *Latin squares*, and we shall meet them again in Section 10.) The student should convince himself that this is just a reflection of the following theorem.

48(a) Theorem

For any given a and b in I_n, there is exactly one x in I_n satisfying the equation $a + x = b$. That is, subtraction is always possible in I_n since this x is just the (unique) number, $b - a$, which must be added to a to get b.

PROOF: See the proof of Theorem 49.

Section 8   Finite Arithmetics

If we look for a similar property in the *multiplication* tables for I_3 and I_5, we shall at first be disappointed. For there is a whole row of 0s and a whole column of 0s in each case. But we note that, except for the row and column of 0s, every number ($\neq 0$) in I_5, say, appears exactly once in every row and every column of the multiplication table. [*Warning:* This is not true for I_n in general. Look at your tables for I_4 and I_6!] This, again, is just another way of stating:

48(b)   Theorem

If n is a prime number p, then for any given $a \neq 0$ and b in I_p, there is exactly one x in I_p such that $ax = b$.

PROOF: See the proof of Theorem 50.

This clock arithmetic is already familiar, in a different form, from Section 5 of Chapter 2 of this book. For the n symbols $0, 1, 2, \ldots, n-1$ of I_n are nothing else than a *complete residue system* of integers modulo n (Chapter 2, Definition 61). Our rules for addition and multiplication in I_n are just a rephrasing of the following: Carry out addition and multiplication in the ordinary way for integers, and then replace this sum or product (if necessary) by that integer (from among $0, 1, 2, \ldots, n-1$) to which the sum or product is congruent modulo n. Theorems 48(a) and (b) are clearly equivalent to Theorems 49 and 50, which we now prove.

49   Theorem

If a and b are among the integers $0, 1, 2, \ldots, n-1$, then there is one, and only one, x (among $0, 1, 2, \ldots, n-1$) such that $a + x \equiv b \bmod n$.

PROOF: There is at least one such x, namely, that integer x in the complete residue system $0, 1, 2, \ldots, n-1$ such that $x \equiv b - a \bmod n$, since certainly $a + (b - a) \equiv b \bmod n$. Moreover, there is only one solution among $0, 1, 2, \ldots n-1$. For, if $a + x' \equiv b \bmod n$, then $x' \equiv b - a \bmod n$. That is, $x' \equiv x \bmod n$. But if two integers x and x' of a complete residue system are *congruent*, then they are *equal*, by Definition 61 (Chapter 2) of a complete residue system. Hence $x' = x$.

50   Theorem

If n is a prime number p and a ($\neq 0$) and b are among the integers $0, 1, 2, \ldots, n-1$, then there is one, and only one, x (among $0, 1, 2, \ldots, n-1$) such that $ax \equiv b \bmod p$.

PROOF: By Corollary 69 of Chapter 2, this unique solution is that x among $0, 1, 2, \ldots, n-1$ for which $x \equiv a^{p-2}b$.

51 REMARK

With reference to Theorem 50, if n is *not* a prime number, say $n = rq$ ($r > 1$, $q > 1$), then there is always some choice of a ($\neq 0$) and b so that $ax = b$ has no solution in I_n. In fact, we have only to note that $rq = 0$ in I_n. (Why?) Hence, in the rth row and qth column of the multiplication table for I_n, there will be a 0. This means that there are at most $p - 2$ nonzero entries in the rth row of the multiplication table. So some element, say b, of I_n does not appear in the rth row. That is, there is no x for which $rx = b$.

52 EXERCISES

1 Find an equation $ax = b$ ($a \neq 0$) in I_4 which has:
  (a) No solution
  (b) Exactly one solution
  (c) More than one solution
2 Do the same in I_6.
3 Use Remark 68 of Chapter 2 to find the reciprocal of 2 in I_{11} (that is, the solution to $2x = 1$) and the reciprocal of 3 in I_7. What is the reciprocal of 2 in I_{11213}?

The important facts about the finite arithmetic I_n can be summed up by saying that I_n forms a *field* if and only if n is a prime number.

53 Definition

A field is any collection F of at least two objects, together with an "addition" and "multiplication" defined in such a way as to have the following properties:

A1 If a and b are in F, so is $a + b$.
A2 $(a + b) + c = a + (b + c)$ for all a, b, c in F.
A3 $a + b = b + a$ for all a, b in F.
A4 The equation $a + x = b$ always has a unique solution x in F. (In particular, the solution to $a + x = a$ is called 0, and the solution to $a + x = 0$ is called $-a$, the "negative of a." The solution to $a + x = b$ is denoted by $b - a$. That is, $b + (-a) = b - a$.)

M1 If a and b are in F, so is ab.
M2 $(ab)c = a(bc)$ for all a, b, c in F.
M3 $ab = ba$ for all a, b in F.
M4 If $a \neq 0$, then the equation $ax = b$ always has a unique solution x in F. (In particular, the solution to $ax = a$ is called 1, and the solution to $ax = 1$ is called $1/a$, or a^{-1}, the "reciprocal" or "inverse" of a. The solution to $ax = b$ $(a \neq 0)$ is denoted by b/a. That is, $b/a = b(1/a) = ba^{-1}$.)

D $a(b + c) = ab + ac$ for all a, b, c in F (and hence also, by M3, $(a + b)c = ac + bc$).

We record some simple properties which all fields have in common with ordinary real numbers:

54 Theorem

(a) *In any field, $1 \neq 0$.*
(b) *In any field, $(-a)b = -(ab)$ for all a, b in the field.*
(c) *In any field, $-(-a) = a$ for all a in the field.*
(d) *In any field, $a0 = 0$ for all a in the field.*
(e) *In any field, if a product ax is 0, then either $a = 0$ or $x = 0$.*

The proofs of these, and other, elementary facts about fields can be found in any elementary abstract algebra text.

We see that a *field* is an algebraic system which has many of the properties which the system of ordinary fractions has, or which the system of *all* ordinary (real) numbers has. Many of these properties will be assumed without proof in what follows. In fact, the system of ordinary numbers is the prototype field. One main difference is in the *order relations:* Given any two ordinary unequal numbers—say, 3 and π or $\frac{2}{3}$ and $-\frac{1}{4}$—we can always say, in a consistent manner, that one is *larger* than the other. But for clock arithmetic, this is not always possible. (Is 6 o'clock earlier or later than 12 o'clock? Of course, I_{12} is *not* a field. But even if the day were only 11 hours long, essentially the same difficulty would arise.)

In fact, it is appropriate to point out that the axioms for a *field* are an incomplete description of ordinary real numbers in the same sense that the axioms of absolute geometry (that is, Hilbert's axioms *minus* the parallel axiom) are an incomplete description of Euclidean geometry. In the case of a field, we can add suitable axioms concerning order relations to get a complete description of ordinary real numbers.

For future reference, we point out that although I_4 itself is not a field, it is possible to construct a field having exactly four elements. In fact, if we denote these four elements by $0, 1, \lambda, 1 + \lambda$, the Tables 5 and 6 for addition and multiplication do indeed define a field. We call this field F_4.

Table 5. Addition table for F_4

+	0	1	λ	$1 + \lambda$
0	0	1	λ	$1 + \lambda$
1	1	0	$1 + \lambda$	λ
λ	λ	$1 + \lambda$	0	1
$1 + \lambda$	$1 + \lambda$	λ	1	0

Table 6. Multiplication table for F_4

	0	1	λ	$1 + \lambda$
0	0	0	0	0
1	0	1	λ	$1 + \lambda$
λ	0	λ	$1 + \lambda$	1
$1 + \lambda$	0	$1 + \lambda$	1	λ

More generally, we record without proof the following facts:

55(a) Theorem

For each prime power $q = p^m$ (p prime), there is one and only one field having exactly q elements. We call this field F_q.

More exactly, if F is a field of q elements $a, b, \ldots$ and F' is a field of q elements, then the elements of F' can be labeled $a', b', \ldots$ in such a way that $(a + b)' = a' + b'$ and $(ab)' = a'b'$. It is also known that these fields are the only fields with a finite number of elements. That is:

55(b) Theorem

If q is not the power of a prime number, then there is no field having exactly q elements. In particular, there is no field having exactly six elements.

The fact that $a + x = c$ and $bx = d$ ($b \neq 0$) always have solutions in a field means, of course, that every *linear equation* (where the variable x appears only in its first power) can be solved in any field. Recall that this is not true in such systems as I_6, where, for example, the linear equation $2x = 3$ has *no* solution (whereas $2x = 2$ has *two* solutions). (See also Problem 1 of Exercises 52.) Can *quadratic* equations also always be solved in a finite field? The answer is *no*, but this should not distresss us, since for ordinary real numbers the answer is also *no*. Such a simple quadratic equation as $x^2 = -1$ has no real solution. (This is just why the so-called "complex numbers" were invented!) More generally, only "half" of the equations $x^2 = a$ can be solved in the real-number field—those for which $a \geq 0$. By looking at your multiplication table for I_5, you will see that something like the same is true there: The equations $x^2 = 0$, $x^2 = 1$, $x^2 = 4$ have solutions in I_5 (What are they?), but the equations $x^2 = 2$, $x^2 = 3$ do not have solutions in I_5. In fact, of course, if $x^2 = a$ ($a \neq 0$) has a solution in I_5, then it has two solutions, one the negative† of the other, just as for real numbers. (The student should verify this.)

56 EXERCISES

1. For which values of a in I_7 does $x^2 = a$ have a solution in I_7? Verify that when there are two solutions, each is the negative of the other.

2. (a) Which of the following have solutions in I_5?

$$x^2 + x + 1 = 0$$
$$x^2 + 2x + 1 = 0$$
$$x^2 + x + 2 = 0$$
$$x^2 + 3x + 2 = 0$$
$$x^2 + 4x + 2 = 0$$

   (b) Which of the first three equations in part (a) have solutions in I_3?

† In the sense that the sum of the two solutions is zero.

*3 Devise a general criterion to determine which equations $ax^2 + bx + c = 0$ (a, b, c in I_5) have solutions in I_5. [Hint: Recall the derivation of the "quadratic formula" by "completing the square." Does this method still work in I_5?]

Section 9 Finite Geometries

Now we show how each finite field can be used to construct a finite analytic geometry, in the same way that the ordinary real numbers are used in ordinary analytic geometry. All the important principles are illustrated if we take the particular case of the field F_5 ($= I_5$), which has exactly five elements. First make a horizontal line and mark five (equally spaced, for convenience) points on it, corresponding to the five elements 0, 1, 2, 3, 4 of F_5, as in Illustration 37(a). Then make a vertical axis (obviously we are assuming this is to be done on a hanging blackboard—otherwise both axes are horizontal!) at 0, and label five points on it in the same way, as in Illustration 37(b). Now complete the grid (at least in imagination) by drawing horizontal and vertical lines at the points marked. We now have the square grid of Illustration 37(c) with exactly $5^2 = 25$ intersection points, each of which can be labeled by a suitable ordered pair (x, y), where x is the number of the vertical line through it and y is the number of the horizontal line. Note that the points on the x axis (that first horizontal line we drew) are now labeled (0, 0), (1, 0), (2, 0), (3, 0), (4, 0) and the points on the y axis (that first vertical line) are now labeled (0, 0), (0, 1), (0, 2), (0, 3), (0, 4). These are the 25 points of our geometry, as shown in Illustration 38. *There are no other points in this geometry.*

Now what do we mean by a *line* in this geometry? We recall that in ordinary analytic geometry, lines turned out to have linear equations, $ax + by = c$, in two variables, x and y. That is, the points (x, y) of a line were exactly those pairs (x, y) which satisfied some particular linear equation $ax + by = c$. Is it reasonable to say that the points satisfying such an equation are the *lines* of our new geometry? Note that the x axis and the y axis, which we certainly want to call lines, do indeed satisfy such equations ($y = 0$ and $x = 0$, respectively). Moreover, each line of the grid satisfies such an equa-

Section 9 Finite Geometries

Illustration 37

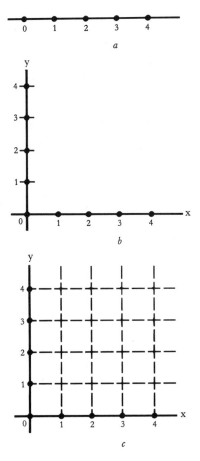

Illustration 38   The 25 points of the analytic geometry constructed from I_5

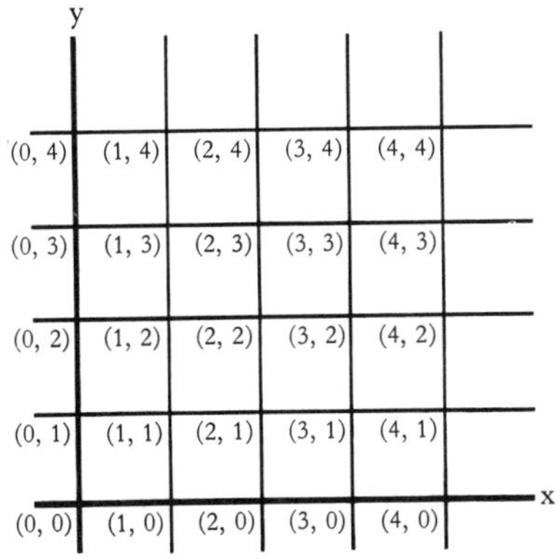

tion. (What are the equations of the horizontal lines? The vertical lines?) Let us try these as tentative definitions:

57 Definition†

A *point* is an ordered pair (x, y) of elements x, y of F_5.

58 Definition†

A *line* is the set of points (x, y) satisfying some particular linear equation $ax + by = c$, where a, b, c are in F_5 and a and b are not both 0.

The reader should note that two coordinate pairs (x_1, y_1) and (x_2, y_2) represent different points whenever they look different, that is, whenever $x_1 \neq x_2$ or $y_1 \neq y_2$. On the other hand, two linear equations which look entirely different may possibly represent the very same line. For example,

59
$$x + 4y = 0$$
$$2x + 3y = 0$$
$$3x + 2y = 0$$
$$4x + y = 0$$

are four different ways of writing the same line. (To see this more clearly, remember that $4 = -1$ and $3 = -2$ in I_5.) This is familiar from elementary algebra, since we know that the solutions to an equation $ax + by = c$ are the same as the solutions to the equation $kax + kby = kc$ obtained from it by multiplying each term by any nonzero constant k. (Or by dividing by any constant $k \neq 0$, which comes to exactly the same thing.)

60 EXERCISES

1 Plot the points satisfying each of the Equations 59.
2 What are $\frac{1}{2}$, $\frac{1}{3}$, $\frac{1}{4}$ in F_5? [Hint: $\frac{1}{2}$ is the number which, when multiplied by 2, gives 1.]

The student should note that a line such as $2x + 3y = 4$ can be "sketched" in very much the same way as in ordinary analytic geometry. We now do this in detail in three ways, to illustrate. [*Warning:* Methods 1 and 2 described below depend on the fact that the field F_5 is the integers modulo 5, I_5. Only method 3 will work if the field is not some I_n. In particular, methods 1 and 2 will not

---

† These are special cases of the more general Definitions 57* and 58* stated at the end of this section.

work for F_4 (Tables 5, 6 in Section 8). This will be seen later in Example 111 and Exercises 112.]

61 Methods of Plotting $2x + 3y = 4$

METHOD 1: If $x = 0$, we have $3y = 4$, $y = \frac{4}{3} = 4 \cdot \frac{1}{3} = 4 \cdot 2 = 3$. Hence $(0, 3)$ is a point on the line. Likewise, if $y = 0$, then $2x = 4$ and $x = 2$; hence $(2, 0)$ is a point on the line. (The student should double-check that each of these actually satisfies the equation $2x + 3y = 4$.) Joining these two points $(0, 3)$ and $(2, 0)$ by a "line" does not give any more points of our geometry, but if we extend our grid, we find that the "line" hits another point on the extended grid of Illustration 39. This point ought to be called $(4, -3)$. (It is, of course, the same as $(4, 2)$.) Now the two points $(2, 0)$, $(4, 2)$ give a more manageable line, which clearly passes through $(3, 1)$ in Illustration 39. We can immediately check that $(3, 1)$ satisfies the equation $2x + 3y = 4$.

Illustration 39 The line $2x + 3y = 4$ cuts the coordinate axes at $(2, 0)$ and $(0, 3)$.

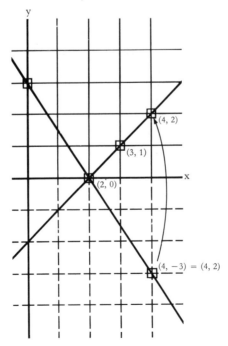

Redrawing this "line" on a grid extended to the right, as in Illustration 40, immediately gives the remaining two points of the line $2x + 3y = 4$, namely, (5, 3) and (6, 4). The latter are of course just (0, 3) and (1, 4) in disguise.

As a final step we can redraw the five points of the line as in Illustration 41. Note that we have essentially used the *intercept form* 42* of the line in this process.

METHOD 2: A modification of method 1 is first to write the line in the so-called *slope-intercept form* 42**; $y = mx + r$. In this case, we solve the equation $2x + 3y = 4$ for y:

$$3y = -2x + 4 = 3x + 4;$$
$$y = \tfrac{3}{3}x + \tfrac{4}{3} = x + 4 \cdot \tfrac{1}{3} = x + 4 \cdot 2 = x + 3.$$

Illustration 40

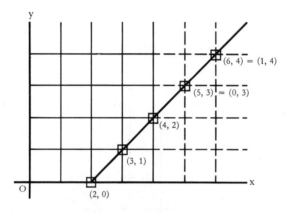

Illustration 41

The five points of the line $2x + 3y = 4$.

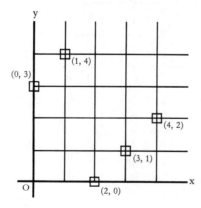

That is, the line $2x + 3y = 4$ is the same as the line $y = x + 3$. From this last equation, we read off immediately that the slope m is 1 and that the y intercept is 3. (See the discussion in Section 7.) Recall that if the slope of a line is m, this means that from any given point on the line, another point of the line can be obtained by moving m units upward and 1 unit to the right. So in this case (where $m = 1$), we move 1 up and 1 right from the point $(0, 3)$ to get another point $(1, 4)$, as in Illustration 42. (We know that $(0, 3)$ is on the line $y = x + 3$ since the y intercept is 3.)

Illustration 43 shows the result of repeating this process from $(1, 4)$, to get $(2, 5)$, that is, $(2, 0)$. Repeating the process two more times gives the remaining points $(3, 1)$ and $(4, 2)$.

Illustration 42

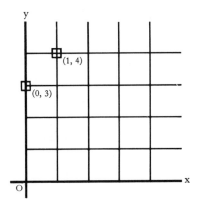

Illustration 43

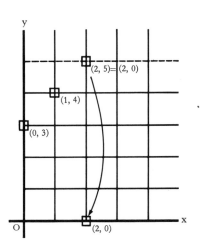

Illustration 44

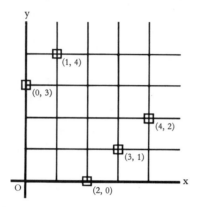

The final result in Illustration 44 is, of course, the same as the result obtained in Illustration 41 by method 1.

METHOD 3: The third method of plotting the points of a line is the easiest to describe but the most tedious to carry out in practice. We simply assign to one of the variables, say x, each of its five possible values 0, 1, 2, 3, 4 and solve the resulting equation for y in each case. This gives Table 7. Plotting these five points gives Illustration 41 or 44 again.

Table 7. Points (x, y) satisfying the equation $2x + 3y = 4$.

x	y
0	3
1	4
2	0
3	1
4	2

62 EXERCISES

(In Problems 1 to 3, the coordinate field is F_5.)

1. Find all points on the lines $x + y = 0$, $3x + 2y = 1$.
2. From the points of the line $y = x + 3$ how can you find the points of $y = x + 4$ painlessly? Of $y = x$, $y = x + 1$, $y = x + 2$? [Hint: These all have the same slope.]

Section 9 Finite Geometries

3 Find the point (or points) which the two lines $3x + 2y = 1$ and $2x + y = 2$ have in common by:
(a) Solving the two equations simultaneously, without any reference to the geometry.
(b) Plotting the two lines and finding the point, or points, which they have in common.

4 Find the points on the lines having the same equations as in Problem 1 but taking the coordinate field to be F_7 (that is, replacing F_5 by F_7 in Definitions 57 and 58).

We already know there are exactly 25 points in our geometry. How many lines are there? We can count them as follows: There are five lines of the form $x = c$. These are the only lines with $b = 0$. All other lines can be written in the form $y = mx + r$, and every choice of m and r gives a different such line (Why?). So there are 25 lines of this form. Hence there are $5 + 25 = 30$ lines altogether.

How many points are there on a line of our geometry? If the line is of the form $x = c$, then it contains exactly the five points $(c, 0)$, $(c, 1)$, $(c, 2)$, $(c, 3)$, and $(c, 4)$. If the line is of the form $y = mx + r$, then each of the five possible values of x gives a unique value of y, so each of these lines also contains exactly five points.

How many lines are there through a given point, say (x_1, y_1)? There is exactly one line of the form $x = c$, namely, $x = x_1$. For the remaining lines, of the form $y = mx + r$, we must have $y_1 = mx_1 + r$. But each of the five possible values of m thus determines r uniquely (since $r = y_1 - mx_1$), so there are exactly five such lines, one for each $m = 0, 1, 2, 3, 4$. Hence there are altogether six lines through each point (x_1, y_1). These facts will be needed in the proof of Theorem 65.

63 EXERCISES

1 Prove that the lines $y = mx + r$ and $y = m'x + r'$ are different lines unless $m = m'$ and $r = r'$.

2 How many lines are there in the plane over F_7? Over F_p, p a prime number?

We have called this system a *geometry*, and we have seen that there are certain objects which can be called "points" and sets of them which can be called "lines" in much the same manner as in Euclidean analytic geometry. But exactly what justification do we have for describing this as a "geometry"? We already have a convenient list—Hilbert's axioms in Section 6—of fundamental properties of Euclidean geometry (from which all others can be derived). How many of these are satisfied by our geometry? In fact, not very many of them. However the axioms of incidence and the parallel axiom hold. More precisely, the following Axioms A1, A2, A3 hold.†

64   Axioms A1, A2, A3

A1   Given two distinct points P, Q, there is exactly one line l containing both of them.

A2   Given a line l and a point P not on l, then there is exactly one line through P not containing any points of l.

A3   There are at least four points, no three of which are collinear.

We propose to show in detail that these three statements are true in our geometry of 25 points. Let us state this formally as:

65   Theorem

The geometry over F_5, described by Definitions 57 and 58, satisfies Axioms A1, A2, and A3.

PROOF: We first prove A1. Let $P_1 = (x_1, y_1)$ and $P_2 = (x_2, y_2)$ be two distinct points. We have to show that (a) there is at least one line through P_1 and P_2, and (b) there is no more than one such line.

(a) Certainly there is at least one line containing P_1 and P_2, for $(y - y_1)(x_1 - x_2) = (x - x_1)(y_1 - y_2)$ is readily verified to be such a line.

(b) We need only show that there cannot be two distinct lines through any given pair of distinct points. We can do this by simply counting the points on the lines through a given point P. There are six lines through P and four points ($\neq P$) on each of them. Therefore, there are 24 points ($\neq P$) on these lines. No point of the plane is overlooked in this count, since we have seen in step (a) that every point is on *some* line through P. Can any point Q have been counted on *two* of the lines through P? (That is, can any two points PQ have two lines through them?) The

† Here A stands for "affine plane," the usual name given to any geometry satisfying these axioms. (Note that although these axioms have the same labels as the axioms of area in Chapter 3, they are not related to those axioms.)

answer is *no*, for if Q had been counted *twice* among the 24 points we counted, then there would really be only (at most) 23 points ($\neq P$) in the plane. But we know from the very beginning that there are, in fact, exactly 24 points ($\neq P$) in the plane. Hence no two lines can contain both P and Q.

This completes the proof of A1. (If the reader is not sure about step (b), he should study it carefully. Arguments of this type will be used often in what follows.)

Now it is a simple matter to prove A2, that is, the existence of a unique line parallel to l at a point P outside l. For if we join P to each of the five points on l, we get five of the six lines through P. The sixth one is the required unique parallel.

The proof of A3—that is, the existence of four points no three of which are collinear—is even simpler. For the four points $(0, 0)$, $(0, 1)$, $(1, 0)$, $(1, 1)$ are in our plane. The (unique) line containing the first two of these is $x = 0$, which certainly does not contain either of the others, and the (unique) line containing the second two is $x = 1$, which also does not contain either of the others. Hence no three are collinear. This completes the proof of Theorem 64.

It is apparent that much of what we have said about F_5 and the geometry defined from F_5 by Definitions 57 and 58 can be said about F_n. More explicitly, let us state:

57* *Definition*
A *point* is an ordered pair (x, y) of elements of F_n.

58* *Definition*
A *line* is the set of points (x, y) satisfying some particular linear equation $ax + by = c$, where a, b, c are in F_n, and a and b are not both 0.

The geometry described by Definitions 57* and 58* is officially called AG(2, F_n), pronounced in full as "the affine two-dimensional geometry over the field F_n of n elements." We shall abbreviate this to AG(F_n) since we are never discussing anything but two-dimensional geometries (i.e., planes). Certainly there are exactly n^2 points in AG(F_n).

A good test of the reader's understanding of the proof of Theorem 65 is to try to prove the following generalization.

66 Theorem

$AG(F_n)$ satisfies Axioms A1, A2, and A3.

67 EXERCISES

1. Can the equation of part (a) in the proof of A1 in Theorem 65 really be written as $ax + by = c$? If so, what are a, b, c?
2. (a) Prove that $AG(F_n)$ has exactly n^2 points and exactly $n^2 + n$ lines.
    [Hint: Compare the remarks preceding Exercises 63.]
    (b) Prove that there are exactly n points on each line of $AG(F_n)$.
3. Use the results of Problem 2 to prove that there are exactly $n + 1$ lines containing each point of $AG(F_n)$. [Hint: The number of points per line, multiplied by the total number of lines, must be equal to the number of lines per point, multiplied by the total number of points.]
*4. Prove Theorem 66. [Hint: Try to follow the proof of Theorem 65, using the results of Problems 2 and 3.]
*5. (a) Find the points of the two parallel lines $y = x$ and $y = x + 1$ in $AG(F_4)$, and mark them on a schematic drawing of the plane $AG(F_4)$.
    (b) Do the same for the points of the two parallel lines $y = \lambda x$ and $y = \lambda x + 1$. What is the point of intersection of $y = x + 1$ and $y = \lambda x$? [Hint: This must be done by method 3. If you try to use one of the other two methods, you will get the wrong points. Use Tables 5 and 6.]

As proved in Problem 2 of Exercises 67, $AG(F_n)$ is a geometry with exactly n points on each line. Hence, because of Theorems 55(a) and (b) (stating the existence of one and only one field F_n for each prime power $n = p^m$), there is an $AG(F_n)$ only if n is a prime power, p^m. But there *are* other finite geometries, not defined over fields, which also satisfy Axioms A1, A2, and A3. It is a major unsolved problem in geometry whether any of these other geometries can have exactly n points on a line if n is *not* a prime power.

For some more possible properties apparently peculiar to a 25-point geometry, the reader should see Howard Eves, *A Survey of Geometry*, Vol. 1, p. 432. For a more detailed discussion of our 25-point geometry the reader may consult the article [10] by H. M. Cundy.

Section 10 Application

Why are we interested in this peculiar geometry? A mathematician is interested because he can get more insight into parts of Euclidean geometry by studying the axioms individually, in somewhat the same way that a biologist gets more insight into the behavior of a frog by dissecting it or in the way a psychologist hopes to get insight into the behavior of higher animals by studying a rudimentary animal such as the flatworm.

But there is another answer to the question about why we study these rudimentary geometries, namely, that they turn out to have applications to the physical world. One such possible application is intuitively apparent to anyone who has heard of the atomic theory of matter. For this theory assumes that blackboards and sheets of paper are not really the solid, continuous objects they seem to be but are simply aggregates of discrete atoms with gaps between them. In this case, a "line" drawn on a sheet of paper will consist only of a finite number of the finitely many atoms which make up the sheet of paper, just as the "lines" of $AG(F_n)$ consist of only finitely many points. The author is not aware of any serious attempt ever being made to apply finite geometries in this way.

However, serious attempts *have* been made to use finite geometries as models of *quantum mechanics*. According to quantum theory, there are discrete physical systems for which a geometric description would again naturally seem to require a *discrete* type of geometry, as opposed to the *continuous* geometry of Euclid. In this case, the gaps are between energy states of particles (rather than between atoms), whereas in classical mechanics there is a continuous allowable distribution of energy states. G. Järnefeldt and P. Kustaanheimo have proposed that a finite geometry very much like our geometry of 25 points is, in fact, the appropriate geometry for quantum mechanics. The interested reader can consult G. Järnefeldt, "Reflections on a Finite Approximation to Euclidean Geometry," *Ann. Acad. Sci. Fenn.*, (A 1) No. **96** (1951); an article by P. Kustaanheimo, *Soc. Sci. Fenn. Comment. Phys.-Math.*, (15)**9** (1959); and other articles by these two authors.

But we propose to describe a much simpler application to physical problems, namely, to the design of experiments. We imagine that tomato plants are to be tested for their reaction to various doses of water and fertilizer in various combinations. To be specific, let us suppose that five different dosages a, b, c, d, e of water are to be tried, and five different dosages A, B, C, D, E of fertilizer. Clearly, if we simply plant 25 tomato plants in a row and apply the 25 different combinations aA, aB, aC, . . . , eE to them, we shall be able to make the test. However, it is not certain that on such a long row the amount of sunshine, the quality of the soil, etc., are constant through the entire length of the row. So it is usual to put the plants out in a square array, like our schematic arrangement of the points of our 25-point geometry $AG(F_5)$, as in Illustration 45.

Now, in order to minimize the influence of a plant's particular position in this array, it is usual to insist that the dosage of water (respectively, fertilizer) be made in such a way that the same dosage does not appear twice in any one row or column of this array of plants. In other words, the dosages a, b, c, d, e of water are supposed to form a *Latin square* (just like the addition table, Table 3, for the arithmetic of F_5). This is easily arranged; simply copy Table 3, replacing 0, 1, 2, 3, 4 with a, b, c, d, e, respectively, as in Illustration 46.

For future reference, we insert here an explicit definition of a Latin square.

68 Definition

A *Latin square of order n* is an $n \times n$ square array of n different symbols, each occurring n times, with the property that all n symbols appear in each vertical column and in each horizontal row of the array.

Of course, the fertilizer dosages must also be arranged as a Latin square. But it is clear that we cannot use the *same* Latin square for this purpose. For then, we would not have all 25 different combinations of water and fertilizers. In fact, it is easy to see that we would only get the five combinations: aA, bB, cC, dD, eE, which already appear along the top row. So the problem is to find two *orthogonal* Latin squares, as defined by:

69 Definition

A Latin square of order n with entries a, b, c, . . . is said to be *orthogonal* to another Latin square with entries A, B, C, . . . if, in

Section 10   Application   275

the juxtaposition of the two squares, each of the n^2 possible combinations aA, aB, aC, ... appears (exactly once).

We shall show how the lines in our 25-point geometry $AG(F_5)$ can help us find such a Latin square orthogonal to the one we already know.

Illustration 45

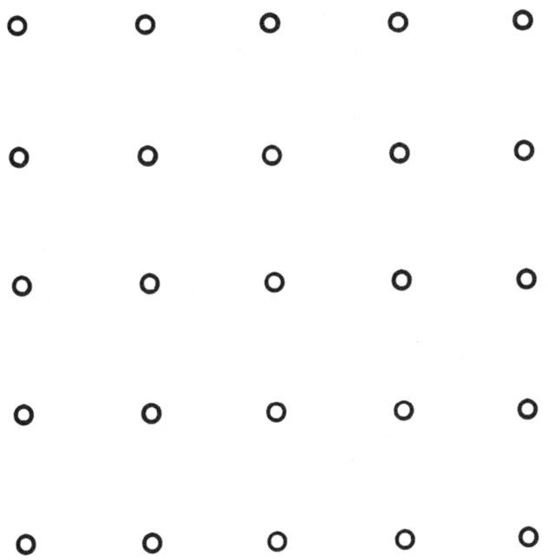

    ○   ○   ○   ○   ○

    ○   ○   ○   ○   ○

    ○   ○   ○   ○   ○

    ○   ○   ○   ○   ○

    ○   ○   ○   ○   ○

Illustration 46

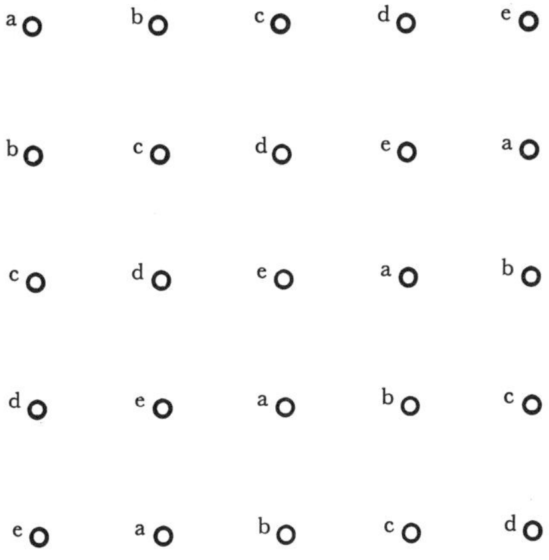

^{a}○ ^{b}○ ^{c}○ ^{d}○ ^{e}○

^{b}○ ^{c}○ ^{d}○ ^{e}○ ^{a}○

^{c}○ ^{d}○ ^{e}○ ^{a}○ ^{b}○

^{d}○ ^{e}○ ^{a}○ ^{b}○ ^{c}○

^{e}○ ^{a}○ ^{b}○ ^{c}○ ^{d}○

We first note that the *a*'s of our Latin square all lie on a single line of $AG(F_5)$. In fact, so do the *b*'s, the *c*'s, the *d*'s, and the *e*'s. (See Illustration 47.) Moreover, these five lines (the line of *a*'s, the line of *b*'s, etc.) are *parallel* to each other, that is, no two of them have a point in common. We say that they form the *parallel class* of lines having slope 1. (In the standard form $y = mx + r$, where *m* is the *slope* of the line $y = mx + r$, these five lines all have $m = 1$. The student should prove this; see Problem 3 of Exercises 70.) This might suggest that we could find another suitable Latin square by taking another parallel class of lines, for example, those of slope 2. Two lines of this family, $y = 2x + r$—namely, $y = 2x$ and $y = 2x + 1$—are shown in Illustration 48.

It is clear that if we label the points of one line with *A*, those of another with *B*, and those of the remaining three with *C*, *D*, and *E*, respectively, then we again get a Latin square. For no line meets any given horizontal row (or any given vertical column) in more than one point, since these rows (and columns) are also lines of $AG(F_5)$ and we know that two lines have at most one point in common.

70   EXERCISES

1 Find a Latin square of order 7.
2 Find two orthogonal Latin squares of order 3.
3 What is the equation of the line of *a*'s in Illustration 46 (or 47)? Of the line of *b*'s? *c*'s? *d*'s? *e*'s?

In fact, of course, each line $y = 2x + r$ ($r = 0, 1, 2, 3, 4$) meets each row (and each column) in *exactly* one point, since $y = 2x + r$ can fail to meet a given row (or column) only if it is *parallel* to that row (or column)— but it isn't. (To put it another way, for any given value of *x*, there is always a value of *y* obtained from the equation $y = 2x + r$. This means that each *column* contains at least one point of the line $y = 2x + t$ since the *x* values are just labels on the columns. Likewise, since for each value of *y* a unique value of *x* is determined from the equation $y = 2x + r$, which is the same as $3y = 3 \cdot 2x + 3r$ or $x = 3y + 2r$, we conclude that each row of the square array contains a point on the line $y = 2x + r$.)

This kind of argument shows that any parallel class of lines in $AG(F_5)$ (except the two classes $x = $ constant and $y = $ constant, which are parallel to the coordinate axes) determines a Latin square.

Section 10   Application

We have only to show that any two different such Latin squares are orthogonal in order to have solved the tomato-plant problem. But it is, in fact, obvious that two such Latin squares are orthogonal. For suppose, if possible, that a pair, say aA, occurred in two different places in the juxtaposed squares. This would mean that the line of a's and the line of A's both pass through those two points of the

Illustration 47

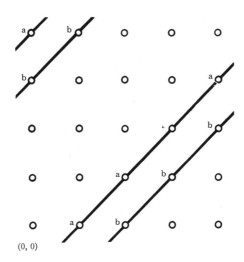

Illustration 48

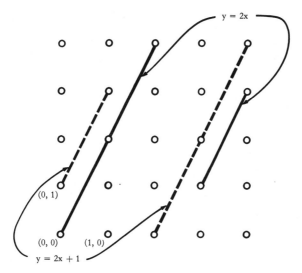

geometry. But this means (since two points determine a unique line) that the line of a's and the line of A's are one and the same line. But this is impossible since, by definition, they are in two entirely different parallel classes. (To put it another way, the two equations $y = m_1 x + r_1$ and $y = m_2 x + r_2$ have exactly one common solution if $m_1 \neq m_2$.)

Illustration 49

aA	bB	cC	dD	eE
○	○	○	○	○

bD	cE	dA	eB	aC
○	○	○	○	○

cB	dC	eD	aE	bA
○	○	○	○	○

dE	eA	aB	bC	cD
○	○	○	○	○

eC	aD	bE	cA	dB
○	○	○	○	○

Illustration 50   *A pair of 10 x 10 orthogonal Latin squares.*

aA	eH	bI	hG	cJ	jD	iF	dE	gB	fC
iG	bB	fH	cI	hA	dJ	jE	eF	aC	gD
jF	iA	cC	gH	dI	hB	eJ	fG	bD	aE
fJ	jG	iB	dD	aH	eI	hC	gA	cE	bF
hD	gJ	jA	iC	eE	bH	fI	aB	dF	cG
gI	hE	aJ	jB	iD	fF	cH	bC	eG	dA
dH	aI	hF	bJ	jC	iE	gG	cD	fA	eB
bE	cF	dG	eA	fB	gC	aD	hH	iI	jJ
cB	dC	eD	fE	gF	aG	bA	iJ	jH	hI
eC	fD	gE	aF	bG	cA	dB	jI	hJ	iH

We record the explicit solution to the tomato-plant problem in Illustration 49. Exactly this same construction can be used to solve any corresponding tomato-plant problem *whenever the number of different dosages is a prime power* $n = p^m$. (Just use the corresponding geometry $AG(F_n)$.) But what if the number of dosages is not of the form p^m? The smallest such number is 6. It was conjectured in 1782 by Euler, and proved more than 100 years later (in 1900) by G. Tarry, that *there are no two orthogonal* 6×6 *Latin squares*. That is, a compact experiment of the type we designed with 25 tomato plants and 5 different dosages (of water and fertilizer) cannot be designed if there are 36 tomato plants and 6 dosages. Euler went further to conjecture that whenever the number of dosages n is of the form $4k + 2$ (for example, $n = 10, 14, 18, 22, \ldots$), then it is also impossible to design such a compact experiment. In 1959 this conjecture was proved to be as wrong as a conjecture could be! In fact, it is now known that such an experiment can be designed for *all n* of the form $4k + 2$, *except for* $n = 6$. An entertaining account of this discovery is given by Martin Gardner in the *Scientific American* of November, 1959. The cover page of that issue shows, in color, the design for $n = 10$, reproduced here as Illustration 50.

Section 11 Circles and Quadratic Equations

In Section 9 we were able to do a certain amount of geometrizing, at least to the extent of defining points and lines in such a way that some of the simplest properties expected of points and lines were valid in our geometry. In the present section we propose to show that analogs of more complicated figures from old-fashioned analytic geometry also can be found in our 25-point geometry and that they have many of the same familiar properties.

Probably everyone would agree that the next simplest geometric object, after point and line, is a *circle*. Euclid defines a circle to be a set of points all at a given distance from a fixed point. Since we haven't defined what the "distance" between two points should be, we can't use this definition directly. But taking a hint from analytic geometry, we might try to say that a circle is the set of all points (x, y) satisfying an equation of the form $x^2 + y^2 = r^2$, where r is a fixed element of F_5. Or, more generally, we might agree to say that $(x - a)^2 + (y - b)^2 = r^2$ is a circle with radius r and center (a, b). However, for F_5, there is a serious objection to calling $x^2 + y^2 = r^2$ a circle. For we would like to be able to call $x^2 + y^2 = 0$ a *point-circle*, that is, it should consist of the single point $(0, 0)$. But, in fact, there are eight other points (x, y) besides $(0, 0)$ for which $x^2 + y^2 = 0$.

71   EXERCISES

1 Which nine points (x, y) of $AG(F_5)$ satisfy the equation $x^2 + y^2 = 0$?

2 Show that if k is not the square of some element in F_5, then $x^2 - ky^2 = 0$ contains only the point $(0, 0)$. [Hint: Solve this equation for k.]

3 (Recall Euclid's construction of an equilateral triangle with given side AB, described in Second Objection 7, Section 1.) Let $A = (0, 0)$ and $B = (1, 0)$ in $AG(F_3)$, and use Euclid's method to construct the third vertex C of an equilateral triangle ABC with side AB. What is peculiar about this "equilateral triangle"? [Hint: The "circle" with center A and radius AB is $x^2 + y^2 = 1$. The "circle" with center B and radius AB is $(x - 1)^2 + y^2 = 1$. Any one of the points they have in common can be called C.]

Problem 2 of Exercises 71 suggests how to remedy the objection just raised. We take, to be specific, $k = 3$ (in F_5) and make the following:

72   Definition

In $AG(F_5)$ a *circle* with center (a, b) is the set of all points (x, y) for which $(x - a)^2 + 2(y - b)^2 = r$.

73   Warning

We do *not* insist that the right-hand side as the "square of the radius," actually be a square. Thus for the circle $x^2 + 2y^2 = 2$, there is no natural "radius" itself, although 2 might be called the "square of the radius."

Section 11 Circles and Quadratic Equations

As we investigate the consequences of Definition 72, some readers are certain to complain that the equation there should be said to represent an *ellipse* (as in Illustration 53) or a hyperbola (as in Illustration 52) rather than a circle. *This complaint is completely justified.* In AG(F_5) it is not really convenient to distinguish between the various kinds of conics. In fact, the properties of "circles" which we make use of (such as the existence of a center, the fact that at most two tangents can be drawn to a circle from any given point, and the distinction between interior and exterior points) are shared by *all* conics. In spite of this legitimate complaint, we retain the name "circle." The reader who chooses to replace this by "ellipse" or "conic" will not go far wrong.

It is relatively obvious that these circles satisfy the requirement that a line can meet a circle in at most two points, since a line has a linear equation, which when substituted into the equation of a circle gives a quadratic equation, which in turn has at most two solutions. Lemma 74 and Theorem 75 spell this out exactly for the general case of F_n.

74 Lemma

The quadratic equation $ax^2 + bx + c = 0$ (a, b, c, in F_n) has no more than two solutions in F_n.

PROOF: Suppose there are three solutions x_1, x_2, x_3 to the equation, so that the three equations E_1, E_2, E_3 all hold:

$$E_1: ax_1^2 + bx_1 + c = 0$$
$$E_2: ax_2^2 + bx_2 + c = 0$$
$$E_3: ax_3^2 + bx_3 + c = 0.$$

We shall show that two of these solutions are necessarily equal. Specifically, we show that the assumption $x_1 \neq x_2$ and $x_2 \neq x_3$ leads to the conclusion that $x_1 = x_3$.

Subtracting $E_1 - E_2$, we get

$$a(x_1^2 - x_2^2) + b(x_1 - x_2) = 0,$$

hence

$$[a(x_1 + x_2) + b](x_1 - x_2) = 0.$$

By Property 5 of Theorem 54, if a product is 0, one of the factors must be. Since $x_1 \neq x_2$, we know $x_1 - x_2 \neq 0$, so that

75(a) $$a(x_1 + x_2) + b = 0.$$

Since $x_2 \neq x_3$, the same reasoning applied to $E_2 - E_3$ leads to

75(b) $$a(x_2 + x_3) + b = 0.$$

Subtracting 75(a) − 75(b) gives

$$x_1 - x_3 = 0,$$

so that $x_1 = x_3$

which completes the proof that there can be at most two distinct solutions to a quadratic equation.

76 Theorem

A line $ax + by = c$ and a circle $(x - A)^2 + 2(y - B)^2 = r$ (both in $AG(F_5)$) have at most two points in common.

PROOF: Solving $ax + by = c$ for y (or for x, if $b = 0$) and substituting into the circle equation yields a quadratic equation in x (or y). This quadratic equation, by Lemma 74, has at most two solutions, x_1 and x_2 (or y_1 and y_2). These are the x coordinates (or y coordinates) of the points which the line and circle have in common. This completes the proof.

Let us see what the circle $x^2 + 2y^2 = 1$ looks like. Since its center is $(0, 0)$, we can make things easier to visualize by changing the representation of the geometry $AG(F_5)$ so that $(0, 0)$ is exactly in the middle of the coordinate system, as in Illustration 51. Then we can find the points on the circle $x^2 + 2y^2 = 1$ by making Table 8. If we

Table 8

x	y
1	0
−1 (= 4)	0
2	1
2	−1 (= 4)
−2 (= 3)	1
−2	−1

plot these points (x, y) as small squares on our graph, we get illustration 52. Actually, of course, the six points of our "circle" would look more "circular" if we pushed the center $(0, 0)$ over to the left, as in Illustration 53. This last figure has the distressing feature that the center of the circle appears to lie "outside" the circle!

Section 11   Circles and Quadratic Equations   283

Illustration 51

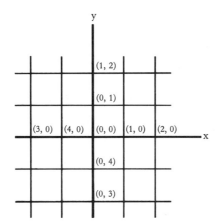

Illustration 52

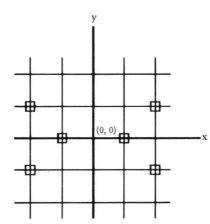

The points of the "circle" $x^2 + 2y^2 = 1$.

Illustration 53

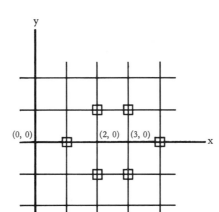

Another way to show the points of $x^2 + 2y^2 = 1$.

This suggests an interesting question: What do we mean by the "outside" and "inside" of a circle in our geometry? Of course, such a question may not have a meaningful answer. The corresponding question, What do we mean by the "interior" and "exterior" of the segment joining two points (0, 0) and (0, 2)? does *not* have a meaningful answer. We propose to show that for our circle $x^2 + 2y^2 = 1$ in $AG(F_5)$ there *is* a meaningful answer. What properties do points on the inside (= interior) of a circle in Euclid's geometry have that the points on the outside (= exterior) of the circle do not have? One such property is that any line through an interior point of a Euclidean circle meets the circle in two points, whereas lines through exterior points may meet a Euclidean circle in no points, one point, or two points. But this property will not define the *interior* of the circle $x^2 + 2y^2 = 1$ in $AG(F_5)$ for us, for it is not hard to see (just by looking carefully at Illustration 53 and remembering what "lines" look like) that *every* point of $AG(F_5)$ which is not on the circle itself has lines through it which do not meet the circle $x^2 + 2y^2 = 1$. In particular, the center of the circle $x^2 + 2y^2 = 1$, namely, (0, 0), might be expected to be an interior point by any reasonable definition of "interior." But the line $x = y$ passes through (0, 0) and misses the circle completely!

77 EXERCISE

Prove that every point (not on the circle itself) of $AG(F_5)$ is on at least one line which does not meet the circle $x^2 + 2y^2 = 1$. (Especially verify this for the points (2, 0) and (3, 0), which seem to be "inside" the circle.)

After this failure, we might try some other definition. We know that through an exterior point of a Euclidean circle there are three kinds of lines: those which miss the circle, those which meet it at exactly one point (these are called *tangents* to the circle), and those which meet it at two points (these are called *secants*, or *chords*, of the circle). Does this property of exterior points (of having three different kinds of lines through them) distinguish exterior points from interior points in our case? If we consider the point (0, 0), we see that it has only two kinds of lines through it—those which do not meet the circle $x^2 + 2y^2 = 1$ at all and those which meet it in exactly two points. That is, through the point (0, 0) there are no tangent lines to the circle. This suggests that we try the following:

78 *Definition*

A point P not on the circle $x^2 + 2y^2 = 1$ in $AG(F_5)$ is called an *interior* point if there are no tangents to the circle through P; otherwise, it is called an *exterior* point.

While we are making definitions, let us also agree to the following abbreviation.

79 *Abbreviation*

In what follows, we shall use the symbol $\mathcal{C}$ to denote the circle $x^2 + 2y^2 = 1$ in $AG(F_5)$.

The remainder of this section is devoted almost entirely to the particular circle $\mathcal{C}$. However, most of the theorems we shall prove about $\mathcal{C}$ will be true about all circles (suitably defined) in all geometries $AG(F_n)$. In fact, as already suggested, these are really theorems about *conic sections*, of which circles are only very special examples. These same theorems will appear in their more general context in Sections 12 to 15.

80 EXERCISE

List all the interior points and all the exterior points of $\mathcal{C}$.

In the case of Euclidean circles, there are exactly two tangent lines through each exterior point. Is this true in our case? We could look at each exterior point and verify that it *is* the case. However, it is better to prove the following general theorem.

81 Theorem

If there is a tangent line to $\mathcal{C}$ through a point P not on $\mathcal{C}$, then there is exactly one other tangent line to $\mathcal{C}$ through P.

We first prove:

82 Lemma

There is exactly one tangent line to the circle $\mathcal{C}$ at each point Q of $\mathcal{C}$.

PROOF: There are exactly six lines altogether through Q. Of these, five are obtainable by joining Q to one of the other points of the circle. The sixth is necessarily a tangent line to Q since it cannot meet the circle at any of these other five points. (See Illustration 54.)

Illustration 54

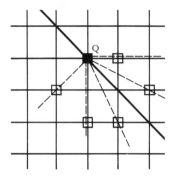

Tangent line at Q

Illustration 55

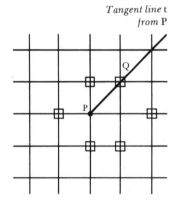

Tangent line t *from* P

Illustration 56

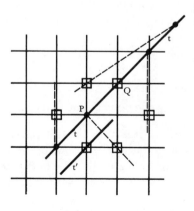

From each of the four points of t ($\neq$ Q), *there is another tangent to* $\mathcal{C}$.

PROOF OF THEOREM 81: We first recall that our circle $\mathcal{C}$ has exactly six points on it. The given tangent line, say t, from P passes through exactly one of these points, say Q. This leaves five points of the circle (see Illustration 55). The lines through P which are not tangent lines each contain either no points of the circle or two points (by Theorem 76). Hence some *odd* number of the remaining lines at P must be tangent lines. That is, there are altogether an even number of tangents to the circle through P.

Of course, there is nothing special about P, and we have actually shown that *through each point of t* (different from the point of tangency, Q), *there are an even number of tangents* (including t itself) to $\mathcal{C}$. We need only show that this even number is *two*.

To do this, first recall that (by Lemma 82) there are exactly six tangents to $\mathcal{C}$ altogether, one at each of its six points. We have just shown that through each of the four points on t (different from Q), there is at least one other tangent to $\mathcal{C}$. The four tangents from these four points meet $\mathcal{C}$ in four points of tangency. Counting the tangent t at Q, this accounts for five of the six tangents to $\mathcal{C}$. (See Illustration 56.)

The tangent at the remaining point of the circle cannot pass through any point R of t since there would then be an odd number (namely, three) of tangents to the circle through R. Hence this remaining line must be *parallel* to t. In particular, it cannot pass through P, and we have shown that there are exactly two tangents to the circle through P.

83 EXERCISE

What are the two tangent lines to $x^2 + 2y^2 = 1$ through $(2, 0)$? Through $(4, 4)$? Notice that these can easily be found by inspection.

Note that we have also proved Theorem 84 as a by-product of the preceding proofs.

84 Theorem

Given a point Q on the circle $\mathcal{C}$ and the tangent line t to $\mathcal{C}$ at Q, there is exactly one other point of $\mathcal{C}$ where the tangent line to $\mathcal{C}$ is parallel to t. (This tangent line is marked t' in Illustration 56.)

This, of course, is also a property of Euclidean circles and helps confirm our feeling that we have chosen an appropriate definition.

We have now proved that *exterior points* are those through which exactly two tangents pass and *interior points* are those through which no tangent lines pass.

We can also note that if we only knew the circle and its tangents, we could find its center, since the *center* of $x^2 + 2y^2 = 1$ is that point O such that for each chord RS through O (R, S on $x^2 + 2y^2 = 1$), the tangents at R and S are parallel. This also agrees with a property of Euclidean circles.

We conclude our discussion of this circle $\mathcal{C}$ by counting the number of exterior and interior points. We know that there are six tangents to the circle and that each pair of them meet (if they are not parallel) in an exterior point. Thus there are $(6 \cdot 5)/2 - 3 = 12$ exterior points. Since there are 25 points in the plane altogether and 6 on the circle, this leaves $25 - 12 - 6 = 7$ interior points.

85 EXERCISES

1 Prove that the center $(0, 0)$ of $\mathcal{C}$ is really an interior point. [Hint: Note that the lines through $(0, 0)$ all have equations of the form $y = mx$, except for the line $x = 0$.]

2 Mark each point on Illustration 53 according to whether it is an interior or exterior point of $\mathcal{C}$, thus verifying that the above calculations are correct.

3 Plot the points of the circle $x^2 + 2y^2 = 4$. Find all exterior and interior points.

We remind the reader again that the theorems we have just proved are not actually restricted to the particular circle $x^2 + 2y^2 = 1$ but are true for all objects which can reasonably be called circles—in fact, for all objects which can reasonably be called conics (such as ellipses or hyperbolas). This will be discussed from a slightly different point of view in the following sections.

Section   12 Finite Affine Planes

In Sections 7 to 11 we have seen how the ideas of Descartes, that is, the ideas of ordinary (real) analytic geometry, can be applied to geometries where a "line" contains only a finite number of "points." Can we also apply the ideas of Euclid to such finite geometries? That is, can we study such geometries from an "elementary" point of view, without the use of coordinates? In the present section we show that the answer is *yes*.

More specifically, we ask what "Euclidean" theorems can be established if we have only Hilbert's axioms, Groups I and IV (the axioms of incidence and the parallel axiom), to work with. Our versions of these axioms have already been recorded as A1, A2, and A3 in Section 9. For convenience, we repeat them now:

A1 Given two distinct points P, Q, there is exactly one line l containing both of them.

A2 Given a line l and a point P not on l, then there is exactly one line through P not containing any points of l.

A3 There are at least four points, no three of which are collinear.

In contrast to Section 9, the meanings of "point," "line," and "a line contains a point" are not now specified. This is one of the powers of the axiomatic method—no specific meaning has to be attached to the things axiomatized except the meaning implicit in the axioms. For example, points and lines can be thought of in terms of tomato plants and fertilizer, as in the example of Section 9, or quite differently, as in Exercises 86.

86 EXERCISES

1 Suppose that by "point" is meant the location of a store where groceries are sold—say Albany, Birmingham, Chicago, or Denver. Suppose that "line" means a particular line of groceries—say apples, bananas, cheese, doughnuts, eels, and figs. Finally, suppose that "a point is on a line" or "a line contains a point" means that a particular line of goods is sold at a particular location. Give an explicit distribution of the six commodities in the four cities so that the Axioms A1, A2, and A3 are satisfied. [Hint: Compare with $AG(F_2)$.]

2 A *point* is defined to be one of the nine symbols $0, 1, 2, \ldots, 8$. The sum of two points is defined to be their sum modulo 8. (Thus $7 + 3 = 2$.) There are two types of *lines:* Eight lines consist of the triples of points of the form $a, a + 1, a + 3$ $(a \neq 0)$. Four other lines consist of the triples $0, 1, 5$; $0, 2, 6$; $0, 3, 7$; $0, 4, 8$. Explain why these points and lines satisfy Axioms A1, A2, A3. Represent the points on a convenient diagram, as in Section 9, to make it obvious which points are collinear.

It is natural to expect that not much can be accomplished on the basis of these meager assumptions alone. However, we shall see that we can, in fact, prove a number of things which at first glance would appear to require many more assumptions than just A1 to A3.

In order to restrict our attention to the case where a line contains only a finite number of points, let us assume the further axiom:

A4   There is at least one line with exactly n (>1) points on it.

These four axioms are the axioms for an *affine plane of order n*. The analytic geometry $AG(F_n)$ (of which we have studied the particular case $AG(F_5)$ in detail) is an example of such a plane, but there are other, essentially different examples. (Since none of the essentially different examples is of order less than 9, we shall not investigate any of them in this book.)

The axioms do not state explicitly that every line contains exactly n points, although we know this to be true for the case $AG(F_n)$. But this can be proved, as Theorem 88. We first prove:

87   Theorem

In any affine plane of order n, each point lies on exactly $n + 1$ lines.

PROOF: Let P be a point of the plane. Axiom A4 guarantees that *some* line, say l, contains exactly n points, say $P_1, P_2, \ldots, P_n$. There are two cases to consider: (1) P is not on l, (2) P is on l. In case (1), exactly n lines through P, namely, $PP_1, PP_2, \ldots, PP_n$, meet l. And exactly one line through P (by A2) fails to meet l. Hence there are exactly $n + 1$ lines through P. In case (2), consider a point Q not on l and a line l' which contains neither P nor Q. (See Illustration 57.) (Why is there always such a line?) By case (1) we know there are exactly $n + 1$ lines through Q. All but one of these meets l', by A2. Hence l' contains exactly n points. But P is not on l', so

Illustration 57

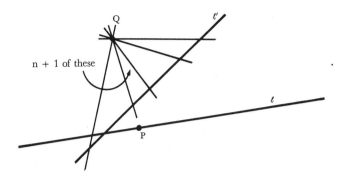

the argument of case (1) shows that there are exactly $n + 1$ lines through P. This completes the proof.

88 Theorem

In any affine plane of order n, each line contains exactly n points.

PROOF: Let l be any line in the plane, and let P be any point not on l. Then of the $n + 1$ lines at P, exactly n (by A2) meet l. These n points of intersection are necessarily the only points on l (Why?). This completes the proof.

89 EXERCISE

Suppose Axiom A4 is replaced by:

A4* There is at least one point which lies on exactly $n + 1$ lines.

Then show that A4 (and hence Theorems 87 and 88) can be proved as a theorem.

90 Definition

If l is a given line, then the lines having no points in common with l, as well as l itself, are said to be *parallel* to l. If l is parallel to l', we write $l \| l'$ or $l' \| l$.

91 Theorem

In an affine plane of order n, each line l has exactly n lines parallel to it.

PROOF: Consider any line s which intersects l, say at P. (Why is there always such a line?) There are exactly n points on s, through each of which

Illustration 58

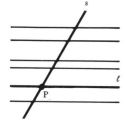

there is exactly one parallel to l. (See Illustration 58.) Hence there are at least n parallels to l. There are no other parallels to l, for if $l'\|l$ and l' fails to meet s, then there are *two* lines at P, namely, s and l, which fail to meet l'. This contradicts Axiom A2.

92 *Definition*

In an affine plane of order n, the n lines parallel to a given line l are called the *parallel class* of lines determined by l.

Thus Theorem 91 says that in any affine plane of order n, each parallel class of line contains exactly n lines.

93 Theorem

In an affine plane of order n, there are exactly n^2 points and exactly $n^2 + n$ lines.

PROOF: The n lines of a parallel class include all points of the plane. (Why?) Since there are n points on each of these n lines, there are $n \cdot n = n^2$ points altogether. Likewise, at each of the n^2 points of the plane, there are $n + 1$ lines. But each of these $n^2(n + 1)$ lines is counted n times, once for each of the n points on it. Hence there are $n^2(n + 1)/n = n^2 + n$ lines altogether, as claimed.

94 EXERCISES

1 How many parallel classes are there in an affine plane of order n?

2 Show, in a different way, that there are exactly $n^2 + n$ lines in an affine plane of order n. [Hint: Consider some fixed line, and count all the lines which intersect it and all the lines which do not intersect it.]

Section 13 Finite Projective Planes

It will be more convenient, in what follows, to look at planes which are slight modifications of the affine planes we have studied so far. The motivation for this change is the lack of symmetry between the roles played by points and lines in the axioms for affine planes. For the axioms require that two points always have one and only one line *containing* both, but two lines do not always *contain* a common point. The following standard axioms for *projective planes* will remedy this lack of symmetry. Any set of points and lines satisfying these four axioms is called a *projective plane of order n*.

P1 Given two distinct points P, Q, there is exactly one line l containing both of them.

P2 Given two distinct lines r, s, there is exactly one point P contained in both of them.

P3 There are at least four points, no three of which are collinear.

P4 There is at least one line with exactly $n + 1$ (>2) points on it.

Of course, the last two axioms are not symmetric, but the corresponding Axioms P3' and P4' which are symmetric to them can be proved as theorems, so that the entire body of theorems is perfectly symmetric. This implies the following:

95 Principle of Duality in Projective Planes

Let S be any statement (or phrase) about projective planes, and let S' be its *dual statement*. That is, let S' be the statement (or phrase) obtained from S by replacing each occurrence in S of the word "point" by the word "line" and each occurrence of the word "line" by the word "point." Then the *principle of duality in projective planes* is: If S can be proved as a theorem for all projective planes, then S' can also be proved as a theorem for all projective planes.

The reason for this principle is clear. In each step of the known proof of S, we just replace "line" by "point" and "point" by "line" to get a proof of the dual theorem S'. Of course, we may have to replace whole expressions in order to avoid awkward terminology. Thus, for example, the dual axiom P3' reads:

P3' There are at least four lines, no three of which pass through the same point.

The word "on" is often used, so that the dual of "a point is on a line" is just "a line is on a point." Or we can say "incident with" to mean the same thing as "on."

96 EXERCISES

1 State P4'.

2 Prove P3'.

3 Prove P4'.

4 In some projective planes, the following famous theorem of Desargues is true: If O, A, A' are three points on a line, O, B, B'

are three points on a second line, and O, C, C' are three points on a third line, then the three points of intersection of lines AB and $A'B'$, lines BC and $B'C'$, and lines CA and $C'A'$ are collinear. Draw a figure in the Euclidean plane to illustrate Desargues' theorem.

5 State the dual of Desargues' theorem, and draw a figure to illustrate this dual theorem.

Of course, just writing down a set of axioms for a projective plane (or for anything else, for that matter) does not guarantee that there is such a thing. However, it turns out that every affine plane can be converted into a projective plane by a simple artifice. This artifice is the familiar one, so richly deplored in high school mathematics, of saying that two parallel lines "meet at infinity." Explicitly, suppose that we have an affine plane A of order n. Then, in A, Axiom P1 already holds. However, there are exceptions to P2 since some pairs of lines do not have a common point, and we must eliminate these exceptions. Which lines do not have a common point with a given line l? Exactly those lines which are parallel to l. In fact, any pair of lines in the parallel class determined by l fails to have a common point. The remedy for this defect is obvious: For each parallel class we add a single point to the affine plane A, and we agree that this point is on each of the lines of the parallel class. These new points are sometimes called *ideal points*.

After this is done, Axiom P2 holds without exception. However, there are now exceptions to P1! For, of course, if we consider two of the added ideal points there is no line joining them. So we also add a new line, which contains exactly the ideal points and no others. This line of ideal points is called the *line at infinity*, or the *ideal line*, and is denoted by l_∞. The ideal points are also sometimes called *points at infinity*. Thus, in this terminology, parallel lines in A "meet at infinity" in the corresponding projective plane.

The net result is that we have added $n + 1$ (ideal) points and a single (ideal) line to the original affine plane A. Consequently, there are $n^2 + (n + 1)$ points and $(n^2 + n) + 1$ lines in the projective plane. Note that the principle of duality is partially confirmed here by the fact that there are exactly the same number, $n^2 + n + 1$, of points as lines. The reader should verify that each of the Axioms P1 to P4 is actually satisfied. In particular, he should verify that an

Section 13 Finite Projective Planes

ordinary point and an ideal point determine a unique line. (Which line?)

This process can also be reversed. That is, if we find a projective plane somewhere and delete one line together with all the points on it, then the resulting system of points and lines is an affine plane. We have to agree, however, that any two lines whose common point (in the projective plane) was on the deleted line will be called *parallel* in the affine plane. Thus our particular construction of a projective plane from an affine plane by adjoining an ideal line and its points is, in a certain sense, the *only* way a projective plane can arise. That is to say, every projective plane which appears to have come from some other source could in fact, have been constructed from a suitable affine plane—namely, from any affine plane obtained by deleting one line, together with all its points, from the projective plane. It should be pointed out, however, that an affine plane obtained by deleting a line and all its points from a projective plane picked at random is not necessarily an $AG(F_n)$.

The reason for saying that a projective plane with $n + 1$ points on a line is of *order n* is that it can be constructed from an affine plane of order n. The student should try to prove, directly from Axioms P1 to P4, the following analogs of the corresponding Theorems 87 and 88 about affine planes. Some of the proofs will be easier than for the corresponding affine theorems because it is no longer necessary to distinguish between "intersecting" and "parallel" lines.

97 Theorem

In any projective plane of order n, each point lies on exactly $n + 1$ lines.

98 Theorem

In any projective plane of order n, each line contains exactly $n + 1$ points.

Note that the analog of Theorem 91 says no more than is already stated by Theorem 97.

99 EXERCISES

1 Prove Theorems 97 and 98.
2 State the analog, for projective planes, of Theorem 91, and explain how it is included in Theorem 97.

Section 14 Ovals in a Finite Plane

Is it possible to define a "circle" in a general affine plane of order n, or rather, in the associated projective plane of order n? We do not yet have any "coordinates" for our points (which may turn out to be tomato plants!), so it is natural to try to use purely geometric properties to define a circle. One of the simplest properties of a circle is that no three of its points are collinear. Of course, this property is shared by other conics, as well as by other oval-shaped curves in ordinary (real) geometry, so we make the following:

100 *Definition*

In a projective plane of order n, any set of $n + 1$ points, no three of which are collinear, is an *oval*.

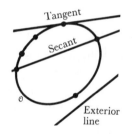

Illustration 59

Although this definition does not in itself assure us that there really are any ovals in a given projective plane, there are certainly ovals in *some* planes, for the circle ℭ discussed in Section 11 has this property. It is a remarkable fact, proved only recently by B. Segre [24], that if n is odd, then every oval in $AG(F_n)$ can be represented by a quadratic equation. But in some other possible projective planes, it is not even certain that there are any ovals at all.

Is it possible to define a *tangent line* to an oval? It is, in the following simple way.

101 *Definitions* (See Illustration 59.)
1. A line l is said to be a *tangent line* to an oval ⊙ if l and ⊙ have exactly one point in common.
2. A line which has exactly two points in common with an oval is called a *secant* of the oval.
3. A line having no points in common with an oval is called an *exterior line* to the oval.

Theorems 102 and 103 show that the tangents to ovals share some important properties with ordinary tangents to ordinary circles.

102 Theorem

At any point Q of an oval ⊙, there is exactly one tangent line.

Illustration 60

PROOF: (Note that the following proof is essentially the same as the proof of Lemma 82.) Suppose we join Q to each of the other n points of the oval, as in Illustration 60. This gives n lines at Q, all of which are secants of $\mathcal{O}$. But we know there are $n + 1$ lines at Q. Hence the $(n + 1)$st line is necessarily a tangent, as required.

14A Ovals in Planes of Odd Order

It turns out that the tangents to ovals behave very differently depending on whether the order n of the plane is even or odd. Throughout Section 14A we shall consider the case where n is odd.

103 Theorem

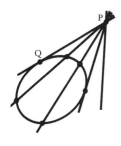

Illustration 61

If a projective plane has odd order n and $\mathcal{O}$ is an oval in the plane, then through each point P not on $\mathcal{O}$ there are either no tangents to $\mathcal{O}$ or exactly two tangents to $\mathcal{O}$.

PROOF: (Compare this with the proof of Theorem 81.) Let P be a point not on the oval $\mathcal{O}$. If there are no tangents to $\mathcal{O}$ through P, then there is nothing to prove. If there is at least one tangent to $\mathcal{O}$ through P, call its point of tangency Q. Now consider the other n lines at P. Each of them either has an even number (0 or 2) of points of $\mathcal{O}$ on it or has an odd number (namely, 1), as in Illustration 61. But there are an *odd* number of points, other than Q, on $\mathcal{O}$. Hence there cannot be an even number of them on each line from P. Hence there must be *at least* one other tangent from P to $\mathcal{O}$. We have only to show that there cannot be *more* than two tangents from any point not on $\mathcal{O}$. To see this, consider the n points of any tangent line t other than the point of tangency. At each of them there is at least one other tangent to the oval (Illustration 62), as we have just proved. This accounts for all $n + 1$ tangents to the oval; hence there are exactly two tangents from each point on t.

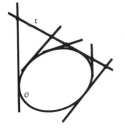

Illustration 62

Note that a consequence of this theorem is that in an *affine plane* of odd order n, if t is a tangent to an oval $\mathcal{O}$ at a point Q on $\mathcal{O}$, then there is exactly one other tangent to the oval *parallel* to t. For at the ideal point on t, there are exactly two tangents to $\mathcal{O}$, t and one other. The other is the required parallel. (Compare with Theorem 84.)

104 EXERCISES

1. A *mini-oval* in a projective plane of order n is a set of n points, no three collinear. Show, by taking $n = 5$ if necessary, that it is not necessarily the case that there is a unique tangent at each point of a mini-oval.

2. If $\mathcal{O}$ is an oval in a projective plane of odd order, prove that every point P in the plane is on at least one secant of $\mathcal{O}$.

3. A *super-oval* in a projective plane of order n is a set of $n + 2$ points, no three collinear. Prove, or disprove, the existence of a super-oval in planes of odd order. [Hint: Any $n + 1$ points of the super-oval form an oval $\mathcal{O}$. Problem 2 shows that the other point of the super-oval lies on a secant of $\mathcal{O}$.]

Theorem 103 now tells us which points in a plane ought to be called "interior points" to a given oval and which points ought to be called "exterior points." For, recalling the properties of an ordinary circle as well as the "circles" of Section 11, we can agree that interior points are those through which no tangent passes and exterior points are those through which at least one (and hence exactly two) tangents pass.

105 Definition

If P is a point in a projective plane of odd order n and $\mathcal{O}$ is an oval in this plane, then P is said to be an *interior* (*exterior*) point of $\mathcal{O}$ if there are no (or exactly two) tangents to $\mathcal{O}$ containing P. Of course, if there is exactly one tangent to $\mathcal{O}$ containing P, then P is a point of the oval.

How many interior points are there to a given oval in a given plane? How many exterior points? Theorem 106 and its corollary answer these questions.

106 Theorem

In a projective plane of odd order n, there are exactly $(n^2 + n)/2$ points exterior to any given oval $\mathcal{O}$.

PROOF: First note that each exterior point E determines a pair of distinct points on $\mathcal{O}$, namely, the points where the tangents from E meet $\mathcal{O}$. Con-

Illustration 63

From an exterior point E, there are exactly two tangents to an oval.

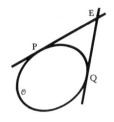

versely, each pair of distinct points P, Q on $\mathcal{O}$ determines a unique exterior point of $\mathcal{O}$, namely, the point of intersection of the two tangents at P and Q. (See Illustration 63.) That is, there is a one-to-one correspondence between the exterior points of $\mathcal{O}$ and the pairs of distinct points on $\mathcal{O}$. Since there are clearly $[(n+1)n]/2$ pairs of distinct points on $\mathcal{O}$ (Why?), there are $[(n+1)n]/2$ exterior points of $\mathcal{O}$.

107 Corollary

An oval $\mathcal{O}$ in a projective plane of odd order n has exactly $(n^2 - n)/2$ interior points.

PROOF: There are exactly $n^2 + n + 1$ points in the plane, of which $n + 1$ are points of $\mathcal{O}$ and $[(n+1)n]/2$ are exterior points. The remaining $(n^2 + n + 1) - (n + 1) - [(n+1)n]/2 = (n^2 - n)/2$ points are interior points, as claimed.

108 EXERCISE

Does this agree with the enumeration of exterior and interior points in Problem 2 of Exercise 85. Explain any discrepancies.

Illustration 64

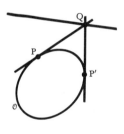

We have already seen that a tangent line to an oval contains only exterior points (except for the point of tangency itself). That is, there are no interior points on a tangent line. This agrees with ordinary (real) Euclidean geometry. But exterior lines behave in a way quite different from that of their Euclidean counterparts. We note first that there are exactly $(n+1)/2$ exterior points on each exterior line l to $\mathcal{O}$. For at each point P of $\mathcal{O}$, there is a tangent which meets l, say at Q. (See Illustration 64.) But then there is another tangent from Q to $\mathcal{O}$, meeting $\mathcal{O}$ at, say, P'. Q is certainly an exterior point of $\mathcal{O}$. This process can be repeated at all points of $\mathcal{O}$, and each of the $(n+1)/2$ pairs of the type P,P' will determine an exterior point on l. Since all the other points of l are interior points we have proved part (b) of Theorem 109. The student should prove part (c) as an exercise.

109 Theorem

If $\mathcal{O}$ is an oval in a projective plane of odd order n, then (a) each tangent line to $\mathcal{O}$ contains no interior points of $\mathcal{O}$, (b) each exterior line to $\mathcal{O}$ contains exactly $(n+1)/2$ interior points of $\mathcal{O}$, and (c) each secant contains exactly $(n-1)/2$ interior points of $\mathcal{O}$.

14B Ovals in Planes of Even Order (Optional)

If n is an *even* number, the arguments used to establish the preceding theorems will not work. But similar arguments can be applied to determine the facts in this case also. The main result is somewhat startling, since when n is even, the ovals do not behave at all like their Euclidean counterparts. We include this striking fact here as Theorem 110 for those who are interested. (However, this section is not needed in the sequel and may be omitted without loss of continuity.)

The definitions of an *oval* $\mathcal{O}$ in a projective plane of even order n and of *tangent lines*, *secant lines*, and *exterior lines* are exactly as in Definitions 100 and 101. And it is still true that there is exactly one tangent line at each point of an oval, for neither the statement nor proof of this fact (Theorem 102) depends on whether n is even or odd. But Theorem 103, which tells us that every point *not* on the oval has either two tangents or no tangents passing through it (and hence enables us to define exterior and interior points of an oval), takes the following very different form in case n is even.

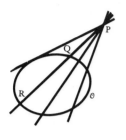

Illustration 65

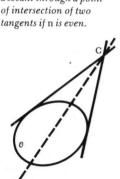

Illustration 66

It is impossible to have a secant through a point of intersection of two tangents if n *is even.*

110 Theorem

If $\mathcal{O}$ is an oval in a projective plane of *even* order n, then all of the $n + 1$ tangents to $\mathcal{O}$ pass through a single point C.

PROOF: Consider a point P, not on $\mathcal{O}$, through which there is a secant line of $\mathcal{O}$. Denote the two points where this secant line meets $\mathcal{O}$ by Q and R, as in Illustration 65. There are an odd number (namely, $n - 1$) of other points on $\mathcal{O}$, so at least one line through P must meet $\mathcal{O}$ in an *odd* number of points. By the definition of oval, this odd number can only be 1. Hence there is at least one tangent line to $\mathcal{O}$ from P. By the same argument, from *each* of the $n + 1$ points (such as P) on a secant line there must be a tangent line to $\mathcal{O}$. But there are only $n + 1$ tangent lines to $\mathcal{O}$ altogether (one at each of the $n + 1$ points of $\mathcal{O}$). Hence there is *exactly* one tangent to $\mathcal{O}$ from each point on a secant line. This means that *two tangents cannot intersect on a secant line*. (See Illustration 66.) Hence, all the $n + 1$ lines through the point of intersection C of two tangent lines must be tangent lines. Since there are only $n + 1$ tangent lines to $\mathcal{O}$ altogether, this completes the proof that all the tangents to $\mathcal{O}$ pass through a single point C.

111 Example

The following example shows how this works in the particular case $n = 4$. We recall from Section 8 that there is a field F_4 consisting of exactly four elements, $0, 1, \lambda, 1 + \lambda$. Addition and multiplication for this field are defined by Tables 5 and 6 in Section 8. We construct $AG(F_4)$ from this field and represent its 16 points in the usual way, as in Illustration 67. Any set of five points, no three collinear, will be an oval in the corresponding projective plane of order 4. In particular, the five points $(1, 0)$, $(0, 1)$, $(1, \lambda)$, $(\lambda, 1)$, (λ, λ) are the points of an oval. One way to verify this is to note that these five points all satisfy the quadratic equation $x^2 + \lambda xy + y^2 = 1$ and to recall that no linear equation ($=$ line) can have more than two solutions ($=$ points) in common with a quadratic equation ($=$ oval). (Recall Lemma 74 of Section 11.)

Illustration 67

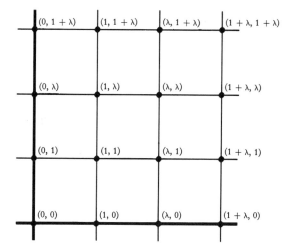

Illustration 68   The oval $x^2 + \lambda xy + y^2 = 1$ and three of its tangents.

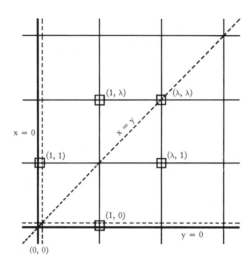

This oval is represented graphically as the boxed points in Illustration 68. Three tangent lines, $x = 0$, $y = 0$, and $x = y$, are obvious by inspection and are marked by dotted lines in the figure. The tangents at the other two points are $y = \lambda x$ and $y = (1 + \lambda)x$. As we already know from the general Theorem 110, these five tangents have a common point, namely, $(0, 0)$.

112 EXERCISES

1. Prove or disprove the existence of a super-oval in projective planes of even order. [Hint: Use Theorem 110.] (Compare with Problem 3 of Exercises 104.)
2. Verify that $y = \lambda x$ and $y = (1 + \lambda)x$ are tangents to the oval $x^2 + \lambda xy + y^2 = 1$ in $AG(F_4)$.
3. Determine the five points on the oval $x^2 + \lambda x + y^2 = \lambda^2$ $(= 1 + \lambda)$ in $AG(F_4)$, the five tangent lines at these points, and their common point of intersection. [Answer: $(0, \lambda)$, $(\lambda, 0)$, $(1 + \lambda, \lambda)$, $(\lambda, 1 + \lambda)$, $(1 + \lambda, 1 + \lambda)$, $x = 0$, $y = 0$, $x = y$, $y = \lambda x$, $y = (1 + \lambda)x$.]

Illustration 69

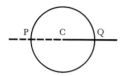

We remark, in passing, that if we think of an oval $\mathcal{O}$ in a projective plane of order 4 as a "circle," then there is some reason to think of the point of intersection of its tangents as the "center" of the circle. For suppose we have a circle with center C. Then, if any line through C meets $\mathcal{O}$ at all, it is unlikely to meet it in two points P, Q. (See Illustration 69.) For, if so, then the "distance" $\overline{CP}$ and the "distance" $\overline{CQ}$ would be equal. But then the "diameter" of the circle would be $\overline{PC} + \overline{CQ}$. But reference to the addition table (Table 5) for F_4 shows that the sum of any two equal elements of F_4 is 0. That is, the diameter of $\mathcal{O}$ is 0, which contradicts our intuition that the diameter of a circle should not be 0. An alternative to this dilemma is that each line through C can meet $\mathcal{O}$ in (at most) one point—which is the case, as shown by Theorem 110.

Section 15 A Finite Version of Poincaré's Universe

The reader will recall the discussion of hyperbolic planes in Section 5, which arose from attempts to prove the parallel postulate. It was eventually realized by Bolyai and Lobachevsky that the parallel postulate was independent of the other assumptions of Euclidean geometry and could equally well be replaced by the non-Euclidean postulate (H) of Section 2, namely:

(H) Given any line l and any point P not on l, then at least two different lines through P do not meet l.

From now on, we shall refer to a non-Euclidean geometry in which this postulate (H) holds as a *hyperbolic geometry* or *hyperbolic plane*.

However, there still remained doubts in the minds of many mathematicians and philosophers about the consistency of such a hyperbolic geometry. Isn't it possible that the assumption of the non-Euclidean postulate about parallels would eventually lead to a contradiction? This fear was dispelled when it was shown by Poincaré and others that if there is a contradiction in hyperbolic geometry, there is also a contradiction in Euclidean geometry itself. What Poincaré did was to find a model of the hyperbolic plane *within* the Euclidean plane. In this way, each theorem about the hyperbolic plane was also a theorem about the Euclidean plane, and any contradictory theorems about the hyperbolic plane would also be contradictory theorems about the model and hence about the Euclidean plane.

The Poincaré model is of interest not only from this strictly mathematical point of view but also because it shows how a non-Euclidean geometry might, in fact, arise from conditions in the physical world. Poincaré [22, Chap. 4] imagines a universe consisting of the interior of a large sphere, say of radius R. He makes three basic assumptions:

1. In this sphere of radius R, the absolute temperature varies from a large value at the center to absolute zero at the surface according to the law that at any point at distance r from the center the temperature is $R^2 - r^2$.
2. All dimensions of all objects in this universe are in direct proportion to the temperature.
3. Every object instantly takes on the temperature of the region in which it is placed.

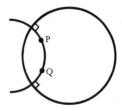

Illustration 70

This universe, which appears bounded to us, will appear *unbounded* to its inhabitants. For, as they walk toward the surface of the sphere, their legs—and hence their steps—become shorter and shorter, so they can never reach the surface. Moreover, it can be shown that the shortest distance between two points P, Q will *not* be along what we, on the outside, would call a straight line between P and Q but will be along an arc of the circle through P and Q which meets the surface of the sphere at right angles. (The only exception to this general rule occurs if P and Q are in line with the center of the sphere, for then this line itself is a "circle" meeting the surface of the sphere at right angles.) That is, it pays a being at point P in this universe, if he is in a hurry to get to point Q, to make a slight detour toward the center of the universe in order to be able to take longer steps, as in Illustration 70. Of course, to him it will not appear to be a detour at all. Thus his "straight lines" will appear to us to be arcs of circles. In much the same way, the shortest plane route from New York to London on a usual flat map of the earth appears to make a detour toward the North Pole. But, in fact, it is not a detour at all.

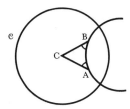

Illustration 71

In Poincaré's universe the sum of the angles of a triangle is less *than two right angles.*

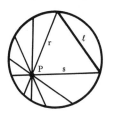

Illustration 72

Poincaré's universe corresponds to the non-Euclidean (hyperbolic) assumption about parallels.

Since we are primarily concerned with *plane* geometry, we can take Poincaré's universe to be the interior of an ordinary circle 𝒞. Then the straight lines are circles meeting the original circle 𝒞 at right angles. (Straight lines through the center are again to be considered as special cases of such circles.) In fact, it is possible to prove that this is an exact model of the hyperbolic plane of Bolyai and Lobachevsky, as described in Section 5. For example, since angles are measured in the usual way in this model, it can be proved that the sum of the angles of a triangle is less than 180°. This is easy to see in the special case where one vertex of the triangle is taken to be the center C of the circle, as in Illustration 71.

There is an even simpler model which we can use to show that at least the *incidence properties* of the hyperbolic plane are as consistent as those of the Euclidean plane. In this model, the *points* are again the interior points of some fixed circle and the *lines* are chords of the circle. Clearly, if P is a point not on line l, there are many lines through P which do not meet l (inside the circle). (See Illustration 72.) It is easy to see which lines are the "right and left parallels" to l through P. They are just the chords (r and s in Illustration 72) through P and the two end points of the chord l.

Is there also a non-Euclidean "hyperbolic" plane with only a *finite* number of points? We could take the following to be two natural axioms for such a plane.

H1 Given two distinct points P, Q, there is exactly one line l containing both of them.

H2 Given a line l and a point P not on l, then there are at least two lines through P not containing any point of l.

However, these incidence relations are not quite enough to ensure that we have a *plane* and not some higher-dimensional space. For example, if we define a *line* to be a line in ordinary Euclidean three-dimensional space, E^3, and a *point* to be a point of E^3, then Axioms H1 and H2 are satisfied. (Note that the corresponding Axioms A1 and A2 are *not* both satisfied.) But it would be unsatisfactory to call such a three-dimensional system a "plane." In order to choose a suitable third axiom to make this really a *plane* geometry, we notice the following property of the Euclidean planes. If we take the three lines determined by any three noncollinear points, then *every* point in the whole Euclidean plane is on some line joining a pair of points on these three lines. (See Illustration 73.) This is the motivation for the third (and final) axiom for the incidence relations in a hyperbolic non-Euclidean plane.

H3 Let three arbitrary noncollinear points P, Q, R be chosen. Consider the points on all lines joining pairs of points on the lines $PQ, QR,$ and RP. Then every point is on a line determined by two of these points.

We have already pointed out that the interior points of an ordinary Euclidean circle, and the chords of such a circle, satisfy Axioms H1 and H2. It is not hard to see that they satisfy H3 also. (The reader should verify that H3 also holds.) The question at hand is whether a *finite* system of points and lines can satisfy these axioms. The Poincaré models show how an *infinite* hyperbolic plane can be found within the ordinary infinite Euclidean plane. This naturally suggests that we look for a *finite* hyperbolic plane in the same way within a finite Euclidean (i.e., affine) plane. As usual, we imagine the affine plane extended to be a projective plane.

All the necessary ideas for this project are already at our disposal. We have defined circles—or rather, *ovals*—in our finite planes, and we know what points are in the *interior* of an oval. Hence we know what a *chord* of an oval $\mathcal{O}$ is, namely, the set of interior points of $\mathcal{O}$ which are also on some line. Thus we could try the following dictionary:

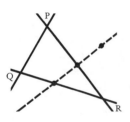

Illustration 73

Hyperbolic plane $\mathcal{H}$	*Projective plane* $\mathcal{P}$ *of odd order* n
1. Point	1. Interior point of an oval $\mathcal{O}$
2. Line	2. Those points on a line in $\mathcal{P}$ which are also interior points of $\mathcal{O}$

To distinguish points and lines of $\mathcal{H}$ from points and lines of $\mathcal{P}$, we shall use capital initials for the former, i.e., Point and Line. And if P and l denote a point and line of $\mathcal{P}$, then we shall denote the corresponding Point and Line of $\mathcal{H}$ by $\bar{P}$ and $\bar{l}$.

Certainly two Points $\bar{P}, \bar{Q}$ determine a unique Line in $\mathcal{H}$, for the corresponding points P, Q in $\mathcal{P}$ determine a unique line l in $\mathcal{P}$

Section 15 A Finite Version of Poincaré's Universe   307

and the corresponding Line $\bar{l}$ contains both $\bar{P}$ and $\bar{Q}$. Thus Axiom H1 is satisfied by this model. Now suppose $\bar{P}$ is a Point not on the Line $\bar{l}$. There are $n + 1$ Lines through $\bar{P}$ since each of the $n + 1$ lines of $\mathcal{P}$ through the point P corresponds to a Line through $\bar{P}$. But we know from Theorem 109 that at most $(n + 1)/2$ of these Lines can meet $\bar{l}$ since l contains at most $(n + 1)/2$ interior points. (In fact, of course, if l is a secant line of $\mathcal{O}$, it contains only $(n - 1)/2$ interior points of $\mathcal{O}$. That is, $\bar{l}$ contains only $(n - 1)/2$ Points.) Hence at least $(n + 1) - (n + 1)/2 = (n + 1)/2 \geq 2$ Lines through $\bar{P}$ fail to meet $\bar{l}$. This proves H2.

We now try to prove H3. Let $\bar{P}, \bar{Q}, \bar{R}$ be three noncollinear Points. Consider the $(n - 1)/2$ Lines joining $\bar{P}$ to some $(n - 1)/2$ Points of Line $\overline{QR}$, as in Illustration 74. Each of these $(n - 1)/2$ Lines contains at least $(n - 1)/2 - 1 = (n - 3)/2$ Points other than $\bar{P}$. Hence there are at least $[(n - 1)/2][(n - 3)/2] + 1$ (counting $\bar{P}$) Points on these Lines. We shall show that each Point of the whole system is on some Line containing two of these Points. This will prove Axiom H3.

Illustration 74   There are $\dfrac{(n - 1)}{2}$ lines at $\bar{P}$, each containing at least $\dfrac{(n - 3)}{2}$ Points different from $\bar{P}$.

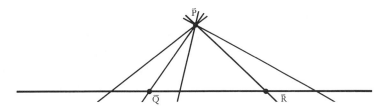

In fact, if $\bar{T}$ is any Point whatsoever, then there are $n + 1$ Lines containing $\bar{T}$. If $[(n - 1)/2][(n - 3)/2] + 1 > n + 1$, then not all the $n + 1$ Lines at $\bar{T}$ can contain exactly one of these Points. That is, some Line through $\bar{T}$ must contain at least two of these Points, which will complete the proof. We need only verify that $[(n - 1)/2][(n - 3)/2] + 1 > n + 1$. This inequality reduces to $n^2 - 8n + 3 > 0$, which is true whenever $n > 7$.

This completes the proof of:

113 Theorem

If $n > 7$ and $\mathcal{O}$ is an oval in a projective plane $\mathcal{P}$ of odd order n, then the interior points of $\mathcal{O}$ and the sets of interior points of $\mathcal{O}$ lying on lines of $\mathcal{P}$ constitute the Points and Lines of a finite hyperbolic plane satisfying Axioms H1 to H3.

In conclusion, we note that a peculiarity of the hyperbolic plane of Theorem 113, in contrast to finite affine planes and projective planes, is that not all Lines contain the same number of Points. Those Lines which come from exterior lines contain $(n + 1)/2$ Points, but those which come from secant lines contain only $(n - 1)/2$ Points. Curiously enough, this peculiarity can be removed if n is *even*. The trick is to use the original orthogonal-circles model suggested by Poincaré. (The mathematical details can be found in [18]. Other interesting properties of the even-order case are discussed in [17].)

APPENDIX TO CHAPTER 4

The definitions, postulates, and common notions of Euclid, as given in T. L. Heath's edition of *The Thirteen Books of Euclid's Elements* [4, pp. 153–155].

Definitions

1. A point is that which has no part.
2. A line is breadthless length.
3. The extremities of a line are points.
4. A straight line is a line which lies evenly with the points on itself.
5. A surface is that which has length and breadth only.
6. The extremities of a surface are lines.
7. A plane surface is a surface which lies evenly with the straight lines on itself.
8. A plane angle is the inclination to one another of two lines in a plane which meet one another and do not lie in a straight line.
9. And when the lines containing the angle are straight, the angle is called rectilineal.
10. When a straight line set up on a straight line makes the adjacent angles equal to one another, each of the equal angles

is right, and the straight line standing on the other is called a perpendicular to that on which it stands.

11 An obtuse angle is an angle greater than a right angle.

12 An acute angle is an angle less than a right angle.

13 A boundary is that which is an extremity of anything.

14 A figure is that which is contained by any boundary or boundaries.

15 A circle is a plane figure contained by one line such that all the straight lines falling upon it from one point among those lying within the figure are equal to one another;

16 And the point is called the center of the circle.

17 A diameter of the circle is any straight line drawn through the center and terminated in both directions by the circumference of the circle, and such a straight line also bisects the circle.

18 A semicircle is the figure contained by the diameter and the circumference cut off by it. And the center of the semicircle is the same as that of the circle.

19 Rectilineal figures are those which are contained by straight lines, trilateral figures being those contained by three, quadrilateral those contained by four, and multilateral those contained by more than four straight lines.

20 Of trilateral figures, an equilateral triangle is that which has its three sides equal, an isosceles triangle that which has two of its sides alone equal, and a scalene triangle that which has its three sides unequal.

21 Further, of trilateral figures, a right-angled triangle is that which has a right angle, an obtuse-angled triangle that which has an obtuse angle, and an acute-angled triangle that which has its three angles acute.

22 Of quadrilateral figures, a square is that which is both equilateral and right-angled; an oblong that which is right-angled but not equilateral; a rhombus that which is equilateral but not right-angled; and a rhomboid that which has its opposite sides and angles equal to one another but is neither equilateral nor right-angled. And let quadrilaterals other than these be called trapezia.

23 Parallel straight lines are straight lines which, being in the same plane and being produced indefinitely in both directions, do not meet one another in either direction.

Postulates

Let the following be postulated:

1. To draw a straight line from any point to any point.
2. To produce a finite straight line continuously in a straight line.
3. To describe a circle with any center and distance.
4. That all right angles are equal to one another.
5. That, if a straight line falling on two straight lines make the interior angles on the same side less than two right angles, the two straight lines, if produced indefinitely, meet on that side on which are the angles less than the two right angles.

Common Notions

1. Things which are equal to the same thing are also equal to one another.
2. If equals be added to equals, the wholes are equal.
3. If equals be subtracted from equals, the remainders are equal.
4. Things which coincide with one another are equal to one another.
5. The whole is greater than the part.

Propositions

The first 29 propositions of Euclid's *Elements*, as given by Heath [4]. Proposition 29 is the first one which requires the parallel postulate for its proof.

1. On a given finite straight line to construct an equilateral triangle.
2. To place at a given point (as an extremity) a straight line equal to a given straight line.
3. Given two unequal straight lines, to cut off from the greater a straight line equal to the less.
4. If two triangles have the two sides equal to two sides, respectively, and have the angles contained by the equal straight lines equal, they will also have the base equal to the base, the triangle will be equal to the triangle, and the remaining angles will be equal to the remaining angles, respectively, namely those which the equal sides subtend.
5. In isosceles triangles the angles at the base are equal to one

another, and, if the equal straight lines be produced further, the angles under the base will be equal to one another.

6   If in a triangle two angles be equal to one another, the sides which subtend the equal angles will also be equal to one another.

7   Given two straight lines constructed on a straight line (from its extremities) and meeting in a point, there cannot be constructed on the same straight line (from its extremities), and on the same side of it, two other straight lines meeting in another point and equal to the former two, respectively, namely each to that which has the same extremity with it.

8   If two triangles have the two sides equal to two sides, respectively, and have also the base equal to the base, they will also have the angles equal which are contained by the equal straight lines.

9   To bisect a given rectilineal angle.

10 To bisect a given finite straight line.

11 To draw a straight line at right angles to a given straight line from a given point on it.

12 To a given infinite straight line, from a given point which is not on it, to draw a perpendicular straight line.

13 If a straight line set up on a straight line make angles, it will make either two right angles or angles equal to two right angles.

14 If with any straight line, and at a point on it, two straight lines not lying on the same side make the adjacent angles equal to two right angles, the two straight lines will be in a straight line with one another.

15 If two straight lines cut one another, they make the vertical angles equal to one another.

16 In any triangle, if one of the sides be produced, the exterior angle is greater than either of the interior and opposite angles.

17 In any triangle, two angles taken together in any manner are less than two right angles.

18 In any triangle, the greater side subtends the greater angle.

19 In any triangle, the greater angle is subtended by the greater side.

20 In any triangle, two sides taken together in any manner are greater than the remaining one.

21 If on one of the sides of a triangle, from its extremities, there be constructed two straight lines meeting within the triangle, the straight lines so constructed will be less than the remaining two sides of the triangle, but will contain a greater angle.

22 Out of three straight lines, which are equal to three given straight lines, to construct a triangle; thus it is necessary that two of the straight lines taken together in any manner should be greater than the remaining one.

23 On a given straight line and at a point on it, to construct a rectilineal angle equal to a given rectilineal angle.

24 If two triangles have the two sides equal to two sides, respectively, but have the one of the angles contained by the equal straight lines greater than the other, they will also have the base greater than the base.

25 If two triangles have the two sides equal to two sides, respectively, but have the base greater than the base, they will also have the one of the angles contained by the equal straight lines greater than the other.

26 If two triangles have the two angles equal to two angles, respectively, and one side equal to one side, namely, either the side adjoining the equal angles or that subtending one of the equal angles, they will also have the remaining sides equal to the remaining sides and the remaining angle to the remaining angle.

27 If a straight line falling on two straight lines make the alternate angles equal to one another, the straight lines will be parallel to one another.

28 If a straight line falling on two straight lines make the exterior angle equal to the interior and opposite angle on the same side, or the interior angles on the same side equal to two right angles, the straight lines will be parallel to one another.

29 A straight line falling on parallel straight lines makes the alternate angles equal to one another, the exterior angle equal to the interior and opposite angle, and the interior angles on the same side equal to two right angles.

The final propositions of Book I of the *Elements* are the celebrated Pythagorean theorem (47) and its converse (48).

47 In right-angled triangles, the square on the side subtending

the right angle is equal to the squares on the sides containing the right angle.

48 If in a triangle the square on one of the sides be equal to the squares on the remaining two sides of the triangle, the angle contained by the remaining two sides of the triangle is right.

References

Sections 1 to 6

1   J. Bolyai, *Theory of Absolute Space*. (See Bonola [2].)
2   Roberto Bonola, *Non-Euclidean Geometry*, Dover Publications, Inc., New York, 1955. (This edition contains also J. Bolyai, *Theory of Absolute Space*, and N. I. Lobachevsky, *Theory of Parallels*.)
3   H. S. M. Coxeter, *Introduction to Geometry*, John Wiley & Sons, Inc., New York, 1961, especially pp. 3–10.
4   T. L. Heath, *The Thirteen Books of Euclid's Elements* (3 vol.), Cambridge University Press, London, 1908 (Dover reprint 1956), especially Vol. 1.
5   D. Hilbert, *Foundations of Geometry*, The Open Court Publishing Company, LaSalle, Ill., 1902.
6   N. I. Lobachevsky, *Theory of Parallels*. (See Bonola [2].)
7   H. Meschkowski, *Unsolved and Unsolvable Problems in Geometry*, Frederick Ungar Publishing Co., New York, 1966.
8   B. L. van der Waerden, *Science Awakening*, John Wiley & Sons, Inc., New York, 1963.
9   H. E. Wolfe, *Introduction to Non-Euclidean Geometry*, Holt, Rinehart and Winston, Inc., New York, 1945.

Sections 7 to 11

10 H. M. Cundy, "25-point geometry," *Math. Gazette*, **36** (1952), pp. 158–166. (A deeper discussion of this same geometry, by W. L. Edge, can be found on pages 113–121 of the same journal in 1955.)
11 Harold L. Dorwart, *The Geometry of Incidence*, Prentice-Hall, Inc., Englewood Cliffs, N.J., 1966.
12 H. Eves, *A Survey of Geometry*, Allyn and Bacon, Inc., Boston, 1963, Vol. 1 (Chap. 8).
13 M. Gardner, "How three modern mathematicians disproved a celebrated conjecture of Leonhard Euler," *Scientific American*, November 1959, pp. 181–188.
14 B. W. Jones, "Miniature number systems," *Math. Teacher*, **51** (1958), pp. 226–231.

15 ——, "Miniature geometries," *Math. Teacher*, **52** (1959), pp. 66–71.

16 H. J. Ryser, *Combinatorial Mathematics*, Carus Mathematical Monograph 14, Math. Assoc. of America, distributed by John Wiley & Sons, Inc., New York, 1963, Chap. 7.

Sections 12 to 15

17 J. W. Archbold, "A metric for plane affine geometry over $GF(2^n)$," *Mathematika*, **7** (1960), pp. 145–148.

18 D. W. Crowe, "The trigonometry of $GF(2^{2n})$, and finite hyperbolic planes," *Mathematika*, **11** (1964), pp. 83–88.

19 P. Dembowski and D. R. Hughes, "On finite inversive planes," *J. London Math. Soc.*, **40** (1965), pp. 171–182.

20 T. G. Ostrom, "Ovals and finite Bolyai-Lobachevsky planes," *Amer. Math. Monthly*, **69** (1962), pp. 899–901.

21 D. Pedoe, *An Introduction to Protective Geometry*, Pergamon Press, New York, 1960.

22 H. Poincaré, *Science and Hypothesis*, Dover Publications, Inc., New York, 1952 (Chap. 4).

23 B. Qvist, "Some remarks concerning curves of the second degree in a finite plane," *Ann. Acad. Sci. Fenn.*, **134** (1952), pp. 1–27.

24 B. Segre, "Ovals in a finite projective plane," *Can. J. Math.*, **7** (1955), pp. 414–416.

EXCURSIONS INTO MATHEMATICS

CHAPTER   5   **GAMES**

BY ANATOLE BECK

While this book was being written, this advertisement appeared in the New York Times Magazine on December 3, 1967. This is a perfect example of the games adults play! (Courtesy of the Lewis Singer Tax Service.)

Section 1 Introduction

The word "games" brings to mind the picture of children at play, of men sitting over a chessboard or a poker table, of athletes at the Olympics. In all of this, one senses the feeling of pastime, of leisure, of the candy and dessert of life. We think of games as frivolous activity, not really worthy of serious attention. We are aware, for example, that military maneuvers are sometimes called war games, but that seems to be a misuse of the word.

Actually, games are often serious business. Much of a person's personality goes into the choice of which games he will play, as well as which he will watch. It is often the case that a game is a reflection of an important phase of the player's environment and that by playing the game, he is exercising his mastery of it. Simple examples include the rough play of lion cubs, who, through frolicking, are also learning the arts of hunting and killing. Little girls "play" at being mothers, and even a cursory examination of the other games our children play will yield a sense of preoccupation with adult problems, or with problems that children might easily mistake for adult problems. In every culture children play at the activities of adults, and in so doing, they learn and master the techniques involved.

And what of the games adults play? In part they are the leftovers of their childhood experience; in part they represent adults seeking the experiences of childhood, just as children seek the experiences of adulthood. But on the other hand, adults still play games which sharpen the skills they find vital. It is no accident that chess has classically been a favorite game among generals and poker among businessmen. One might well ponder the significance of the fact that today's generals seem to prefer poker to chess in some parts of the world.

In recent years, games have been devised to help people understand the submarine war in the North Atlantic (1940–1945), the sharing of profits among partners, and the conflicts of strategy in the Nuclear Test Ban Treaty of 1963.† Some games, like Nim (see, for example, *Last Year at Marienbad*),‡ have been completely ana-

---

† See, for example, [1] and [9].
‡ See [11, pp. 38–40, 54–57, 143–146].

lyzed, and all the winning strategies are known. For these games, the spice of uncertainty has been removed. Other games, like chess and Go, still defy analysis, while the game of Hex is half-analyzed. These last games still fascinate mathematicians.

We shall discuss only games of mental (rather than physical) prowess, and we shall divide them into a few large categories.

Section 2 Some Tree Games

We shall look first at some games, well known and otherwise, which we shall call *tree games* for reasons which will become apparent later. The first is rather simple. Twelve matches are laid on a table, and the first player (whom we call Mr. White) can take one, two, or three of them. Then the second player (Mr. Black) takes one, two, or three; then Mr. White, etc. The object is to obtain the last matchstick. Try this game for awhile and convince yourself that Black can always win, no matter what the play by White. Suppose we had started with 130 matches. Would Black still win? 2,048 matches? In fact, it is easy to see that Black can win exactly if the number of matches is divisible by 4 and that White wins otherwise. How do you show this formally?

Now let us change the game somewhat. The 12 matches are set in a straight line, evenly spaced. The removal of any number of matches together leaves us with either one or two groups of matches. If another group of matches together is removed, we have one, two, or three groups left, etc. Under the new rules, we must always remove a group of matches together, the group being one, two, or three in number. In this new game, White can win, and he can win no matter what counterstrategy Black employs. In fact, White will win if the original number of matches is 45 or 128 or anything else. Another game which is similar is played in the film "Last Year at Marienbad." The game starts with matches in four piles containing, respectively, 1, 3, 5, and 7 matches (or cards or dominoes). Each player, in turn, chooses as many matches as he wishes *but always from one pile*. The one who takes the last match wins. This game goes by the name of Nim, which is an anglicized form of the German word meaning "Take!" Actually, in the Marienbad game, the one who took the last match (card, domino) was the *loser*. This rule creates some difficulty, but the difficulty is not intrinsic to the analysis. We shall go along with the rules as we have stated them rather than with the movie version.

1   EXERCISES

1   In the game of Nim, let us substitute for the numbers 1, 3, 5, 7 the numbers
    (a) 1, 2, 3   (b) 2, 4, 4
    (c) 2, 4, 6   (d) 1, 4, 5
    (e) 1, 2, 3, 4, 5   (f) 1, 7, 8
    Who wins in each case, Black or White?

2   A chessboard is 8 by 8 inches, consisting of sixty-four 1-inch squares. A domino is 1 by 2 inches. Each player in turn covers two adjacent squares with a domino, and no square can be covered twice. The game is played until no further move is possible. Last player to move is the winner. Who wins?

3   Show that Tic-Tac-Toe is a draw when properly played.

2   *Definition*

A game is a *tree game* if:

1   At each turn, the number of moves that the player who moves can choose from is finite.
2   There are two players, and they alternate turns.
3   The game lasts at most a certain number of turns, and by the time that number is reached, the outcome is determined.
4   Each player knows the moves of the other player.

In addition, we often take as a convention:

5   White moves first.

The convention saves some time later on, but on the other hand, we also sometimes need to consider games satisfying:

5'   Black moves first.

These, for technical reasons, are called *inverted tree games*.

We shall now examine a particular tree game, which is actually played on a "tree." Observe the diagram in Illustration 1. White goes first and chooses one of the three "branches," 1, 2, or 3. If he chooses 2, for example, then Black must choose one of the "branches" d, e, f, or g. If Black chooses g, then White takes one of the four branches available to him, and the game is a victory for White, a victory for Black, or a draw, according as the letter at the end of that branch is W, B, or D. Before you read our analysis of this game, try playing it yourself for awhile.

Illustration 1

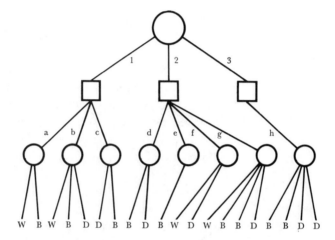

Illustration 2

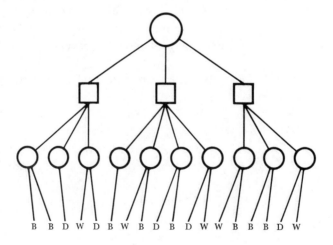

3 Analysis of the Game

If White plays alternative 1 on his first move, then Black is faced with *a*, *b*, or *c* as his choices. If he plays *a*, then White will be able to win. If he plays *b*, then White will also be able to win. If he plays *c*, then White will be able to draw at best. Thus, the worst best that Black can give to White is to play alternative *c*; White will then get a draw. We thus conclude that if White plays 1 on the first move, then we can expect a draw.

Suppose White chooses 2 as his first move. Then we see that if Black takes *d*, White gets a draw; if Black takes *e*, then Black wins; while if Black takes *f* or *g*, White can win. We can thus anticipate that Black would choose *e* if White chose 2, and Black would win.

If White were to choose 3, Black would have no choice but *h*, and White, faced with a Black win or a draw, would presumably choose a draw. Thus, White can manage a draw if he chooses 1 or 3 and will give the game to Black if he chooses 2. It makes sense, then, that he would choose 1 or 3, and we refer to a game like this as a *natural draw*. To be more general, this game has a *natural outcome*, which, in this case, is a draw. We are interested in knowing which tree games have natural outcomes, and what they are. In order to look further into this question, let us look at the two games shown in Illustrations 2 and 3.

Illustration 3

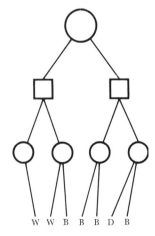

4   EXERCISE

Show that the game in Illustration 2 is a *forced win for Black* and that the game in Illustration 3 is a *forced win for White*. By this, we mean that there is a way for Black to win the first game against any play by White, and conversely in the second game.

5   *Definition*

If a tree game has the property that it is possible for White to win every time, regardless of the play by Black, then we say that "*White to win*" is the *natural outcome* of the game. We define similarly the *natural outcome* "*Black to win.*" If it is within Black's power to categorically prevent White from winning, no matter how he plays, and if White can similarly prevent Black from winning (e.g., Tic-Tac-Toe), then we say that the game is a *natural draw* or that the *natural outcome is* "*Draw.*"

Suppose we were to set out to prove that every tree game has a natural outcome. We would consider White's first move. White has a choice of finitely many moves, and for each possible choice of a first move, he *presents Black with an inverted tree game*. Suppose that each of the inverted tree games so created has a natural outcome. What then? Well, if one of these games has the outcome "White to win," then White can force a win in the original game by choosing the corresponding branch for his first move. Suppose, on the other hand, that every one of the inverted games has the natural outcome "Black to win." In that case, no matter what original move White makes, Black can win no matter what subsequent moves White makes. It follows immediately that "Black to win" is the natural outcome of the original game.

Now suppose that neither of these cases holds. Then none of the inverted games is a win for White, and not all are wins for Black. Since each of them is assumed to have a natural outcome, some must be natural draws. Thus White has a choice among inverted games that Black can win and games that can be drawn, but White cannot force a win in any of them. We take as natural his choice of a draw, and then we see that the original game is a natural draw.

The inverted games are "smaller," in some sense, than the original game. This should make us think of induction and of the relationship of inverted games to ordinary games.

6 Definition

If G is a game, then G' is the game with the roles of Black and White reversed. G' is an inverted game and is called the *inverse* of G. G is also called the inverse of G'.

7 Lemma

Let G be an ordinary game, and let G' be its inverse. Then G' has a natural outcome just if G does.

PROOF: It is clear that if some strategy will win for White in G, then the corresponding strategy will win for Black in G', and conversely. In the same way, if Black has forced a win in G, then White has one in G', and conversely. Finally, if the natural outcome of either game is a draw, then the inverse game is also a natural draw. Thus the lemma is proved.

8 Theorem

Every tree game has a natural outcome.

PROOF: We prove this theorem by induction. The *parameter of induction* is the length of the game. For each game G, we say that G *is of rank n*, where n is some natural number, if the longest possible sequence of moves in G is of length n. It should be clear that every game of rank 1 has a natural outcome, for in such a game, White makes one move and chooses the result. If he can win, he does so. If he cannot win, he draws, if that is possible. Otherwise, he has no choice, and Black wins. This grounds our induction.

Suppose now that every tree game of rank $1, 2, \ldots, n-1$ has a natural outcome, and we are given a tree game G of rank n. Let the first move of White be a choice among k alternatives. Then if White chooses alternative j the remaining moves constitute an inverted game G_j. Since the inverted game G_j can last at most $n-1$ moves (remember that G can last only n moves at most), its inverse G'_j has a natural outcome. Thus G_j also has a natural outcome for each j from 1 to k. Thus, if G_j has the outcome "White to win" for at least one value of j, then G also has that outcome. If G_j has the outcome "Black to win" for all values of j, then again G also has the outcome "Black to win." If some of the G_j are "Black to win" but some are draws, then White can ensure a draw by choosing one of these, and we see that G is also a natural draw. This proves the theorem.

We thus see that all tree games are solved in principle, each having its natural solution. It is interesting to note that the proof, as given, only asserts the existence of a natural outcome, without saying how to determine it. We can find such a way, perhaps, if we return to Illustration 1. Note that the points of choice in this figure are marked with circles or squares according as they represent a White move or a Black one. These points will be called *nodes* of the tree. The top circle is called a node *of rank* 1, the squares in that diagram are all nodes of rank 2, and the remaining circles are nodes of rank 3. Now let us start with the nodes of rank 3 and write in each circle the letter of the best outcome White could obtain if he were to reach that node. Please do so now.

Reading from left to right, the best White can obtain for himself at each node is WWD; DBWW; D. Black, at the previous move, would play to give White the *worst best*. This worst best, or minimax, we write in the appropriate square, and it comes out as D; B; D. Now White on his first move will arrange for the best worst best; i.e., he will choose 1 or 3 and take the draw.

Thus, if the tree of the game is of the simple form given above, with a reasonably small number of nodes, we can use this procedure to obtain a natural strategy. This is what we call the *method of exhaustion* (so called because we exhaust the possibilities, not because we exhaust ourselves). In a simple game like Tic-Tac-Toe, however, the number of nodes of rank 9 is $9 \cdot 8 \cdot 7 \cdot 6 \cdot 5 \cdot 4 \cdot 3 \cdot 2$, which is roughly a third of a million, and the nodes of rank 8 are some 40,000 more. This would surely be the method of self-exhaustion! Instead, as in many cases, we use an analysis employing certain kinds of symmetry to show certain facts about the game. For example, it is quite easy to show that the natural outcome of Tic-Tac-Toe is not "Black to win." The proof proceeds from this strange lemma:

9   Lemma

If "Black to win" is the natural outcome of Tic-Tac-Toe, then "White to win" is also.

REMARK

Since "Black to win" and "White to win" cannot both be natural outcomes, this lemma would show that "Black to win" is impossible.

PROOF: We assume (contrary to the actual fact) that Black can force a win. That means that for every position in which Black might find himself, he has a move which guarantees that by the ninth move he will have won. That is, for each situation in which Black might find himself, he has an indicated move, and if he always follows this method (which we shall call the *Answering Strategy*), then by our assumption, he will win. Consider White's position. He places the first mark. Then Black places a mark. At this point, if White could erase his first mark, he could then act as second player and win by using the Answering Strategy. However, White cannot remove his first mark; he is stuck with it. In some games this can be a major disadvantage, but not in Tic-Tac-Toe. White merely plays according to the Answering Strategy as though the first mark were not there. We are now really talking about two games: one is actually being played, and the other is in White's mind. The difference is that the second game has one white mark fewer. And the two games could go to completion, side by side, but for the fact that White, using the Answering Strategy, might, at some time, be called upon to put a mark into the square that actually holds his first mark. The rules do not allow two marks in the same square. But for this little difficulty, the Answering Strategy would give a win for White. At the point when White, in his fantasy game, imagines he has won, he actually has in the real game. Can we eliminate the difficulty?

Yes, we can. If the Answering Strategy requires White to fill the square occupied by the extra marker, he merely ceases to ignore it. Now he has the board as it should be at the end of his move, and he hasn't even moved. So now he must move. He puts his mark into any vacant square and *ignores that one instead*. The fantasy game now squares with the Answering Strategy, and White is ignoring a different mark. Thus White can always keep the board just as it would be if he were playing second, plus an extra white mark, which he ignores. Black can't ignore it, of course, but then White isn't interested in why Black makes the moves he does; White can win against any Black moves (in the fantasy game). Now, whenever White has three marks in a row in the fantasy game, he has them also in the real game.† Thus the Answering Strategy, used (or misused) in this way, would yield a win for White. This proves the lemma.

This disposes of the possibility that "Black to win" could be the natural outcome of Tic-Tac-Toe. Note that this proof does not yield any clue to how the game is to be played. However, we do know that the key question we need to answer is: "Can Black always obtain a draw?" If so, the game is a natural draw. Otherwise, White always wins.

† Note that the converse of this statement is not true.

When we analyze Black's possibility for a draw, we use the natural symmetry of the situation to reduce our task. For example, White seems to have nine moves at the start. In fact, he has only three that are essentially different from each other. If he chooses the center for his first move, then Black really has only two possible replies: a corner, which draws, or a side, which lets White win. Analyze this game to show that Black can draw.

It often happens in mathematics that one can show that something can be done without knowing how to do it. Similarly, we can sometimes show that something is impossible without examining all the cases. Thus, we showed there was no winning strategy for Black at Tic-Tac-Toe without going into the one-third of a million possible games.

10   EXERCISES

1 We play a game with a penny and a chessboard. The penny is put down on one of the squares bordering White's edge of the chessboard and then moved in alternate turns by Black and White. Each move is either forward toward Black's edge or sideways. It is always forbidden to make a move which returns the penny to a position occupied earlier in the game. The object is to reach the last row. The winner is the one who pushes the penny into that row. White places the penny originally; Black makes the first move.
   (a) Is this a tree game?
   (b) What is its natural outcome?

2 Suppose the game is played instead on the vertices of the board (the corners of the squares).
   (a) Is this a tree game?
   (b) What is the natural outcome?

3 White and Black take turns placing pennies on an elliptical table until it is impossible to continue. No penny may overlap another penny. Last one to place a penny successfully wins.
   (a) Is this a tree game?
   (b) What is the natural outcome?

4 Suppose the table in Problem 3 were rectangular. How would the analysis differ? Suppose it were triangular. What property of the table is in question? Generalize!

Section 3   The Game of Hex

Section 3   The Game of Hex

Our next object of study is a truly remarkable game which mathematicians and others have been playing for about 20 years. It goes by the name of Hex and has been variously attributed to Piet Hein, the Danish mathematician, to John Nash, the American mathematician, and to others. Mr. Martin Gardner [3] credits Hein with the game and Nash with an excellent proof that the natural outcome is "White to win," as we see below.

Hex is played on a board whose size is n by n, where n is an integer. If n is odd, then the first move seems always to be in the center. Thus players prefer n even, and we shall take $n = 14$, as it was in the Yale Mathematics Common Room in 1952. The game is played with round markers, such as Go stones; Black and White each have 98 of these. Now let us imagine 196 hexagons, each just large enough to accommodate one marker, arranged in a large rhombus, as in Illustration 4.

Illustration 4

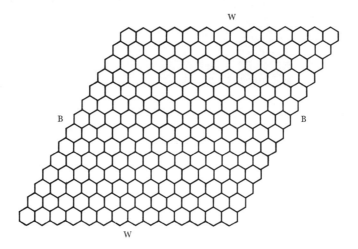

Piet Hein (1905-1996) is a Danish engineer, poet, and intellectual jack-of-all-trades. As a close friend of many distinguished professional mathematicians he has applied mathematics to both architecture and games. In the photograph he is writing the equation of the "super-ellipse" which he suggested to solve a Stockholm problem of putting a traffic circle into a rectangular space. He is credited with the invention of the game of Hex in 1942. (Photograph by Henri Dauman.)

Section 3 The Game of Hex

White plays first and sets a White marker somewhere on the board. Wherever he puts it, there it will remain, as in the case of Tic-Tac-Toe. Then Black puts a marker, then White, then Black, etc. White's purpose is to join the top edge of the board to the bottom with an unbroken string of White markers adjacent to one another. Black's purpose is to join the two sides in a similar manner. We now set the student to prove the following lemmas. (Our own proofs will be given later. Don't peek!)

11 Lemma

If the Hex board is completely covered with White and Black markers, then either there is a White chain joining the top to the bottom or there is a Black chain joining the left to the right.

12 Lemma

The game of Hex never ends in a draw.

13 Lemma

Either White can force a win or Black can force a win.

14 Lemma

If Black can force a win, then White can force a win.

15 Theorem

White to win at Hex.

The fascinating thing about this game of Hex is that no one knows a winning strategy. It is easy to see that going first confers an advantage, and any player of course wants the first move. But no one can realize from this advantage its full benefit, an assured victory. Good players beat weaker ones against this advantage.

Until recently, a commercially marketed game called Bridge-it shared this characteristic [4]. Now a simple strategy is known whereby one can force a win there [5].

A number of mathematicians have attempted to devise an explicit strategy for Hex. The question commands some interest, and whoever solves it will achieve thereby a worldwide, if fleeting, renown.

The proofs of the lemmas are as follows:

PROOF OF LEMMA 11: Assume that the board is completely covered with White and Black markers. Assume that there is *no* chain of Black markers joining the left side to the right side. Let D consist of all the Black markers which can be joined by a chain of Black markers to the left side of the board. D is then a collection of markers. If there are no markers at all in D, then, for example, every space on the left-hand edge of the board is occupied by a White marker. In that case, of course, White wins. If D is not empty, then let D_0 consist of all the markers not in D which touch markers in D or touch the left side of the board. It is clear from the definition that every marker in D_0 is white; otherwise it would have been included in D. Now it is easy to see, but not so easy to prove, that D_0 contains a chain of White markers joining the top of the board to the bottom. That would prove the lemma. Try to prove it!! We shall come back to it later.

PROOF OF LEMMA 12: The rules specify that the game is played until there is a winner or the board is full (n^2 moves). When the board is full, there is a winner. Therefore, there is a winner within n^2 moves.

PROOF OF LEMMA 13: If White cannot force a win, then either Black can force a draw or Black can force a win, by Theorem 8, since Hex is a tree game. Since a draw is impossible, the assumption that White cannot force a win implies that Black can force a win.

PROOF OF LEMMA 14: The proof here is exactly the same as in the discussion of Tic-Tac-Toe, above. If Black can win at Hex, then White can win at Hex', the game inverse to Hex. White's maneuver is thus to put down a marker in an arbitrary place and then ignore it, pretending to himself that he is playing Hex', which White wins, by assumption. If at some move, he finds he needs the space occupied by the "extra" marker, he puts a marker somewhere else, anywhere else, and makes *that* the "extra" marker. Continuing in the same way, White wins.

PROOF OF THEOREM 15: Either "Black to win" or "White to win." But "Black to win" implies "White to win." Therefore, "Black to win" cannot be the case. Therefore, "White to win," which is what we were to prove.

Now we come to the part that mathematicians find so fascinating. We have shown that White can win if he plays properly, but we have no idea what proper play is. Our proof does not tell us what moves to make, what moves to avoid, what constitutes a winning or a losing position. In fact, it is well known that strong players playing

Section 3   The Game of Hex

Black regularly beat weaker players playing White. Until recently, it was not even known whether White's winning was dependent on his first move. It is conceivable, after all, that every possible first move belongs to some winning strategy for White. The proof of "White to win" does, in fact, seem to show just this, but a closer examination will show that it only proves that on the false hypothesis "Black to win." Here we introduce a fact well known to mathematicians but to few others, namely, that any conclusion whatsoever can be derived from a false hypothesis.

16   EXERCISE

1   Some of the statements below follow from others; some have been proved. State which follow from others and which have been proved.
   (a) Black to win or White to win.
   (b) If Black to win, then White to win.
   (c) If Black to win, then White to win with any first move.
   (d) "Black to win" is false.
   (e) White to win with any first move.
   (f) White to win.

We can now ask the following question: Suppose we alter the rules of Hex to allow Black to dictate White's first move (note again the principle of the worst best). Will this change in the rules (we shall call the modified game *Beck's Hex*) be enough to allow Black to win, or will White be able to win with any first move, however bad? The answer is to be found in the following material.

+ ## 17 Lemma

If White wins Beck's Hex, then Black wins Beck's Hex.

PROOF: We note first that White wins Beck's Hex exactly if he wins true Hex regardless of his first move. Suppose then that that is the case (i.e., suppose the hypothesis holds). Then Black, under the rules of Beck's Hex, plays White's first marker into an acute corner of the board and puts a Black marker next to it, in the manner shown in Illustration 5. Given the hypothesis, it is now the case that Black can win. We see this by imagining that the corner marker is also Black, so that Black plays as though the initial

Illustration 5

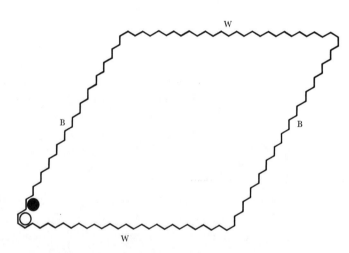

Illustration 6

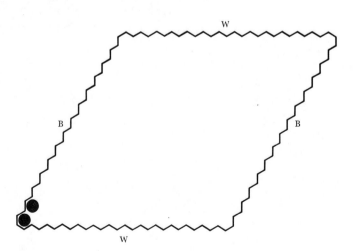

Section 3   The Game of Hex   333

position were as in Illustration 6, with White to move. Let us compare the relative strengths of the position in Illustration 6 and the situation which faces White with only one marker in the acute corner and Black to move. Because the game is symmetric along the long diagonal of the rhombus, Black is *at least* as strong in Illustration 6 (White to move) as White is with only one marker in the corner (Black to move). Thus if White has a winning strategy with this beginning position, then Black can force a win out of the position of Illustration 6, no matter how White plays. Suppose now that Black and White play out the game, with Black laboring under the delusion that the corner marker is black and not white. Black follows White's supposedly winning strategy and, at some move, suddenly arrives at the point where he imagines he has won. At that point, the illusion is no longer necessary, for we shall show that he actually has won.

It may be that Black has several chains which he finishes simultaneously. In that case, look at all the chains that Black imagines win for him, and select from among these one with the minimum number of markers. You should be able to show that such a minimal chain has exactly one marker on the left edge of the board. If that marker is not in the corner, in the space numbered 1 in Illustration 7, then all the markers in the chain are black and Black has really won. If, however, the chain does contain a black marker in space 1, then it must also have one in the space numbered 2, since it cannot have another marker on the left edge. Let us look, then, at the chain made by substituting the marker in space number 3 (which really *is* black) for the marker in space number 1. This is also a winning chain and is made up of genuine black markers. Thus, from the assumption that White can force a win at Beck's Hex, we derive the conclusion that Black can force a win at Beck's Hex, which is what we were to prove (in Latin, *quod erat demonstrandum*).

Illustration 7

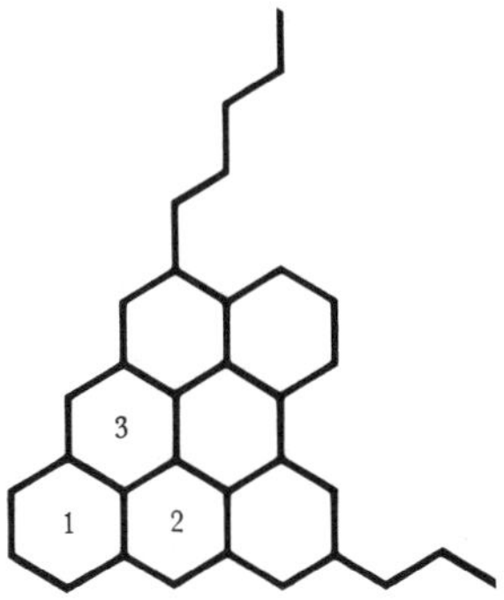

\+ **18 Theorem**

"Black to win" is the natural outcome of Beck's Hex.

PROOF: This is clear from the lemma. As one wit† has put it, this "wrecks Beck's Hex."

There is an interesting corollary to the method of this theorem:

\+ **19 Corollary**

If White, playing Hex, places his first marker in the acute corner, then he has lost his advantage, and Black can force a win.

20 EXERCISES

1 Some of the statements below follow from others; some have been proved. State which follow from others and which have been proved.

(a) If White can win at Hex with any first move, then White to win at Beck's Hex.

(b) If White to win at Beck's Hex, then White can win at Hex with any first move.

(c) If White can win at Beck's Hex, then Black can win at Beck's Hex when he places the first White and Black markers in spaces 1 and 3, respectively.

(d) If White can win at Hex when he places his first marker in an acute corner, then Black can win at Beck's Hex.

(e) Black to win at Beck's Hex.

(f) Black can win at Hex if White puts his first marker into an acute corner.

(g) Black can win at Beck's Hex if he puts the first White and Black markers in spaces 1 and 3, respectively.

2 Prove Corollary 19.

3 Suppose at Hex that Black places White's first move and then White places Black's first move. After that, each plays for himself. What is the natural outcome of *this* game?

We have not yet proved Lemma 11. Here is a sketch of a proof:

We consider the collection D of black markers which are attached to the left side of the board by a chain of black markers. Let us now imagine that

† Prof. M. N. Bleicher.

Section 3 The Game of Hex   335

Illustration 8

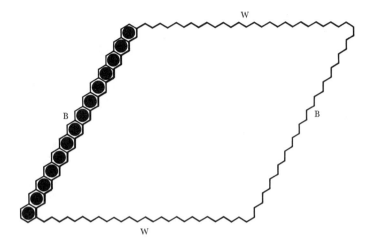

we add a row of black markers to the board, as shown in Illustration 8. The board is now 15 rows wide and 14 rows high, or $n + 1$ wide by n high.

We now start at the upper right most marker in D, and beginning with the white marker adjacent to it on the right, we follow the edge of D as far as we can. It will turn out that we do not run in circles but, in fact, that we come to a point where we cannot continue. We shall show that this point can only occur at an edge. Since we have covered the left edge with black markers, which are now in D, it is not the left edge. Since D does not reach to the right edge of the board, it is not the right edge. We also see that it cannot be the top edge, since then the white markers would form a chain cutting off the first black marker from the left-hand edge, which is impossible. Thus, the process stops at the lower edge, and White has a winning chain.

Illustration 9

a

b

Of course, the previous paragraph is not a proof. But it is the outline of a proof, if only we could make it more definite. In making it formal and rigorous, we shall introduce the concept of a *pair* of markers. A pair of markers consists of two markers, one black and one white, which are adjacent. We can imagine for each pair of markers an arrow running along the line dividing their two spaces and having the black marker on *its own right*, as in Illustration 9. The marker at the head of the arrow is called the *forward marker indicated by the pair*. The marker at the tail of the arrow is called the *rearward marker indicated by the pair*.

If a pair of markers indicates a forward marker, then we can make a new pair by combining the forward marker with the marker of the opposite color in the given pair. We call this pair the *next pair after* the given pair. In the same way, if the pair indicates a rearward marker, then we can make a new pair by combining the rearward marker with the marker of the opposite color in the pair. This pair will be called the *previous pair before* the given pair. It is an elementary fact, which we give as an exercise, that if a pair P_1 is the next pair after a pair P_0, then P_0 must be the preceding pair before P_1, and conversely. Also, it is clear that if P_1 is the next pair after P_0, then P_1 and P_0 have a marker in common, and the other two markers are of the same color and adjacent.

Now let us return to the Hex board, which is full of black and white markers and to which we have added a row of black markers on the left edge. We choose the rightmost marker from D in the top row and the white marker adjacent to it on the right. These form a pair, and we see at once that this pair is not the next pair after some other pair, since this pair does not indicate a previous pair (Why?). We now make a list, into which we put the pair we mentioned, and then the next pair after it, and then the next pair after that, etc., until we either come to a pair we had before or come to a pair which has no next pair after it.

When we have proved (as we shall) that we can never repeat a pair, then we shall know that we must come to a pair which has no pair after it, since the process must come to an end. The reason it must come to an end is that there can be only finitely many pairs on the board. [In the exercises at the end of the proof, we shall consider how many.] In order to have no next pair, a pair must lie against one side of the rhombus. We observe at once that the black markers in the listing form a chain from the original black marker to the final one, and the white markers have this same property. Thus when we are able to show that the final pair lies at the bottom of the board, it will follow that White has a winning chain.

Let us designate the first pair by P_1, the next by P_2, etc. It is clear that we never have P_1 the same as P_i, for any i, since that would require P_1 to be the next pair after P_{i-1}, and as we have observed, P_1 has no rearward marker and thus no previous pair. We see also that P_2 cannot be the same as P_i if $i > 2$, since that would require P_1 to be the same as P_{i-1}, which we just showed is impossible. In fact, we now show easily, by the method of induction, that P_j cannot be the same as P_i if $j < i$. (The student is to write out the induction.) Since no pair can be repeated and there are only finitely many pairs, let $\overline{P}_m$ be the pair which has no next pair after it. That is, the forward marker indicated by P_m is not a marker at all but a blank space off the board. We see then that both markers of P_m lie on one edge of the board. It cannot be the left edge, since that is now made up entirely of black markers. If it were the right edge, then there would be a chain of black markers joining the original black marker to the final one. Since the original black marker was chosen in D, it is connected to the left-hand side. Thus every black marker in the chain would be in D, and there would be a black chain joining the left and right sides, contrary to the hypothesis.

Illustration 10

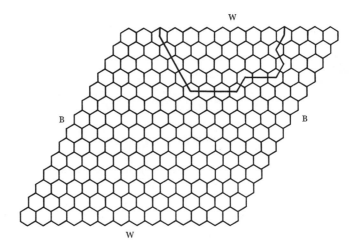

Now let us consider the possibility that the last pair lies against the *top* edge of the board. Since the forward marker is not on the board, the black marker in P_m must lie to the right of the white marker. Since the black marker in P_1 is the rightmost marker in D in the top row and since the black marker in P_m must be in D, the white marker in P_m lies to the left of the black marker in P_1. At this point, I would like to be able to say that it is obvious that the black marker in P_1 is cut off by the chain of white markers from the left side of the board. It is so clear visually that I am tempted not even to mention that it requires proof. One technique for disposing of such a problem is to make the assertion and precede it by the words "It is obvious that. . . . " This technique was forever ruined for me by my professor in college,† a man of impeccable honesty and keen perception, who observed that the operational definition of "obvious" was "difficult or impossible to prove." I shall digress here to remark that in the intervening years, I have found it wise counsel when looking for the flaw in an argument, whether mathematical, political, economic, etc., to look first and hardest in the neighborhood of the word "obvious."

So we proceed to a proof of this "obvious" assertion. We know that there is a chain of white markers joining the white markers in P_1 and P_m. There might be other such chains. Choose the one with the smallest number of markers. In that case, we can show that no marker is repeated in the chain. (Is this "obvious"?) Look at the spaces occupied by the markers of the chain and connect each to the next by a line segment joining their centers. Also connect the centers of the first and last spaces to the upper edge of the board by a vertical line segment (see an example in Illustration 10).

† Prof. Walter Prenowitz, of Brooklyn College.

Now let us take a coping saw and *cut* along this line. The line passes only through spaces occupied by white markers and never along any edge. When we have cut through the line, we shall have divided the board into two pieces. You might ask: "How do you know?" I would have to say that the answer to the question is either very simple or very, very hard. In fact, we shall say that you are entitled to know that if you saw through a piece of wood in this manner, you get two pieces. If not, and you are looking for a principle of geometry, you must look to the Jordan Curve Theorem, which is very deep.

Let us now lift slightly the piece of the board containing the black marker of P_1, as in Illustration 11. We assert that if we have two black markers adjacent, then both will be picked up or neither will, for if one is raised, then the whole space which it occupies is raised, since the cut does not go through any space occupied by a black marker. Thus, if one is raised and the other not, then the cut must include the edge between them, as in Illustration 12. But the cut never runs along any edge. Thus, no pair of adjacent black markers is so affected. We see then that if we take the lifted portion away, say across the room, we do not break any adjacent pair. Thus, there can be no chain joining the left edge of the board, which is fixed, to the black marker of P_1, which is taken away. This is a contradiction, since that black marker was taken to be in D.

Only one possibility remains. The pair P_m lies against the bottom of the board, so that the white markers form a chain linking the top to the bottom. If we now remove the black markers we added, the truth of the previous sentence remains unchanged. This proves our lemma. Q.E.D. [*Quod erat demonstrandum.*]

Illustration 11

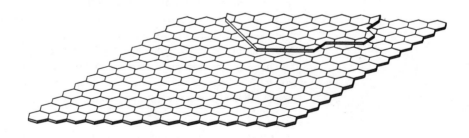

Illustration 12

21 EXERCISES

The problems numbered 1 to 6 have for their locale an augmented Hex board, i.e., a Hex board 14 by 14, or n by n, to which a row of black markers has been added at the left edge, as in Illustration 8. The rest of the board is filled with black and white markers, as noted below.

1. If the original Hex board was 14 by 14, then the number of pairs (of adjacent black and white markers) is no more than 210×196.
2. If the original board was n by n, then the number of pairs is no more than $n^2(n^2 + n)$.
3. If the original board was 14 by 14 and had equally many black and white markers, then the number of pairs is no more than 112×98.
4. If the original board was n by n and similarly filled (with one additional white marker if n is odd), then the number of such pairs is no more than $\frac{1}{2}n^2(\frac{1}{2}n^2 + n)$.
5. If the original board was n by n, then the number of pairs is no more than $3(n^2 + n)$, no matter how the board is filled.
6. If D is defined as in the proof of Lemma 11 and if the pairs in question are to have their black markers in D, then there can be no more than $\frac{5}{2}(n^2 + n)$ pairs.
7. In the proof of Lemma 11, show that the number m satisfies the inequality $m \leq \frac{5}{2}(n^2 + 1)$.
8. If P and P' are two pairs on a Hex board and if P is *not* the preceding pair before P', then P' is not the next pair after P.
9. Show that if a Hex board has on it a winning chain for White, then it cannot also have a winning chain for Black.

Section   4   The Game of Nim

In direct contrast to Hex, we shall now examine the classical game of Nim, in which the winning strategies are known. That is to say, not only is it known who can win, but it is also known how that player can ensure this victory. First, let us examine the game in some detail. There are actually infinitely many games of Nim, one for each of the possible ways of setting out the original matchsticks. Thus (1, 2, 3, 4, 5) is one game of Nim and (1, 3, 5, 7) (Marienbad) is another, while (1, 4, 5) is a third. We shall classify each such game as a *winning game* or a *losing game* according as the player who goes first can or cannot force a win.† We say that the "game" of Nim in which no pile contains any matches is a "losing game." It isn't really a game at all, but it will fit the purposes of our notation to call it a *losing game* or a *lost game*. Mathematicians are full of little tricks like this. Notice that the player who goes first, by removing some of the matchsticks, creates for the other player *another* game of Nim, and he has some choice in what this other game will be. In Marienbad, for example, he takes away something from (1, 3, 5, 7) and thus can leave for his opponent one of the following games:

(3, 5, 7)	(1, 3, 4, 7)	(1, 3, 7)	(1, 3, 5, 3)
(1, 2, 5, 7)	(1, 3, 3, 7)	(1, 3, 5, 6)	(1, 4, 5, 2)
(1, 1, 5, 7)	(1, 3, 2, 7)	(1, 3, 5, 5)	(1, 3, 5, 1)
(1, 5, 7)	(1, 3, 1, 7)	(1, 3, 5, 4)	(1, 3, 5)

If we knew that every one of these games is a winning game, we would then know that (1, 3, 5, 7) is a losing game.

In a similar way, we could *learn* that (1, 3, 5, 6), for example, is a winning game if we knew that (3, 5, 6), (1, 2, 5, 6), and (1, 3, 4, 6) are losing games.

22 *Definition*

We now define a way of making a game out of two games by playing them together. We can then say something about the new game. If, for example, (a, b, c) and (d, e, f, g) are games of Nim, we can look at the game (a, b, c, d, e, f, g), which we shall call their *union* or *join*. It is worthwhile noticing that:

† It should be observed that notations vary. The definition by Hardy and Wright [6, pp. 117–120] of "winning game" and "losing game" is exactly opposite.

Section 4   The Game of Nim

The game of Nim is important in the film "Last Year at Marienbad," and is played several times. The text below introduces the game (in the movie) at the card table. The picture shows another occasion, when the game is played with dominoes.

The beginning of the conversation between X and M takes place offscreen, as the image shifts toward their group.

M's Voice: No, not now . . . I have another game to suggest instead: I know a game I always win . . .

X's Voice: If you can't lose, it's not a game!

M's Voice: I can lose (*Short pause; M appears on the screen at this moment; it is he who is talking.*)

M (continuing): . . . But I always win.

X: Let's try.

M (laying out the cards in front of X): It takes two people to play. The cards are arranged like this. Seven. Five. Three. One. Each player picks up cards in turn, as many cards as he wants, on condition that he takes from only one row each time. The man who picks up the last card has lost. (*A brief pause, then pointing to the cards he had laid out*): Would you like to start?

M, standing quite rigidly, evidently in the same position as in the preceding shot, has laid out the cards in front of X according to the following diagram:

```
0 0 0 0 0 0 0
 0 0 0 0 0
  0 0 0
   0
```

They play quickly and in silence, without music. After a second's reflection X takes one card from the row of seven. M very quickly takes one card from the row of five. X reflects for three seconds and picks up the rest of the row of seven. M, still without pausing to think takes two cards from the row of five. X takes a card from this same row. M takes two cards from the row of three. X thinks for a few seconds, smiles as if he realizes he has lost, takes one of the cards that remain (row of five). M takes one of the others (row of three). A single card remains; since all the cards have been laid out face down, it is not seen what card this one is. Yet the camera has moved close to the table during the game and remains momentarily fixed on this remaining card, as if it had some meaning. The image is interrupted only at the moment of A's laughter (A is invisible), which is repeated just as before, after the last card has remained for X. This laughter lasts to the end of the shot and a little beyond. (Excerpts from the film "Last Year at Marienbad." Courtesy of Terra films and Grove Press, Inc.)

23 Lemma

The join of two losing games is a losing game.

PROOF: Intuitively, each move you make is a move in one of the two games. Whatever move you make is answered *in that game*. Thus you are actually playing the two games simultaneously, and with proper play, your opponent gets the last matchstick in each game. It is clear that he gets the last matchstick in the join.

Formally, we use a course-of-values induction on the number of matchsticks in the join of the two games (the number $a + b + c + d + e + f + g$ in the example above). It is clear that the removal of any number of matchsticks from any pile will leave your opponent with the join of a winning game and a losing game. If he chooses the proper move in the winning game, he makes it into a losing game and presents you with your new situation, the join of two losing games. (Do you see why we call the "game" with no matchsticks a "losing game"?) However, the total number of matchsticks is smaller, and thus by the hypothesis of the induction, the join is a losing game. Since he had a winning game regardless of your first move, you had a losing game to start with. Q.E.D.

24 Corollary

The join of a winning game and a losing game is a winning game.

PROOF: Prove it yourself.

25 EXERCISES

1 Is the join of two winning games ever a losing game? Is it ever a winning game? Show!
2 How much of the analysis above applies to games other than Nim? Which games?

26 Lemma

The join of any game of Nim with itself is a losing game.

PROOF: Prove it yourself.

27 EXERCISES

Which of the following are winning games?

1 (1, 2, 3) 2 (1, 3, 4) 3 (1, 4, 5)
4 (1, 5, 6) 5 (1, 6, 7) 6 (1, 7, 8)

Prove this lemma:

28 Lemma

$(1, n, n + 1)$ is a winning game if and only if n is odd.

We can look at Lemma 28 in this way: The game (n, n) is a losing game (How do you know?). If we combine it with the game $(1, 1)$ to form either $(1, 1, n, n)$ or $(n + 1, n + 1)$, then we get another losing game. If, however, we combine the two to give $(1, n, n + 1)$, then this gives a losing game exactly if n is even.

Prove this lemma:

29 Lemma

$(1, m, m + n)$ is a winning game if n is an integer greater than 1.

30 EXERCISES

Which of the following are winning games?

1 $(2, 3, 4)$	2 $(2, 3, 5)$	3 $(2, 4, 6)$
4 $(2, 9, 10)$	5 $(2, 9, 11)$	6 $(2, 10, 12)$
7 $(2, 12, 14)$	8 $(2, 12, 18)$	9 $(2, 17, 18)$

Prove this lemma:

31 Lemma

$(k, n, n + 1)$ is a winning game if n is even and $k \geq 2$.

At this point, we hope to see that a losing game is a game in which the numbers are in some kind of equilibrium.

32 EXERCISES

1 Is $(1, 2, 3, 4, 5)$ a winning game or a losing game?
2 If (a, b, c) is a losing game and your opponent chooses some matchsticks from the third pile, then you must choose from the first or second pile if you are to make it a losing game again. Prove this assertion.

Suppose (a, b, c) is a losing game. We shall call the player who goes first L and the other W. We might imagine a new game, in which the numbers of matchsticks are $2a$, $2b$, and $2c$ and in which each player must take an *even* number of matchsticks at each turn. It should be clear that this game is the same as (a, b, c), for everything is doubled. Thus L, who still goes first, must lose this game just as surely as he loses (a, b, c) if W plays properly. If we require only L to choose an even number of matchsticks at each turn, without restricting W, that of course will not hurt W. He can still force a win and, in fact, can do so in a way which will always have him taking an even number of matchsticks.

Now let us consider what happens if we *remove* the restriction from L. So long as L takes an *even* number of matchsticks, W has him beaten by the strategy above. However, if he takes an *odd* number, we can think of the resulting position as a winning position in the game of even moves, *with an extra matchstick*. It would be nice for W if he could always get the game back to a position which is losing in the even-move game. However, he cannot always do that. Perhaps he can counter the odd matchstick in another way, by balancing it with *another odd matchstick*. It is always within W's power to ensure that the number of odd matchsticks is even (in this case, 0 or 2) by properly adjusting his moves.

As a matter of fact, the secret of the game of Nim is hidden in the previous two paragraphs, but it is well hidden, and it will take some work to extract it. Those paragraphs are not meant to *prove* anything but are only meant to indicate why we should imagine that a theorem like the next one might be true. Actually, the discussion does contain the seed of the proof, and this theorem is the kernel of the theory of the game of Nim.

33 Theorem

If (a, b, c) is a losing game, then so are $(2a, 2b, 2c)$ and $(2a + 1, 2b + 1, 2c)$.

PROOF: Before starting on this proof, it is important that you have done Problem 2 of Exercises 32, just previous. Also, please note that a game is a winning or losing game regardless of the order of the piles. This elementary note will be of importance later.

The proof will go by induction, and the parameter of the induction will be the sum $a + b + c$. To ground our induction, we let $a + b + c = 0$, that is, $a = b = c = 0$. This is a losing game, and so are $(2a, 2b, 2c) = (0, 0, 0)$ and $(2a + 1, 2b + 1, 2c) = (1, 1, 0)$. Here we see a use of the convention that $(0, 0, 0)$ is a losing game. To do the "induction step" of the course-of-values induction, we shall assume that the theorem holds for all cases where $a + b + c < m$, a given number. We wish to show that it holds also when $a + b + c = m$.

The situation here is illustrative of a common situation in mathematics. We really have *two* theorems:

A: If (a, b, c) is a losing game, then $(2a, 2b, 2c)$ is a losing game.
B: If (a, b, c) is a losing game, then $(2a + 1, 2b + 1, 2c)$ is a losing game.

In proving either of these theorems, we would wish to know the other as well for all smaller values of the parameter of induction. Thus, by combining the results into a single theorem, we make it *easier* to prove. Now that we know both results for all lesser values of the parameter, we prove each of our two results. The method of proof is the same in the two theorems. We assume that a player plays the new game and makes some move. We shall demonstrate an answering move, and we shall use the hypothesis of the induction to show that the new game after the two moves is a losing game. It will follow, then, that every initial move makes a winning game, and that will prove the theorem. Result A is easier to prove and will demonstrate the method.

PROOF OF PART A: This part of the theorem [i.e., that $(2a, 2b, 2c)$ is a losing game] is itself divided into two cases:

Case A1: The player who goes first chooses an even number of matchsticks.
Case A2: The player who goes first chooses an odd number of matchsticks.

Proof of Case A1 Let us assume the first player chooses $2k$ matchsticks from the first pile. (It does not matter which pile he chooses from, since he can switch them around.) Now let us consider for a moment the game (a, b, c). If we remove k matchsticks from the first pile here, we get $(a - k, b, c)$, *which must be a winning game.* Thus there is a countermove which makes it into a losing game. That countermove involves taking away j matchsticks from some *other* pile, say the second, to make $(a - k, b - j, c)$, which will then be a *losing* game. Since

$$(a - k) + (b - j) + c = a + b + c - (k + j) < a + b + c = m,$$

the hypothesis of the induction tells us that $(2(a - k), 2(b - j), 2c)$ is a losing game. Therefore, if we take $2j$ from the second pile, we make a losing game.

Proof of Case A2 Suppose the first player takes an odd number of matchsticks, again assumedly from the first pile. We designate that odd number as $2k - 1$ for some positive integer k. Then $(a - k, b, c)$ is again a winning game, and again we can choose a j so that $(a - k, b - j, c)$ is a losing game. In that case, the hypothesis of the induction assures us that $(2(a - k) + 1, 2(b - j) + 1, 2c)$ is a losing game. Thus we can make $(2a - (2k - 1), 2b, 2c)$ into a losing game by taking $2j - 1$ matches from the second pile.

Let us review the proof of A, since the proof of B is more complicated but basically the same. We have shown that no matter what number of matchsticks we take from $(2a, 2b, 2c)$, we get a winning game. It doesn't matter whether the number of matchsticks is even (Case A1) or odd (Case A2). We prove that the game is a *winning* game by showing that some move makes it into a "losing game." However, Part A will not be actually proved until Part B is proved, since it relies on Part B with smaller parameters.

PROOF OF PART B: Part B also divides into two cases, one of which will be divided into subcases:

Case B1: The player who goes first (in the game $(2a + 1, 2b + 1, 2c)$) takes an even number of matchsticks.
Case B2: The player who goes first takes an odd number of matchsticks.

Proof of Case B1 The first player removes $2k$ matchsticks from some pile (we don't say which). Let us consider what happens in (a, b, c) if we remove k matchsticks from that pile. We get a winning game. Thus the removal of j matchsticks from some other pile will make it a losing game again. Let us call this new losing game (a_1, b_1, c_1). Since

$$a_1 + b_1 + c_1 = a + b + c - (k + j) < a + b + c = m,$$

we know that $(2a_1 + 1, 2b_1 + 1, 2c_1)$ is a losing game. Suppose the answer to the removal of the $2k$ matchsticks, then, is a removal of $2j$ matchsticks from the correct pile (the one corresponding to the pile where j matchsticks are chosen). This answer will make a losing game $(2a_1 + 1, 2b_1 + 1, 2c_1)$. Thus, the removal of an even number of matchsticks makes

$$(2a + 1, 2b + 1, 2c)$$

into a winning game.

Proof of Case B2 This case divides into two subcases:

Case B2a: The odd number of matchsticks is taken from the even pile.
Case B2b: The odd number of matchsticks is taken from an odd pile.

Proof of Case B2a Let the odd number be denoted as $2k - 1$, so that the new game is $(2a + 1, 2b + 1, 2(c - k) + 1)$. We wish to show it is a winning game. We know that $(a, b, c - k)$ is a winning game, since (a, b, c) is losing. Thus we can make $(a, b, c - k)$ into a losing game by removing some number j of matchsticks from some pile to make (a_1, b_1, c_1). If we were to remove $2j$ matchsticks from the same pile in

$$(2a + 1, 2b + 1, 2(c - k) + 1),$$

we would get $(2a_1 + 1, 2b_1 + 1, 2c_1 + 1)$, which would be a losing game, under the hypothesis of the induction, if only one of the piles were missing a matchstick. But that is easily arranged: we remove $2j + 1$ matchsticks instead.

Proof of Case B2b What would you make of the fact that we again divide into two subcases?

Case B2bi: The odd number removed from the odd pile is 1.
Case B2bii: The odd number removed from the odd pile is not 1.

Proof of Case B2bi If 1 is removed from an odd pile, the removal of 1 from the other odd pile will give $(2a, 2b, 2c)$, which is a losing game, as shown in Part A.

Proof of Case B2bii Here we write the number removed as $2k + 1$, and we can suppose it is taken from the first pile (Why?). We then have the game $(2(a - k), 2b + 1, 2c)$. We again look at the game $(a - k, b, c)$, which is a losing game, and there is an answer, namely, taking j matchsticks from either the second pile or the third. This gives us yet two more subcases:

Case B2biiα: The reply is in one of the first two piles.
Case B2biiβ: The reply is in the third pile.

Proof of Case B2biiα If the removal of j matchsticks from the second pile leaves $(a - k, b - j, c)$ and this is a losing game, then the removal of $2j + 1$ matchsticks from the second pile in $(2(a - k), 2b + 1, 2c)$ will leave $(2(a - k), 2(b - j), 2c)$, which will also be a losing game, by the hypothesis of the induction.

Proof of Case B2biiβ If the removal of j matchsticks from the third pile leaves $(a - k, b, c - j)$, which is a losing game, then the removal of $2j - 1$ matchsticks from the third pile in $(2(a - k), 2b + 1, 2c)$ will yield

$$(2(a - k), 2b + 1, 2(c - j) + 1),$$

which will be a losing game, by the hypothesis of the induction.

Let us recapitulate the proof: In Part A, we show that the removal of any number, even or odd, from the game $(2a, 2b, 2c)$ can be answered to yield a losing game. Thus $(2a, 2b, 2c)$ is itself a losing game. Similarly in Part B. The situations can be outlined thus:

A The game $(2a, 2b, 2c)$
   Case 1: An even first move
   Case 2: An odd first move
B The game $(2a + 1, 2b + 1, 2c)$
   Case 1: An even first move
   Case 2: An odd first move
        $2a$: Taken from the even pile
        $2b$: Taken from an odd pile
           $2b$i: The odd number is 1.
           $2b$ii: The odd number is not 1.
               $2b$iiα: The answer is in one of the first two piles.
               $2b$iiβ: The answer is in the third pile.

This enormous regression into cases and subcases and sub-sub-subcases proves the induction step, and now, by induction, the theorem is proved.
Q.E.D.

34 Corollary

If (a, b, c) is a winning game, then so are $(2a, 2b, 2c)$ and $(2a + 1, 2b + 1, 2c)$.

PROOF: Prove this corollary and the following theorem yourself.

35 Theorem

The following games are either all losing games or all winning games, where a, b, and c represent the same numbers throughout.

$$(a, b, c) \quad (2a, 2b, 2c) \quad (2a + 1, 2b + 1, 2c)$$
$$(2c + 1, 2b, 2c + 1) \quad (2a, 2b + 1, 2c + 1)$$

36 Theorem

$(2a + 1, 2b + 1, 2c + 1)$ is a winning game.

PROOF: If (a, b, c) is a losing game, then so is $(2a + 1, 2b + 1, 2c)$, which we can obtain by taking one matchstick from the last pile. If (a, b, c) is a winning game, then there is a move which makes it a losing game, say taking k from a. Then $(a - k, b, c)$ is a losing game. Thus

$$(2(a - k), 2b + 1, 2c + 1)$$

Section 4   The Game of Nim

is also a losing game, and we can obtain this game from
$$(2a + 1, 2b + 1, 2c + 1)$$
by taking $2k + 1$ matchsticks from the first pile. Thus, in either case, $(2a + 1, 2b + 1, 2c + 1)$ can be made into a losing game in one move. Therefore, it is a winning game. Q.E.D.

37   Corollary

$(3, 5, 7)$ is a winning game.

38   Theorem

For any values of a, b, and c, $(2a + 1, 2b, 2c)$ is a winning game.

PROOF: If (a, b, c) is a losing game, then taking one matchstick from the first pile makes $(2a, 2b, 2c)$, which is a losing game.

If (a, b, c) is a winning game, then there is a number $k > 0$ such that $(a - k, b, c)$ or $(a, b - k, c)$ or $(a, b, c - k)$ is a losing game. If $(a - k, b, c)$ is a losing game, then taking $2k + 1$ matchsticks from the first pile of $(2a + 1, 2b, 2c)$ will give $(2(a - k), 2b, 2c)$, which is also a losing game. If $(a, b - k, c)$ is a losing game, then taking $2k - 1$ matchsticks from the second pile of $(2a + 1, 2b, 2c)$ will give $(2a + 1, 2(b - k) + 1, 2c)$, which is also a losing game. If $(a, b, c - k)$ is a losing game, then so is
$$(2a + 1, 2b, 2(c - k) + 1)$$
which we get from $(2a + 1, 2b, 2c)$ by taking $2k - 1$ matchsticks from the third pile.

We have shown that the removal of an appropriate number of matchsticks from $(2a + 1, 2b, 2c)$ will make it a losing game, regardless of whether (a, b, c) is a winning or losing game. Thus $(2a + 1, 2b, 2c)$ is a winning game. Q.E.D.

In the next theorem, we encounter the distinction between "if" and "only if." A number can be a multiple of 4 *only if* it is even. However, not every even number is a multiple of 4, so that we cannot say that "a number is a multiple of 4 *if* it is even." We say that being even is *necessary to* being a multiple of 4 but is *not sufficient* for that. The next theorem is an even better example.

39   Theorem

A game (a, b, c) is a losing game ONLY IF the number of odd piles is even.

PROOF: This is a combination of previous theorems. Can you find which theorems?

40 Definition

If (a, b, c) is a game G, then we can look at a game G^* which is made from G by removing one matchstick from each odd pile. The game G^* is called the *trim* of G.

Prove these three theorems:

41 Theorem

$(2a, 2b, 2c)$ is a losing game IF AND ONLY IF (a, b, c) is.

42 Theorem

If $G = (a, b, c)$ is a game with an even number of odd piles, then G is a losing game IF AND ONLY IF G^* is.

43 Theorem

Let (a, b, c) be a losing game. Then $(2a, 2b, 2c)$ and $(2a+1, 2b+1, 2c)$ are losing games, and every losing game with three piles can be so represented, except for the order of the piles.

44 Definition

We now introduce a concept of *reducing* a game. We reduce a game by trimming it (removing a matchstick from each odd pile) and halving each remaining pile. The reduced game is called $\hat{G}$.

45 Theorem

A game $G = (a, b, c)$ is a losing game IF AND ONLY IF both:

1 G has an even number of odd piles.
2 $\hat{G}$ is a losing game.

PROOF: This is already proven. Can you see where and how?

46 PROBLEM

Show that $(182, 147, 37)$ is a losing game.

SOLUTION: $(182, 147, 37)$ has two odd piles and is therefore a losing game if and only if $(91, 73, 18)$ is. This also has two odd piles, and it is a losing game if and only if $(45, 36, 9)$ is. This is a losing game if and only if $(22, 18, 4)$ is. $(22, 18, 4)$ has no odd piles and is a losing game if and only if $(11, 9, 2)$ is. $(11, 9, 2)$ has two odd piles and is a losing game if and only if (5, 4, 1) is, (5, 4, 1) is a losing game if and only if (2, 2, 0) is, and (2, 2, 0) *is* a losing game.

We might reduce the verbiage by writing a column of games, each the reduction of the preceding game, and noting next to each the number of odd piles. Thus:

(182, 147, 37)	2
(92, 73, 18)	2
(45, 36, 9)	2
(22, 18, 4)	0
(11, 9, 2)	2
(5, 4, 1)	2
(2, 2, 0)	0
(1, 1, 0)	2
(0, 0, 0)	0

Each game will eventually come down to (0, 0, 0). If, in this process, all the entries in the right-hand column are even, then the game is a losing game. Otherwise, one of the intermediate games has an odd number of odd-numbered piles, and then every game from that point upwards is a winning game, including the original game.

47   EXERCISES

Which of the following are losing games?

1 (1, 40, 57)   2 (7, 40, 57)
3 (12, 15, 19)   4 (2, 17, 19)

Now we know which of the possible games of three-pile Nim are winning games and which are losing games. Thus, if you are confronted with a game of Nim, say the game (a, b, c), and if (a, b, c) is a winning game, then one of the moves open to you will make the game into a losing game. How can you find the right move? One way would be to look at all the possibilities and decide whether the resulting game is a winning game or a losing game. There are exactly $a + b + c$ possible moves (Can you show this?). Depending on your point of view, there are *only* $a + b + c$ moves or *all of* $a + b + c$ moves. That is to say, the number of moves is large but within reasonable bounds (unlike the number of games of Tic-Tac-Toe, for example). Perhaps we can reduce this number by some careful side considerations.

For example, if we know that the winning game in which you are to move is obtained from a losing game by making one move, *then we know that any proper answering move will be in a different pile from that move*. How do we know this?

Furthermore, consider the possibility that the number of odd piles is even. Then, since the number of odd piles in the new (losing) game must be even, we cannot make any even pile odd or any odd pile even. That is, we cannot take an odd number of matchsticks if the resulting game is to be a losing game.

On the other hand, if the number of odd piles is odd, then we must make it even if the resulting game is to be a losing game. Note that this is a *necessary* condition, not a *sufficient* one. Thus the move must be an *odd* one. In fact:

48 Theorem

If (a, b, c) is a winning game, then it can only be made into a losing game by an even move if $a + b + c$ is even or by an odd move if $a + b + c$ is odd.

PROOF: The previous discussion is a proof. There is also a shorter proof. Can you find it?

Theorems like Theorem 48 can decrease the number of possible moves to be investigated. Actually, we can find a better way to figure out the winning moves. We shall rewrite our original column of game reductions, noting not only how many piles are odd at each stage but also which. An odd pile is marked 1, an even pile 0.

(182, 147, 37)	0 1 1
(91, 73, 18)	1 1 0
(45, 36, 9)	1 0 1
(22, 18, 4)	0 0 0
(11, 9, 2)	1 1 0
(5, 4, 1)	1 0 1
(2, 2, 0)	0 0 0
(1, 1, 0)	1 1 0
(0, 0, 0)	

In this array, the 1s stand for something very specific. Those in the top row indicate that the pile is odd. The first 1 in the second row shows that *half* of 182 is odd. That is, 182 is of the form $4k + 2$,

Section 4   The Game of Nim

and the 1 under that shows that k is itself odd. If $k = 2m + 1$, then 182 is of the form $8m + 4 + 2$. The next 0 down shows that m is even—say, $m = 2n$—so that in actuality, $182 = 16n + 4 + 2$. Continuing in this way, we see that $182 = 128 + 32 + 16 + 4 + 2$. In fact, the significance of the array of 0s and 1s is made clear in the next chart:

(182, 147, 37)	0	1	1	0	1	1
(91, 73, 18)	1	1	0	2	2	0
(45, 36, 9)	1	0	1	4	0	4
(22, 18, 4)	0	0	0	0	0	0
(11, 9, 2)	1	1	0	16	16	0
(5, 4, 1)	1	0	1	32	0	32
(2, 2, 0)	0	0	0	0	0	0
(1, 1, 0)	1	1	0	128	128	0
(0, 0, 0)				182	147	37

If we assign each 1 in the top row a value of 1, each 1 in the second row a value of 2, and each 1 in the sixth row a value of $32 = 2^5$ etc., then the sum of the numbers in each column adds up to the original size of that pile. A game is a losing game if and only if this sort of scheme gives us each power of 2 an even number of times. Let us now examine a winning game and see what this tells us:

(15, 37, 38)	1	1	0	1	1	0
(7, 18, 19)	1	0	1	2	0	2
(3, 9, 9)	1	1	1	4	4	4
(1, 4, 4)	1	0	0	8	0	0
(0, 2, 2)	0	0	0	0	0	0
(0, 1, 1)	0	1	1	0	32	32
(0, 0, 0)				15	37	38

In this case, we can restore the "balance" by eliminating one of the 4s and the 8:

1	1	0	1	1	0
2	0	2	2	0	2
4	4	4	0	4	4
8	0	0	0	0	0
0	0	0	0	0	0
0	32	32	0	32	32
15	37	38	3	37	38

The removal of 12 from the first pile makes (15, 37, 38) into the losing game (3, 37, 38). Sometimes it is a little harder, as in (11, 37, 34):

(11, 37, 34)	1	1	0	1	1	0
(5, 18, 17)	1	0	1	2	0	2
(2, 9, 8)	0	1	0	0	4	0
(1, 4, 4)	1	0	0	8	0	0
(0, 2, 2)	0	0	0	0	0	0
(0, 1, 1)	0	1	1	0	32	32
(0, 0, 0)				11	37	34

We have an extra 8 and an extra 4. Since they are in different piles, we *cannot* get rid of *both*. But we *can* change the 8 to a 4.

1	1	0	1	1	0
2	0	2	2	0	2
0	4	0	4	4	0
8	0	0	0	0	0
0	0	0	0	0	0
0	32	32	0	32	32
11	37	34	7	37	34

Removing four matchsticks from the first pile makes (11, 37, 34) into the losing game (7, 37, 34).

49 EXERCISES

Make these into losing games:

1 (11, 42, 45)   2 (9, 42, 45)
3 (9, 34, 37)   4 (25, 34, 37)

In general, we notice that for any number, if we take out of it any power of 2 and then put into it any sum of distinct, smaller powers of 2, we diminish the number. This is because, for every power 2^k of 2, we have the inequality

$$2^k > 2^{k-1} + 2^{k-2} + \cdots + 4 + 2 + 1.$$

(Prove this inequality!!!)

To make a winning game of Nim into a losing game, decide which column contains the largest unbalanced power of 2. A power of 2 is called unbalanced if there are an odd number of them. Re-

Section 4   The Game of Nim

move from that column *all* the unbalanced powers of 2 which occur there:

(25, 34, 37)	1	0	1
(12, 17, 18)	0	2	0
(6, 8, 9)	0	0	4
(3, 4, 4)	8	0	0
(1, 2, 2)	16	0	0
(0, 1, 1)	0	32	32
(0, 0, 0)	25	34	37

Now add to the same column those powers of 2 which are still unbalanced:

1	0	1	1	0	1
0	2	0	2	2	0
0	0	4	4	0	4
0	0	0	0	0	0
0	0	0	0	0	0
0	32	32	0	32	32
1	34	37	7	34	37

Changing 25 to 7 makes the game into a losing game. We note that this change can be accomplished because $16 + 8 > 4 + 2$. In fact, $16 + 8 > 16 > 8 + 4 + 2 + 1 > 4 + 2$.

Sometimes, we have a choice about our course of action if the largest unbalanced power of 2 occurs more than once, e.g. (25, 50, 53):

1	0	1
0	2	0
0	0	4
8	0	0
16	16	16
0	32	32
25	50	53

1	0	1	1	0	1	1	0	1
2	2	0	0	0	0	0	2	2
4	0	4	0	4	4	0	0	0
0	0	0	8	8	0	8	0	8
0	16	16	16	0	16	16	16	0
0	32	32	0	32	32	0	32	32
7	50	53	25	44	53	25	50	43

We can make (25, 50, 53) into (7, 50, 53), (25, 44, 53), or (25, 50, 43), all of which are losing games.

50 EXERCISES

Solve these games:

1 (157, 204, 45)   2 (192, 183, 15)
3 (104, 108, 110)   4 (89, 209, 136)
5 (51, 52, 53, 54)   6 (32, 33, 34, 35)

We now consider the possibility that the game of Nim has more than three piles, as in Problems 5 and 6 of Exercises 50 or in (1, 2, 3, 4, 5), a well-known game. A natural conjecture would be that the same theory of balancing will do for these games. It is, of course, not clear whether the technique we used is tied to the fact that there are just three piles. Perhaps with four piles we should use powers of $3 (= 4 - 1)$ or powers of some other integer. Powers of 2 have two special properties shared by no other number. These properties are expressed in the following:

51 Lemma

Every positive whole number can be expressed as the sum of distinct nonnegative powers of 2, and in only one way.

PROOF: The first half of the proof goes by induction. We note that if $k = 2^{a_1} + 2^{a_2} + \cdots + 2^{a_{n-1}} + 2^{a_n}$ is the sum of distinct powers of 2, then

$$2k = 2^{a_1+1} + 2^{a_2+1} + \cdots + 2^{a_{n-1}+1} + 2^{a_n+1}$$

and $\quad 2k + 1 = 2^{a_1+1} + 2^{a_2+1} + \cdots + 2^{a_{n-1}+1} + 2^{a_n+1} + 1$

are both the sums of distinct powers of 2 since 2^{a_i+1} and 2^{a_j+1} are different just if 2^{a_i} and 2^{a_j} are and $2^{a_i+1} \neq 1$ for all nonnegative a_i. Now, we are ready for the proof. Certainly 1, which is a power of 2, can be written trivially as a sum of distinct nonnegative powers of 2. Suppose that the result holds for all positive integers less than m. If m is even, then $m = 2k$, where $k < m$, and thus k is the sum of distinct nonnegative powers of 2, by the hypothesis of the induction. Our prior comment now assures us that $m = 2k$ also has this property. If m is odd, then $m = 2k + 1$, where $k < 2k < m$. Again, k is a sum of the required kind, by the hypothesis of the induction, and again, $m = 2k + 1$ is of the desired form by the previous comment. Thus m is, in either case, of the desired form, and we

Section 4   The Game of Nim

have proved by induction that every positive integer can be represented as a sum of distinct nonnegative powers of 2.

Suppose that

$$m = 2^{a_1} + 2^{a_2} + \cdots + 2^{a_n} = 2^{b_1} + 2^{b_2} + \cdots + 2^{b_m}.$$

We can subtract from both sides any power of 2 which appears on both sides, which would give us an equation $2^{c_1} + \cdots + 2^{c_s} = 2^{d_1} + \cdots + 2^{d_t}$, where 2^{c_1} is the highest power of 2 showing and no power appears twice. Then

$$2^{c_1} + \cdots + 2^{c_s} \geq 2^{c_1} > 2^{c_1-1} + 2^{c_1-2} + \cdots + 2^2 + 2^1 + 2^0$$
$$\geq 2^{d_1} + \cdots + 2^{d_t},$$

since $d_1, d_2, \ldots, d_t$ are distinct and less than c_1. This gives us a contradiction. Thus m does *not* have two different representations.   Q.E.D.

52   EXERCISE

Prove, by induction or otherwise, that

$$1 + 2 + \cdots + 2^{n-1} + 2^n = 2^{n+1} - 1.$$

Let us now examine a game of Nim and reduce all its piles to their representation as the sums of distinct powers of 2. This is the same as writing them out "in base 2." Do you see that? We shall therefore refer to these as their *binary representations*. If every power of 2 appears an even number of times, then we call the game *balanced*. We now wish to show these two lemmas:

53   Lemma

If a game G of Nim is balanced, then every possible move leaves it unbalanced.

54   Lemma

If a game G of Nim is unbalanced, then some move will leave it balanced.

If we can prove these assertions, then it is clear that the player who leaves his opponent a balanced game can force a win, since his opponent must always give him an unbalanced game and the game $G_0 = (0, 0, \ldots, 0)$ is not an unbalanced game. So let us prove the lemmas.

PROOF OF LEMMA 53: Suppose we have a game $G = (a, b, \ldots, c)$, which is balanced, and we make a move—say we remove k from the first pile. Then $a > a - k$, and if we write a and $a - k$ in their binary representations there must be a power of 2 in the representation of a which does not occur in the representation of $a - k$. Then that power of 2 occurs an even number of times in $(a, b, \ldots, c)$ and once less in $(a - k, b, \ldots, c)$. Thus, it appears an odd number of times in $(a - k, b, \ldots, c)$, so this is not a balanced game. Q.E.D.

PROOF OF LEMMA 54: If $(a, b, \ldots, c)$ is an unbalanced game, let 2^{m_0} be the largest power of 2 which appears an odd number of times. We can assume, for reasons of symmetry, that 2^{m_0} occurs in the binary representation of a. Let $2^{m_0}, 2^{m_1}, \ldots, 2^{m_s}$ be all the powers of 2 which occur an odd number of times in the game and occur in a, while $2^{n_1}, 2^{n_2}, \ldots, 2^{n_t}$ are all the *other* powers of 2 which occur an odd number of times in the game. Then

$$2^{m_0} + 2^{m_1} + \cdots + 2^{m_s} \geq 2^{m_0} > 2^{m_0-1} + 2^{m_0-2} + \cdots + 2^1 + 1$$
$$\geq 2^{n_1} + 2^{n_2} + \cdots + 2^{n_t}.$$

Let $k = 2^{m_0} + 2^{m_1} + \cdots + 2^{m_s} - (2^{n_1} + \cdots + 2^{n_t}) > 0$. We now claim that $(a - k, b, \ldots, c)$ is balanced. Certainly, the binary representation of $a - k$ is made by removing $2^{m_0}, 2^{m_1}, \ldots, 2^{m_s}$ from the binary representation for a and putting in $2^{n_1}, 2^{n_2}, \ldots, 2^{n_t}$. Thus every power of 2 which appeared an even number of times in the game $(a, b, \ldots, c)$ still appears the same number of times, while every power which appeared an odd number of times has had the number of its appearances either decreased by 1 (if it is $2^{m_0}, 2^{m_1}, \ldots$ or 2^{m_s}) or increased by 1 (if it is $2^{n_1}, 2^{n_2}, \ldots$ or 2^{n_t}). Therefore, every power of 2 appears an even number of times in $(a - k, b, \ldots, c)$. Q.E.D.

With these lemmas, we have therefore *exhibited* a winning strategy for Nim, namely, that you should give your opponent a balanced game to contend with at his next move. If you can do it once, then you can certainly continue to do so.

In the game of Nim played in the movie "Last Year at Marienbad," the person who takes the last matchstick *loses*. The following exercises will show a winning strategy under these rules.

55 EXERCISES

1 If G is a game of Nim and every pile has at most one matchstick, then G is a winning game under the new rules if and only if it is a losing game under the old rules.

2   If G is a game of Nim and exactly one pile has more than one matchstick, then G is a winning game under either set of rules.
3   If G is a game of Nim and G has at least one pile with more than one matchstick, then G is a winning game under one set of rules if and only if it is a winning game under the other set.
*4 State a winning strategy under the new rules.
*5 Construct a winning strategy for this game of Nim: You have a certain number of piles (as before), and at each step you draw from one *or two* piles.

Section 5   Games of Chance

The simplest game of chance is the tossing of a coin. I toss a coin, and you call for heads or tails. There is no way of knowing in advance which way it will come up. If it comes up as you "predicted," then you win a certain stake; if not, then I win. We all feel intuitively that the possible gain should equal the possible loss in this circumstance, for reasons of symmetry.

Suppose we now throw a die. You call the face which will be uppermost when the die comes to rest on the floor. If you "predict" correctly, you win; otherwise, I do. It would seem that you should get more if you win than I get if I win. In fact, in a situation like this, I can expect to win about five times as often as you, and so if I could persuade you to take odds of less than 5 to 1 (i.e., your winnings, when you win, would be less than five times as great as your losses when you lost), then I would be ahead. It seems so intuitively, and in fact, it is supported by both theory and experience.

The basis of this belief is that we have six possible outcomes, and there is no cause for believing that one is more or less likely than any of the others. Following a line of reasoning based on symmetry, we imagine six players, of whom you are one. You choose your face, and the other five divide up the remaining faces. Each has the same chance of winning, it seems, and the winner collects a fixed stake from each of the losers. Basically, your situation in this game is not changed if all the other players are actually one person. Thus, your gain, when you win, should be five times as great as your loss, when you lose.

Similarly when you play roulette. The wheel has 36 numbers and also one or two places marked in green and numbered 0 or 00 (one in Monte Carlo, two in Reno). If the zeros were absent, then a player at a sound roulette wheel could expect to win about 1 time in 36, and by the reasoning of symmetry, he should have odds of 35 to 1, with a rationale similar to that in the previous paragraph. However, he wins only 1 time in 37 or 38, depending on where he goes for his action, and so he comes up a loser at the prevailing odds of 35 to 1.

Not all chance situations, however, lend themselves to such considerations of symmetry. The corridor outside my office is tiled in squares which are 9 inches on each side. Suppose I throw my key, which is an irregularly shaped object, down the hall and go to see whether it lands on one of the lines between adjacent tiles. Suppose that I receive a "payoff" of size x if the key lands on a line and receive a "payoff" of size y if it does not. We allow the possibility that x or y or both are negative, which accounts for my using the word "payoff" in quotation marks. What can I expect of such a situation?

If we knew, for instance, that the key would fall on a line about one time in six, then we could look to the analogy of the die and conclude that out of $6n$ tries (where n is some large integer), the key would fall on a line about n times, leaving $5n$ times when it did not. Thus, I would *expect* about $n \cdot x + 5n \cdot y$ as my total "payoff" for $6n$ tries. I divide the payoff by the number of tries and assert that I have been getting an *expected*

$$\frac{nx + 5ny}{6n} = \frac{1}{6}x + \frac{5}{6}y$$

for each of the $6n$ tries. Since the key falls on a line about one time in six, we denote $\frac{1}{6}$ as the *probability* of that occurrence. If the probability of the key falling on the line were, instead, $\frac{12}{37}$, then the *expectation* of this game would be

$$\tfrac{12}{37} \cdot x + \tfrac{25}{37} \cdot y.$$

In general, if p represents the probability of the key falling on the line, then $1 - p$ represents the probability that it does not, and the *expectation* of the game is $px + (1 - p)y$. The game is denoted as being "favorable," "unfavorable," or "fair" according as $px + (1 - p)y$ is greater than, less than, or equal to 0.

It is worth noting that not all probabilities are fractions. In general, they are real numbers lying between 0 and 1. In a certain sense, they are almost certain to be among those numbers (like $\pi - 3$, for example) which cannot be represented by any fraction. If, instead of using the 9-inch tiles outside my office and my metal key, I had used a mathematical model consisting of a 9-inch grating of lines in the plane and if I had substituted for the key a 9-inch line segment, then the probability would have been $3/\pi$. Ideal probabilities are easier to measure than real ones.

If we imagine a chance event, i.e., one which may or may not occur, we can imagine a probability attached to it as a measure of its likelihood. Thus we might attach a probability $\frac{1}{2}$ to a coin coming up heads, unless the coin is loaded, in which case the probability could be different. We shall not go into the question of how we determine these probabilities or what the philosophical meanings of the numbers are, except to say that an event with a higher probability is more likely to occur, in some sense.

We shall actually be more interested in the expectations. A common view of these expectations is as follows: When I play a game with expectation z, I consider that I am being given an amount z (positive, negative, or zero), and then I am playing a fair game. The game, being fair, does not enrich or impoverish me (in the long run, in a certain sense), and my relation to the game is basically determined by the size of z. This is a rather simple view and does not take into account many of the actual complexities of actual people actually gambling.† However, in the absence of a technical course in probability, we shall take this as an assumption.

56 Definition

A *game of chance* consists of a chance event to which we attach certain *payoffs*. We call the possible outcomes of the chance events $\omega_1, \omega_2, \ldots, \omega_n$, where n is the number of outcomes, and we assume that

† See [2].

the probabilities of these outcomes are $p_1, p_2, \ldots, p_n$, respectively. We require that $0 \leq p_i \leq 1$, for all $i = 1, 2, \ldots, n$, and that $p_1 + p_2 + \cdots + p_n = 1$. We take for the payoffs certain real numbers $x_1, x_2, \ldots, x_n$, which are positive, negative, or zero. We take the sum $x_1 p_1 + x_2 p_2 + \cdots + x_n p_n$, the weighted average of the payoffs, as the *expectation* of the game. The expectation is the measure we shall use of the *value* of the game to the player.

Until now, the games of chance have lacked one of the important aspects of games, namely, the interaction of the players. A man plays "against" the gods or the Fates, and although we assume that he has an agreement with someone to give him the payoff and make that player's losses into his own gains, the second party does not in any real sense take a part in the game. In fact, neither does the first.

Now let us get the first player into the game. Mr. House, the second player, runs a casino where Mr. Mark, the first player, must play. Mr. House's casino offers two games, and since Mr. House is in business for profit, both games have negative expectations for Mr. Mark. His only choice is which game to play. One game has for its payoffs x and y, with probabilities p and q (where $p \geq 0, q \geq 0$, and $p + q = 1$). The other has payoffs w and z, with probabilities r and s. We assume that Mr. Mark will play the first game or the second according as $px + qy$ is greater than or less than $rw + sz$. Thus, we know Mr. Mark's attitude. Now we give Mr. House a part to play in this game. He is allowed to choose p, q, r, and s, but we require that $p = r$ and $q = s$. $x, y, w,$ and z are given; he cannot change them. How should Mr. House play this game?

Case 1: $x \geq w, y \geq z, x \geq y$ In this case, no matter what Mr. House chooses for p and q, it will always be the case that $px + qy \geq pw + qz$. Thus Mr. Mark will *always* play the first game since it yields the greater return. Mr. House wants to minimize this return, so he chooses $p = 0, q = 1$. The value of this game is now y. If Mr. House chooses $p > 0$, then the value of the game to Mr. Mark increases, unless $x = y$.

Case 2: $x \geq w, y \geq z, x \leq y$ The same analysis will show that Mr. House sets $p = 1, q = 0$, and the value of the game is x.

Note that Mr. Mark always takes the better of the choices available to him and that Mr. House tries to arrange to keep this best as bad as possible. This is again the principle of the *worst best*, or *minimum maximum*, which we shall see frequently in this section.

Case 3: $w \geq x$, $z \geq y$, $w \geq z$ As in Case 1, $p = 0$, $q = 1$, and the value of the game is z.

Case 4: $w \geq x$, $z \geq y$, $w \leq z$ As in Case 3, $p = 1$, $q = 0$, and the value of the game is w.

Case 5: $x > w$, $y < z$ This is the interesting case, together with Case 6, which is symmetric with it. If $x \leq y$, then clearly $p = 1$, $q = 0$, and $v = x$.† If $w \geq z$, then $p = 0$, $q = 1$, and $v = z$. If $x > w$, $y \geq z$, $x > y$, and $w < z$, then we have some calculating to do. For some choices of p and q,

$$xp + yq > wp + zq$$

for instance, if $p = 1$, $q = 0$. Now as q grows, $p = 1 - q$ decreases. Using our inequalities, we see that $xp + yq$ decreases while $wp + zq$ increases. By the time $q = 1$, $xp + yq = y < z = wp + zq$. Let p_0 be the value satisfying

$$x(1 - p_0) + yp_0 = w(1 - p_0) + zp_0.$$

Then, if $0 \leq p < p_0$, we have

$$\begin{aligned} x(1 - p) + yp &> x(1 - p_0) + yp_0 \\ &= w(1 - p_0) + zp_0 \\ &> w(1 - p) + zp. \end{aligned}$$

Similarly, if $p_0 < p \leq 1$, we have

$$\begin{aligned} x(1 - p) + yp &< x(1 - p_0) + yp_0 \\ &= w(1 - p_0) + zp_0 \\ &< w(1 - p) + zp. \end{aligned}$$

Thus, if $p < p_0$, the greater of $x(1 - p) + yp$ and $w(1 - p) + zp$ is greater than $x(1 - p_0) + yp_0$, and the same holds if $p > p_0$. Thus the choice of p_0 by Mr. House gives Mr. Mark the smallest possible

---

† We use v to denote the value of the game.

maximum, i.e., the worst best, or *minimax*. This can be shown on a graph (see Illustration 13). Let x and w be marked off on the vertical axis, and let the horizontal axis be the p axis. Let z and y be marked on the line $p = 1$. Then for each value of p, the vertical line through that point intersects the line joining x to y in $x(1 - p) + yp$ and intersects the line joining w to z in $w(1 - p) + zp$. The greater of these two values for each possible p is shown by the heavy line. The low point of that line occurs at p_0, and that is the minimax.

Case 6: This case is exactly like Case 5, except that $w > x$ and $y > z$. The analysis is also similar, *mutatis mutandis*.

Thus we see that House always chooses his probabilities so as to give Mark the minimax. If we have three sets of payoffs instead of two, namely, x_1 and y_1 or x_2 and y_2 or x_3 and y_3, then again House gives Mark the minimax. For each choice of p, the expectation of the game having payoffs x_i and y_i is $(1 - p)x_i + py_i$, and the worst best of these is shown in Illustration 14 for two cases.

Illustration 13

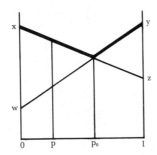

Illustration 14

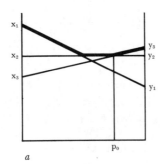

a

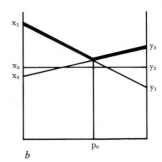

b

We might also consider games of chance having more than two outcomes, e.g., roulette. In such games, a similar analysis can be made, but the presentation requires higher-dimensional graphs, which we choose not to deal with in this treatment.

57 EXERCISES

Calculate the strategies of House and Mark for the following values of the payoffs:

1. Two games—one with payoffs 0 and 1, the other with payoffs -1 and 2.
2. Two games—one with payoffs 0 and 1, the other with payoffs 2 and -1.
3. Two games—one with payoffs 0 and 1, the other with payoffs 0 and -1.
4. Three games—one with payoffs 0 and -1, the second with payoffs $+3$ and -5, the third with payoffs -5 and $+3$.

Section 6 Matrix Games

Our next category of games has a deceptively simple description In spite of this, we shall soon show that we can use these games to describe some interesting situations in the "real world." The most elementary matrix game is called *Matching Pennies*, and in this game, we have two players, A and B. Each puts down a penny, either heads side up or tails side up. If both are Heads or both are Tails, then A takes B's penny. If there is one of each, then B takes A's penny. Simple enough. It was in this game that Tom Sawyer could always beat Huckleberry Finn, because he could follow Huck's thinking well enough to guess what strategy Huck would use. Let us see what we can do to help Huck understand this game, and games like it.†

Huck is the loser, and so we shall assume that all he wants out of the game is his "fair share." It is clear that we cannot create a strategy which will give each more than his fair share since, by

† The author is indebted to Prof. T. A. A. Broadbent for the observation that this game (or rather one just like it) is discussed not by Mark Twain, but by Edgar Allen Poe, in *The Purloined Letter*.

definition, Tom gets more than his share only if Huck gets less, and vice versa. Matching Pennies is what we call a *zero-sum game*, i.e., *A* wins what *B* loses, and vice versa. There are, of course, significant games in which the *sum* of the *payoffs* is not zero. In some such games, the sum can be positive; in others, negative. The most important game in which the sum of the payoffs can be positive is Production, sometimes known as Life. A game in which the sum of the payoffs is always negative is War (also sometimes called Life), and there one player might win, but always less than the other player loses.†

Let us modify the game of Matching Pennies to the following: If both pennies are Heads, then Tom gets a penny from Huck; if both are Tails, then he gets 3¢ from Huck. On the other hand, if there is a Head and a Tail, Huck gets 2¢ from Tom. We represent this game by the matrix in Illustration 15a. A *matrix* is a rectangular array of objects, usually real numbers. The matrix we show here is the *matrix of payoffs to Tom*, and in this section, since we deal with zero-sum games, the payoff to Huck is the negative of the payoff to Tom. Thus, if the coins show up HT, then Tom "gets" −2¢, i.e., he *pays* 2¢, which Huck gets, of course. The original game of Matching Pennies has the matrix shown in Illustration 15b, and Huck's problem is how to outwit Tom Sawyer and come out even in this game. We see from basic assumptions of symmetry that a fair outcome should offer the two players the same expectation of gain in Matching Pennies.

Illustration 15

	Tom	
	H	T
Huck H	1	−2
T	−2	3

a

	H	T
H	+1	−1
T	−1	+1

b

† Of course, nothing is simple in this Brave New World of ours. From 1941 to 1945, the United States took part in a game of War Production, and it is hard to calculate whether the net effect was positive or negative; responsible economic opinions exist on both sides.

Here is Huck's strategy: He tosses his coin in the air and lets it come up Heads or Tails, as it will. We assume it is a good coin, so that it will come up Heads and Tails about equally often in the long run. Now if Tom picks Heads, then he is actually choosing to play a game of chance in which he gets $+1¢$ for Heads and $-1¢$ for Tails on Huck's coin. If he picks Tails, then he is choosing to play the game which pays $-1¢$ for Heads and $+1¢$ for Tails. In either case, Huck has transformed his inferior position (inferior only because Tom can "figger" him) into a fair game, in which his gains or losses are tied only to the whims of fortune.

Please note that Huck has, in a sense, put himself in the position of House in Section 5. He gives Tom a choice of two games to play, and in this case, the expectations are both zero. It is interesting to see what we can make of the other game for Huck. Suppose he once more tosses his coin. Then, if Tom plays Heads, he can expect a payoff of $1¢$ or $-2¢$ with equal likelihood. If Tom plays Tails, then he gets a game of chance with equally likely payoffs of $-2¢$ and $3¢$. Since the Heads game has an expectation of $-\frac{1}{2}¢$ and the Tails game has $+\frac{1}{2}¢$, Tom would clearly play Tails at every turn and win in the long run.

It might occur to Huck to play his Heads and Tails with unequal probabilities. For example, he might choose to play tails only twice for every three times he plays Heads. Once again, he plays the role of Mr. House of Section 5 and gives Tom two games to play, namely, a Heads game, in which the payoffs are 1 or -2 according as Huck's coin comes up Heads or Tails, and a Tails game, in which the payoffs are -2 and $+3$. Note that in these games, the probabilities of Heads are equal, namely, $\frac{3}{5}$; and the probability of Tails is $\frac{2}{5}$ in both games. Tom's Tails game would have an expectation of $\frac{3}{5} \cdot (-2) + \frac{2}{5} \cdot 3 = 0$. Tom's Heads game, on the other hand, would have an expectation of $\frac{3}{5} \cdot 1 + \frac{2}{5} \cdot (-2) = -\frac{1}{5}$. Thus Tom would play the Tails game, which gives him a greater average payoff, namely, 0. Thus Huck comes out more or less even in what at first seemed a disadvantageous situation. Actually, he can do better than that, as we shall see below. The question of how to determine the proper attack for games like these is our next subject of study. We shall use an approach very much like the one we used for House and Mark.

Although John von Neumann (1903–1957) was born in Budapest, he spent his student years in Switzerland, and his later life in the United States, where he was considered to be one of the greatest contemporary mathematicians. The invention of game theory *as a mathematical discipline is due to him (in a paper published when he was 25), but he did equally important work in applied mathematics, functional analysis, and quantum mechanics. He was an active participant in the development of the atomic bomb and acted as scientific consultant and advisor to many U. S. government agencies in the years after World War II. (Courtesy of Mrs. Marion Whitman.)*

58 Definition

A *matrix game* is a rectangular array of real numbers, like the one in Illustration 16. Tom selects a column, and Huck selects a row. The matrix shows the payoff to Tom. The payoff to Huck is in each case the negative of the payoff to Tom.

What can we determine by examination? In the first place, it is clear that Tom *never* has any need to play the second column. He can always do at least equally well by playing the first column (or the third or the fourth) no matter how Huck plays. Thus the second column is superfluous; we say it is *dominated by* the first column. If we remove the second column, as in Illustration 18, then we see that the second row is greater, term for term, than the fourth. Since the matrix shows the payoff to Tom and Huck is trying to *minimize* this payoff, he will *never* play the second row. Thus the second row is dominated by the fourth. Note that in the original matrix, the second row is *not* dominated by the fourth row. Now, in the new matrix (second column and second row removed), we can again remove a column by domination, this time the first, which is dominated by the second. Now we have the game in Illustration 20, which is, in some sense, the heart of the original game since we

Section 6   Matrix Games

Illustration 16

1	−2	3	5	−4
3	1	2	4	3
−4	−5	0	−2	4
2	2	2	2	2

Illustration 17

1	−2	3	5	−4
3	1	2	4	3
−4	−5	0	−2	4
2	2	2	2	2

Illustration 18

1	3	5	−4
3	2	4	3
−4	0	−2	4
2	2	2	2

Illustration 19

1	3	5	−4
−4	0	−2	4
2	2	2	2

Illustration 20

3	5	−4
0	−2	4
2	2	2

have concluded, for one reason or another, that no one would ever play the missing rows or columns.

At this point, we introduce a new concept, the concept of a *mixed strategy*. (The formal definition will come later.) Suppose Huck adds a new row to the matrix, and suppose that row is the outcome of a little game of chance. Huck tosses a coin and chooses either the first row or the second with equal likelihood. This new alternative is called a *mixed strategy*. Tom has no choice about whether he wants to play this game if Huck chooses this strategy. No matter which column he picks, this game is included. Following the method of our previous analysis, we write down the expectation of the game of chance in each column, below the third row, giving us Illustration 21. As the game stands now, the third row is dominated by the fourth, so we remove it, giving us the game we see in Illustration 22.

One way of approaching the game we now see is to mix the three remaining strategies. Let us assign the three rows the probabilities p_1, p_2, and p_3, where $p_1 + p_2 + p_3 = 1$. Then, depending on the column which Tom chooses, he gets to play a game with expectation $3p_1 + 1\frac{1}{2}p_3$, $5p_1 - 2p_2 + 1\frac{1}{2}p_3$, or $-4p_1 + 4p_2$. Thus we have Huck once again playing House to Tom's Mark. Let us now apply the minimax reasoning. We can simplify this analysis by

Illustration 21

	3	5	−4
	0	−2	4
	2	2	2
	$1\frac{1}{2}$	$1\frac{1}{2}$	0

Illustration 22

p_1	3	5	−4
p_2	0	−2	4
p_3	$1\frac{1}{2}$	$1\frac{1}{2}$	0

Section 6 Matrix Games

observing that a decision to play the third row with probability p_3 is actually a decision to play the first row with an additional probability of $\frac{1}{2}p_3$, and the second row likewise. Thus we are actually offering Tom the choice of three games with payoffs of $3(p_1 + \frac{1}{2}p_3)$, $5(p_1 + \frac{1}{2}p_3) - 2(p_2 + \frac{1}{2}p_3)$, and $-4(p_1 + \frac{1}{2}p_3) + 4(p_2 + \frac{1}{2}p_3)$. We now simplify these formulas by setting $p_1 + \frac{1}{2}p_3 = q_1$ and $p_2 + \frac{1}{2}p_3 = q_2$. Then $q_1 + q_2 = 1$, and the payoffs are $3q_1$, $5q_1 - 2q_2$, and $-4q_1 + 4q_2$, respectively. Tom would, of course, choose whichever is greatest, and so Huck contrives to make that maximum as small as possible. Let us note that $q_2 = 1 - q_1$, so that $5q_1 - 2q_2 = 7q_1 - 2$, while $-4q_1 + 4q_2 = -8q_1 + 4$. If we now look at the numbers $3q_1$, $7q_1 - 2$, and $-8q_1 + 4$ to determine which is largest, we see that it depends on what q_1 is. If we plot the three quantities against the possible values of q_1, as in Illustration 23, then the maximum for each value of q_1 is shown by the heavy broken line. The problem now becomes the same as the problem of Messrs. House and Mark in Section 5. We shall be looking for the minimax again, as you will see. Actually, the scale of Illustration 23 is somewhat distorted to allow you to see what is going on. The picture should actually be much narrower. Please note the value of q_1 denoted as q. There the greatest expectation for the three games

Illustration 23

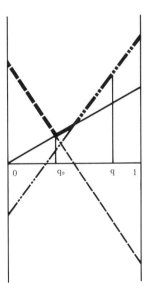

$7q_1 - 2$
$+8q_1 + 4$
$3q_1$

is $7q - 2$. The smallest value taken on by this maximum—that is to say, the *worst best*, or *minimax*—is assumed at the point denoted q_0, where the dotted line and the solid line meet. At point q_0, we have $3q_0 = -8q_0 + 4$, or $q_0 = \frac{4}{11}$. Thus, for example, if Huck should mix his strategies in our original game in such a way as to play the first row at random $\frac{4}{11}$ of the time, the third row $\frac{7}{11}$ of the time, and the other two rows not at all, then Tom would be confronted with a choice of five games of chance, whose expectations are, respectively, $-\frac{24}{11}$, $-\frac{43}{11}$, $\frac{12}{11}$, $\frac{6}{11}$, and $\frac{12}{11}$. In this case, Tom would always choose the third or the fifth column and make an *expected* $\frac{12}{11}$ for the game. We have now shown that Huck can so operate his options that he can hold Tom to an expectation of $\frac{12}{11}$ for each play of the game. Can he, perhaps, do even better in some other way? The answer here is, in a real sense, *no*.

Let us assume that Tom wishes to be sure of getting at least an expected $\frac{12}{11}$. He would then choose the third column $\frac{8}{11}$ of the time and the fifth column $\frac{3}{11}$ of the time.† The expected values of the rows, considered as games of chance for Huck, give expectations of $\frac{12}{11}$, $\frac{30}{11}$, $\frac{12}{11}$, and $\frac{22}{11}$. Since Huck is trying to minimize Tom's gain, he chooses the first or third row, or some combination of them, and Tom gets an expected $\frac{12}{11}$.

Thus Huck has a mixed strategy which will prevent Tom from realizing more than an expected $\frac{12}{11}$ on the game regardless of which mixed strategy Tom employs, while Tom has a mixed strategy which assures him at least an expected $\frac{12}{11}$ against any mixed strategy of Huck's. We say that $\frac{12}{11}$ is the *value* of the game in question, and that will certainly be the outcome between two players who know their game theory.

Now, is this an accident? Is this a special property of this particular game? No, Virginia, there is always a Santa Claus. Every matrix game has a value, in this sense, a guaranteed amount that each side can have for its expectation, and no worse, if it plays properly. You may remember that we denoted as the minimax the worst best, the amount that Huck can hold Tom to. Similarly, we denote as the *maximin* the "best worst" that Tom can extract from Huck. In the game just described, these were equal, and in fact,

---

† The analysis which yields these numbers is too complicated to show here. The reader is merely invited to observe that these numbers do work.

they are equal for all matrix games. This is the famous *Minimax Theorem*, which we shall investigate and prove specifically for 2 by 2 games and which we state below for other games.

59 Definition

Let $\mathcal{G}$ be a 2 by 2 matrix, which we designate as shown in Illustration 24, where the rows and columns of the matrix are denoted by the letters H and T and the entries are real numbers. The payoffs shown are those to Tom, who chooses a column; Huck chooses a row. Tom's mixed strategy consists of (p, q), the probabilities with which he plays H and T, respectively, and Huck's mixed strategy is denoted (r, s). Of course, $p + q = 1 = r + s$. We assume that the two players behave *independently*, which means that the probability that Tom will choose H and Huck choose T is $p \cdot s$. In that case, we have the expected value of the game:

$$E = x \cdot p \cdot r + y \cdot q \cdot r + w \cdot p \cdot s + z \cdot q \cdot s$$

We now assert the following theorem.

60 Theorem

There exists a real number v and a pair of mixed strategies (p_0, q_0) and (r_0, s_0) such that

$$xp_0r + yq_0r + wp_0s + zq_0s \geq v$$

for all mixed strategies (r, s), while

$$xpr_0 + yqr_0 + wps_0 + zqs_0 \leq v$$

for all mixed strategies (p, q).

PROOF: There are two important cases:

Case 1: One row or column dominates another.
Case 2: There is no domination.

Case 1 We can assume that column H dominates column T and that $x \leq w$; all other cases are symmetric to this one. Then we have $x \geq y$, $w \geq z$, $x \leq w$. We assert that $p_0 = 1$, $q_0 = 0$, $r_0 = 1$, and $s_0 = 0$ and that $v = x$. Clearly, for any (r, s),

$$xp_0r + yq_0r + wp_0s + zq_0s = xr + ws$$
$$\geq xr + xs$$
$$= x(r + s) = x = v.$$

Illustration 24

	H	T
H	x	y
T	w	z

Conversely, for any (p, q),

$$xpr_0 + yqr_0 + wps_0 + zqs_0 = xp + yq$$
$$\leq xp + xq$$
$$= x(p + q) = x = v.$$

All this calculation merely shows what is obvious, namely, that both Tom and Huck always play Heads in such a game. There are seven other possibilities of domination, and they all look like this.

Case 2 Since there is no domination, either $x > y$ or $x < y$. If $x > y$, then we must have $w < z$. If $y \geq z$ or $x \leq w$, we would have domination. Thus we have $y < x, w < x, y < z, w < z$.

If we had started with $x < y$, we would have had the same analysis with the signs reversed or, equivalently, with the rows interchanged. Therefore, we need only to analyze this one set of inequalities. Let us assume that Tom chooses to play the columns with probabilities p and $q = 1 - p$. Then if Huck chooses row H, his expectation will be

$$px + qy = px + (1 - p)y$$

If we plot this value against that of p, where $0 \leq p \leq 1$, we obtain the graph in Illustration 25. If we do the same for row T, we obtain a similar graph, except that it is a declining graph. The conditions of nondomination ensure that the two graphs actually do intersect, as in Illustration 26. We call the value of p for which they intersect p_0. For each choice, the worst Huck can allow Tom is the lesser of the two points of intersection, and then only if he plays a pure strategy. If he plays a mixed strategy, the expectation will be somewhere between. By setting $p = p_0$, Tom maximizes this minimum. p_0 is the solution of the linear equation

$$px + (1 - p)y = pw + (1 - p)z$$

Let us denote this common value as v_T. Solving the equation,

$$p_0(x - y - w + z) = z - w,$$

or

$$p_0 = \frac{z - w}{x - y - w + z}.$$

Let us suppose that Huck uses the mixed strategy (r, s). Then the expected value of the game is

$$E = xp_0r + y(1 - p_0)r + wp_0s + z(1 - p_0)s$$
$$= [xp_0 + y(1 - p_0)]r + [wp_0 + z(1 - p_0)]s$$
$$= v_Tr + v_Ts = v_T \quad \text{(since } r + s = 1\text{).}$$

Section 6　Matrix Games　　375

Now we can do the same thing with respect to Huck. If he chooses a mixed strategy (r, s), then Tom can secure $xr + ws$ by playing the H column and $yr + zs$ by playing the T column. If he plays these columns with a mixed strategy, he gets something between the two. The minimax occurs at the intersection of the two lines shown in Illustration 27, and we

Illustration 25

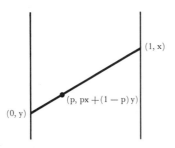

Illustration 26

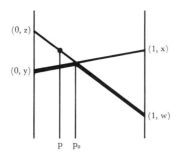

Illustration 27

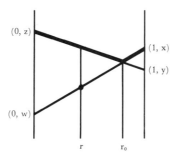

denote the point as (r_0, v_H). If Huck uses the mixed strategy $(r_0, 1 - r_0)$ and Tom uses any mixed strategy (p, q), then

$$\begin{aligned} E &= xpr_0 + yqr_0 + wp(1 - r_0) + zq(1 - r_0) \\ &= [xr_0 + w(1 - r_0)]p + [yr_0 + z(1 - r_0)]q \\ &= v_H p + v_H q = v_H \quad \text{(since } p + q = 1\text{).} \end{aligned}$$

Thus Huck can ensure that Tom gets no more than v_H, no matter what mixed strategy he uses. By examining

$$xp_0 r_0 + y(1 - p_0)r_0 + wp_0(1 - r_0) + z(1 - p_0)(1 - r_0),$$

we see that this quantity is simultaneously equal to v_T and v_H. Thus $v_T = v_H$, and we denote it v.

Recapitulating, we find that in either Case 1 or Case 2, the theorem holds. Q.E.D.

61 Definition

The number v, which appears in Theorem 60, is called the *value* of the game, and the two mixed strategies are called *optimal strategies*. The use of the word "optimal" is meant to reflect only the meaning of the theorem itself, i.e., that no mixed strategy can *guarantee* a better *expectation*. "Guarantee" here means against the best possible play by the opponent. In certain cases, the optimal strategies are not unique. There may be more than one, and in that case, there will be infinitely many. We shall not be looking at such games. A *solution* of a game consists of the value together with the optimal strategies.

Illustration 28

$$\begin{array}{ccccc} x_{11} & x_{12} & x_{13} & \cdots & x_{1n} \\ x_{21} & x_{22} & x_{23} & \cdots & x_{2n} \\ x_{31} & x_{32} & x_{33} & \cdots & x_{3n} \\ \cdots & \cdots & \cdots & & \cdots \\ x_{m1} & x_{m2} & x_{m3} & \cdots & x_{mn} \end{array}$$

62 Theorem (Minimax Theorem)

If $\mathcal{G}$ is the matrix game shown in Illustration 28, then there exists a real number v and mixed strategies $(p_1^0, \ldots, p_n^0)$ and $(q_1^0, \ldots, q_m^0)$ such that the sum

$$x_{11}p_1^0 q_1 + x_{12}p_2^0 q_1 + \cdots + x_{1n}p_n^0 q_1 +$$
$$+ x_{21}p_1^0 q_2 + x_{22}p_2^0 q_2 + \cdots + x_{2n}p_n^0 q_2 +$$
$$+ \cdots +$$
$$+ x_{m1}p_1^0 q_m + x_{m2}p_2^0 q_m + \cdots + x_{mn}p_n^0 q_m \geq v$$

for any mixed strategy $(q_1, \ldots, q_m)$, while

$$x_{11}p_1 q_1^0 + x_{12}p_2 q_1^0 + \cdots + x_{1n}p_n q_1^0 +$$
$$+ x_{21}p_1 q_2^0 + x_{22}p_2 q_2^0 + \cdots + x_{2n}p_n q_2^0 +$$
$$+ \cdots +$$
$$+ x_{m1}p_1 q_m^0 + x_{m2}p_2 q_m^0 + \cdots + x_{mn}p_n q_m^0 \leq v$$

for any mixed strategy $(p_1, \ldots, p_n)$.

We have shown that every matrix game with two rows and two columns has a solution. As we have commented before, every matrix game has a solution, no matter how many rows and columns the game has, as long as these are finite in number, but we did not prove this assertion in general.

63 EXERCISES

Find the solutions for the following games:

$$1 \begin{pmatrix} 1 & -2 \\ -2 & 3 \end{pmatrix} \qquad 2 \begin{pmatrix} 1 & 0 \\ -1 & 0 \end{pmatrix}$$

$$3 \begin{pmatrix} 1 & -2 \\ -2 & 4 \end{pmatrix} \qquad 4 \begin{pmatrix} 1 & -10 \\ -10 & 20 \end{pmatrix}$$

$$5 \begin{pmatrix} 1 & -a \\ -a & b \end{pmatrix}$$

Section 7   Applications of Matrix Games

One of the most famous examples of a matrix game arose during the Second World War. Airplanes were used by the Allies for hunting German submarines in the North Atlantic. Each such flight had a certain expectation of finding and destroying a submarine. It is a little difficult to measure this expectation, and we shall not concern ourselves with the mechanism or the reliability of the measuring technique.

At one point during the war, the search planes began to use radar, and this expanded the radius within which submarines were visible. The result was that the "expectation of kill" was markedly increased, and for a period of time, the airplanes were showing much more effectiveness. The next technological advance was made by the Germans, who discovered that the radar impulse could be detected even from below the horizon, and a submarine using such a detector could often avoid the searching plane by "running away" in some suitably chosen direction. Unfortunately, the use of this radar detector required that the sub lie close to the surface, and if an airplane should come along with its radar shut off, then the submarine would be easier to spot than it would be if it lay deeper.

We now abstract the game and imagine that it is being played between a single airplane and a single submarine. The pure strategies for the airplane are to search visually or by radar; for the submarine, they are to lie deep or to use the radar detector. The game is interesting because, as a qualitative judgment, there is no dominance. In order to have some idea of how the game should be played, we need some estimates on the expectation of kill, and we include some such estimates in the matrix of Illustration 29. It now follows immediately that there is an optimal strategy for each side, *i.e.*, a best "mix" to maximize (or minimize) the expectation of kill.

Illustration 29

	Visual	Radar
Deep	0.4	0.7
Radar detector	0.6	0.2

64 EXERCISE
Solve the North Atlantic Game.

Another famous war game concerns the defense of two passes. It is greatly simplified in order to permit any analysis at all. The generic problem concerns an army which must defend a city against an enemy army. The natural points of defense are a number of mountain passes, one of which must be negotiated by the enemy if he is to reach the city. The pass is gained or held according as the attackers do or do not have a preponderance of force; if the forces are equal, the defenders win. In the simplest possible interesting case, there are two passes, and each "army" consists of one man. Then if the attacker attacks through the defended pass, the defender wins. Otherwise, the attacker wins. Clearly, this is our familiar game of matching pennies. Let us score 1 for the attacker if he succeeds and 0 if he does not. Then the expected value of this game is $\frac{1}{2}$, and either side can ensure this expectation by playing his choices with equal probabilities. A more interesting game is observed if each "army" has two men. Let us designate the strategies of each side as 2-0, 1-1, 0-2, where these denote the number of men assigned to each pass. Then the matrix is as shown in Illustration 30. Unfor-

Illustration 30

	1-1	2-0	0-2
1-1	0	1	1
2-0	1	0	1
0-2	1	1	0

Illustration 31

	0-2	1-1
0-2	$\frac{1}{2}$	1
1-1	1	0

tunately, this is a 3 by 3 game, and we have not discussed 3 by 3 games. However, we can consider that each side has a choice of two strategies, namely, to divide or to consolidate the two men. If both divide the two men, then the defenders win. If only one of them divides, then the attackers win, since the attackers always have a preponderance at one of the passes. If neither side divides its men, then they play the previous game, which is known to have a value of $\frac{1}{2}$. Thus, we can take the matrix of Illustration 31 as a reasonable description of the game.

Illustration 32

	2-0	0-2
2-1	0	1
1-2	1	0

Illustration 33

	3-0	2-1
3-0	?	1
2-1	1	?

Illustration 34

	0-3	2-1
0-3	½	1
2-1	1	½

Illustration 35

65 EXERCISE
Solve the game in Illustration 31.

It is interesting to observe that giving the defense three men will not ensure its winning, though giving it four will do so if the offense has only two. Let us consider the game with three men. The offense can still be thought of as having the same two strategies, 2-0 and 1-1, while the defense has the strategies 3-0 and 2-1. Since two soldiers will hold any pass against the full force of the enemy, the defense 3-0 makes no sense. Since the only reasonable defenses are 2-1 and 1-2, the attack 1-1 makes no sense. Thus the game is equivalent to matching pennies, as we see in Illustration 32.

A more interesting variant gives each side three soldiers. Then the attack or defense deployments are 3-0 or 2-1. If the choices are mixed, then the attack succeeds (see Illustration 33). If both sides elect 3-0, then the game is matching pennies, which has a value of ½. If both elect 2-1, then the game is again matching pennies. Thus, we can make up a new game, which is, in some sense, essentially the same and is a 2 by 2 game. We see that the value of this game is ¾, and the optimal strategies are obtained by flipping coins, first to determine the distribution of forces (3-0 or 2-1) and then to decide which force goes where.

66 EXERCISES
1. Solve this game for
    (a) Four defenders, three attackers
    (b) Five defenders, three attackers
2. Let v_n be the value of the game above, with n attackers and n defenders. Show that the value of v_{n+2} can be calculated as the value of the game in Illustration 35.
3. Show that $v_{n+2} = \dfrac{2 - v_n}{3 - 2v_n}$ whenever $n > 2$.
4. The values of v_n for $n = 1, 2, 3, 4$, respectively, are ½, ⅔, ¾, and ⅘. What is the obvious conjecture? Is it true?
5. Find optimal strategies in the game of n against n.
6. How many defenders can stand against five attackers and have an even chance of winning?
7. If there are six attackers and n defenders, what are the chances of success for the attackers, for all values of n from 1 to 20.

Section 8   Positive-sum Games

Section 8   Positive-sum Games

It may sometimes happen that one player gains more (or less) than the other player loses. If that is the case, then the game is no longer zero-sum, and we must represent the payoff as a double matrix, i.e., a matrix in which the entries are *pairs* of real numbers, representing the payoffs to the two players. In Illustration 36, we have such a

Illustration 36

	H	T
H	+1, −1	−1, +1
T	−1, +1	+2, −1

game. The game looks like Matching Pennies, except that there is a bonus to Tom if TT comes up. How might one analyze such a game? One way to play the game might be for the two players to agree to play Tails at every try and then share the gains thus created (+1 at each play) according to some scheme. What scheme they might use will be discussed later. This technique is known as a *side payment*, and once we allow side payments, it is clear that both players should play Tails at every turn. Tom must, of course, give Huck the penny to pay his losses each time, and also some share of the profit. Any other arrangement merely represents a loss to the two of them together of the profits potentially available.

Let us assume, then, that side payments are impossible because of some circumstance. It is clearly to Tom's advantage to get Huck to select the second row (Tails) as often as possible, preferably all the time. He cannot, of course, do this if he chooses Tails himself all the time. Suppose he chooses Tails 49% of the time and Heads 51%. If Huck knows this and if he is seeking his best interest, then he will play Tails, which will give him an average gain of 0.02, while he will make a gain of −0.02 by choosing Heads. Will he be "wise" enough to play Tails every time? If so, Tom will obtain an expected $2(0.49) - 0.51 = 0.47$. The numbers 49% and 51% are not crucial. What we have shown is that Tom can obtain an expectation as near to $\frac{1}{2}$ as he likes by giving Huck a little more than the 0 which is his minimax.

In mathematical writing, we often write ϵ, the Greek letter epsilon, for a small quantity, especially for an arbitrary small quantity. If Tom plays H with probability $\frac{1}{2} + \epsilon$ and T with probability $\frac{1}{2} - \epsilon$, then Huck has a maximum expectation of 2ϵ, which he obtains by playing Tails every time. Tom then obtains $2(\frac{1}{2} - \epsilon) - 1(\frac{1}{2} + \epsilon) = \frac{1}{2} - 3\epsilon$. In the previous example, we took $\epsilon = 0.01$.

It happens that Huck can use the same approach as Tom, however. If he plays H and T with probabilities q and $1 - q$, respectively, and if $q = \frac{1}{2}$, then he gets an expectation of 0, while Tom, if he can count on Huck's behavior, will always play T, taking an expected $\frac{1}{2}$ for his trouble. This favoring of T by Tom emboldens Huck to increase q. If he increases q just a little, Tom will continue to play T and Huck will obtain more than 0 for his expectation. Tom's minimax occurs at $q = \frac{3}{5}$, at which point he is indifferent to H and T. Thus, if Huck increases q above $\frac{1}{2}$ but not as far as $\frac{3}{5}$, then Tom will, by this analysis, continue to prefer T to H. If ϵ is again a small but positive quantity, so that $\frac{3}{5} - \epsilon$ is actually less than $\frac{3}{5}$ but not by much, then let us consider what happens if Huck plays the mixed strategy $(\frac{3}{5} - \epsilon, \frac{2}{5} + \epsilon)$. In that case, if Tom plays his best answer, which is T, then Tom gets an expectation of $\frac{1}{5} + 3\epsilon$, while Huck expects $\frac{1}{5} - 2\epsilon$.

Note that the House (in this case, the House is benevolent, as in the game called Production) is willing to give away as much as 1 unit per try to the players together. If they cannot find any means of making side payments and if Tom calls the tune, they can get just under $\frac{1}{2}$, of which almost all goes to Tom. If Huck calls the tune, then they can get just over $\frac{2}{5}$ per try, and Tom will get a bare majority of it. If Tom and Huck both try to call the tune, then Tom will play $(\frac{1}{2} + \epsilon, \frac{1}{2} - \epsilon)$, while Huck will play $(\frac{3}{5} - \epsilon, \frac{2}{5} + \epsilon)$. The result will be that Tom will obtain just somewhat more than $\frac{1}{5}$ (the exact expectation will be $\frac{1}{5} + \frac{9}{10}\epsilon - 5\epsilon^2$), while Huck will obtain a small negative expectation $(-\frac{2}{5}\epsilon + 4\epsilon^2)$.

67 EXERCISE

What conclusions can one draw in this example concerning the economic doctrine that all will be for the best if everyone pursues his own best interest? What conclusions can one draw about antitrust laws?

Section 9 Sharing

In a certain sense, we now know that Tom's mimimax is $\frac{1}{5}$ and that Huck's is 0. That is, each can obtain this much no matter what the other does. Suppose we accept as a ground rule that each player must obtain at least his minimax. Each player will be restricted to a mixed strategy; Tom will play his options with probabilities p and $1 - p$, while Huck will play with probabilities q and $1 - q$. The payoffs to the two players, respectively, will be

$$2 - 3p - 3q + 5pq$$

and

$$2p + 2q - 1 - 4pq.$$

The shaded area in Illustration 37 represents all the possible payoffs, with the x value being the payoff to Tom and the y value

Illustration 37

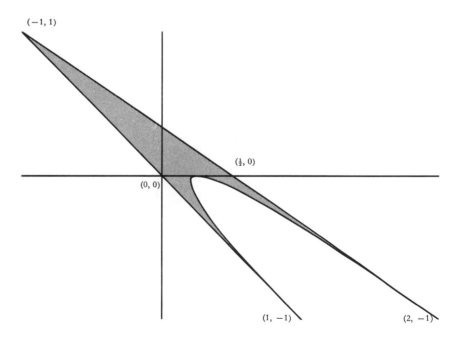

the payoff to Huck.† The requirement that each player receive at least his minimax leaves us the shaded triangle in Illustration 38, and a simple analysis shows that we need only consider the line segment which is the hypotenuse of that triangle. Note that the worst that the players can do in this triangle is the point $(\frac{1}{5}, 0)$, which represents their minimaxima.

Of the possible points on the line segment in question, now, current theory takes no preference; they are all equally good. There is a certain amount of feeling that the midpoint is the "fairest," but then why isn't $(\frac{1}{5}, \frac{1}{5})$ even "fairer"? In order to resolve this issue, we shall try to isolate the phenomenon. As we indicated at the beginning of the chapter, one way to deal with a phenomenon you

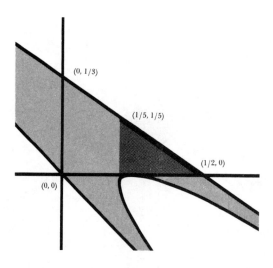

Illustration 38

† The curve in Illustration 37 has for its equation
$$16x^2 + 40xy + 25y^2 - 8x - 6y + 1 = 0.$$
Its derivation is too difficult to include here.

wish to understand is to make a game of it, and we shall discuss such games briefly in the next section.

Before we do, however, let us return to the case in which side payments are allowed. In order to obtain the maximum benefit the game allows, Tom and Huck must agree to play tails at every try, and clearly this requires, as we said before, that Huck's losses must be paid out of Tom's winnings and also that he must get something of Tom's profit. How much is not clear, and questions of this sort are the material for the theory of cooperative games, which follows.

Section 10 Cooperative Games

This section deals with games which are the subject of current research. One of the most frequent sources of cooperative games arises from the theory of positive-sum games, as in the last two sections. Cooperative games for two players have certain limitations, as we shall see, and the most interesting cases are for more players. Such theory as does exist is quite difficult, far beyond the scope of this book. However, I am including this section so that you can see what is being worked on.

In a simple cooperative game, we have a number of players who can obtain certain payoffs in a certain situation by forming coalitions. The simplest such game is similar to the situation in the last section. Tweedledum and Tweedledee are offered 100 by the Red Queen if they can agree on how to share it. "All right," says Tweedledum, "50-50." "That's all very well for you," says Tweedledee, "but in fact, I'll take 60 or I won't play." "Then you'll get nothing," says Tweedledum. "You, too," says Tweedledee. "Do you want the 40 or don't you?" Does it sound farfetched? Not at all. The game is played all over the world, a million times a day, as part of the game called Production.

68 EXERCISE

Find four distinct instances of "Tweedledum and Tweedledee" in real economic life.

As we said before, the game of Tweedledum and Tweedledee really has no answer, and you now understand it about as well as anyone. An interesting variant involves Winken, Blinken, and

Illustration 39

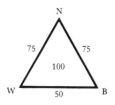

Nod, any two of whom can have 75 if they agree on how to share it. If they can all three come to an agreement, they get 100. No one can join two coalitions, and anyone standing alone gets nothing. Here we can argue by symmetry that they should share the 100 equally, since they are all in it symmetrically. But suppose we have Winken and Blinken together making 50, while either makes 75 with Nod and all together make 100. This scheme is shown in Illustration 39. Is it clear that Winken and Blinken get 25 each, while Nod gets 50? Suppose Winken and Blinken make 60 together, and everything else remains the same. What if they make 40? 100? 0?

69 EXERCISES

Investigate these games:

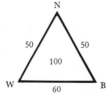 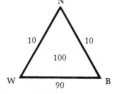

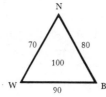 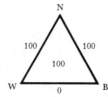

References

1. M. Dresher, "A sampling inspection problem in arms control agreements: A game theoretic analysis," *Memorandum RM-2972-ARPA*, The RAND Corporation, 1967.
2. Lester E. Dubins and Leonard J. Savage, *How To Gamble If You Must*, McGraw-Hill Book Company, New York, 1965.
3. Martin Gardner, "Mathematical Games" (column), *Scientific American*, **197** (1957), pp. 145ff.
4. *Ibid.*, **199** (1958), pp. 124ff.
5. *Ibid.*, **205** (1961), pp. 148ff.
6. G. H. Hardy and E. M. Wright, *An Introduction to the Theory of Numbers*, Clarendon Press, Oxford, 1965.
7. H. W. Kuhn and A. W. Tucker, *Contributions to the Theory of Games*, Princeton University Press, Princeton, N.J., 1950.
8. R. Duncan Luce and Howard Raiffa, *Games and Decisions*, John Wiley & Sons, Inc., New York, 1957.
9. Michael Maschler, "A Price Leadership Method for Solving the Inspector's Non-constant-sum Game," *Naval Research Logistics Quarterly*, **13** (1966), pp. 11–23.
10. John Dennis McDonald, *Strategy in Poker, Business, and War*, W. W. Norton & Company, Inc., New York, 1950.
11. Alain Robbe-Grillet, *Last Year at Marienbad*, Grove Press, Inc., New York, 1962.
12. John von Neumann and Oskar Morgenstern, *Theory of Games and Economic Behavior*, 3d ed., Princeton University Press, Princeton, N.J., 1963.
13. John Davis Williams, *The Compleat Strategyst*, rev. ed., McGraw-Hill Book Company, New York, 1965.

EXCURSIONS INTO MATHEMATICS

CHAPTER   6   **WHAT'S IN A NAME?**

BY MICHAEL N. BLEICHER

Section   1   Introduction

Most of us take our arithmetic and the real-number system for granted. We know the rules of the game: The sum (product) of 2 and 3 is the same as the sum (product) of 3 and 2, or the sum of 2 and 3 added to 7 yields the same result as adding 2 to the sum of 3 and 7. (These and other basic facts can be found in Begle [1].) We have handy methods (*algorithms*) for efficient addition, subtraction, multiplication, etc., without really worrying about why or if the methods work. For example, why do we borrow when we do subtraction or what is really going on when we do long division?

Turning to a different aspect of the question, how do we add the number $1.8989898989\cdots$ to the number $1.976976976\cdots$? Worse yet, how can we add $1.89889888988889888889\cdots$ to $1.898899888899988889999\cdots$? Perish the thought that we should be asked to multiply them! Do those decimals really stand for numbers? Are there simpler decimals for the same numbers? Can we write these numbers as fractions?

On the other hand, you may have asked: Can you find the decimal expansions of any of $\sqrt{2}$, $\sqrt{9}$, $\sqrt{-2}$, $\sqrt{-11}$, $\log_{10} 27$, π, $\sin 15°$, $\sin 30°$? Do they all have decimals? How many? Are decimals worth the trouble anyway, or can we find a better system? Some of these questions will be answered here. Answers to others will be hinted at; while some are beyond the scope of this chapter.

If we are given two decimals, it is not difficult to tell which is larger. But can you tell which is the larger of $\frac{12,347}{12,348}$ and $\frac{49,901}{49,905}$ or of $\sqrt{2}$ and $\frac{577}{408}$ and $\frac{1,393}{975}$?

We have noted before, in the chapter on perfect numbers, that arithmetic problems which we now consider very elementary (say, junior high school level) were once beyond the capabilities of the best mathematicians simply because they hadn't appropriate notations for stating and working them.

We will examine various methods of naming the numbers, and we will discover some of the advantages and disadvantages of each. We begin with methods of naming the positive whole numbers or natural numbers, as they are sometimes called. In presenting these systems, we will take certain liberties in altering both chronologically and conceptually some of the numeration methods discussed. The reader who desires more detail and accuracy should consult one of Cajori's books [4, 5, 6] or that of Ore [17].

Section 2 Historical Background

The first and perhaps most primitive way of recording positive whole numbers, or natural numbers, is to place as many marks or stones as the number calls for, e.g.,

$$3 \leftrightarrow ///$$
$$12 \leftrightarrow ////////////$$
$$47 \leftrightarrow ///.$$

This system was undoubtedly used by many early human cultures. It has a certain simplicity, and in fact, without the user knowing the names of the numbers or being literate, numbers can be recorded, compared, added, and subtracted in this system. On the other hand, large numbers get completely out of hand, and multiplication is very tedious.

1   EXERCISES

1. How would you perform addition and subtraction in this system?
2. How would you perform multiplication and division?
3. How would you compare two numbers? (Remember, you can't count if you don't have names for numbers and ways to tell from the names which is bigger.)

Probably one of the first simplifications to occur was the grouping of the marks together into uniform bunches. The most common sizes for the bunches on this planet seem to be 5 (one hand), 10 (both hands), or 20 (hands and feet). However, the Babylonians used 60, vestiges of which remain with us in 60-minute hours, 60-second minutes, etc. Some African cultures have used systems based on 3 and 4; and certain North American Indian tribes had systems based on 3, 4, and 8. (See Eells [8].)

Section 2   *Historical Background*   393

Let us for a moment suppose we take bunches based on 5; then our numbers above might appear as

3 ↔ ///
12 ↔ ///// ///// //
47 ↔ ///// ///// ///// ///// ///// ///// ///// ///// ///// //

or perhaps

47 ↔ ///// ///// ///// ///// /////
     ///// ///// ///// ///// //.

Now if we want to compare two natural numbers, instead of comparing the strokes we can compare bunches of strokes to see which is bigger. Of course, someone might have felt that ///// was clearer than /////, which led to a system one can still find in common use today. As notation developed, someone might have wanted to replace ///// by something simpler, say by not making the first three strokes. Thus, instead of /////, we might write \\, or more symmetrically, V.

Thus 47 might be written as

$$47 \leftrightarrow \begin{matrix} VVVVV \\ VVVVII. \end{matrix}$$

Of course the symbol $\begin{smallmatrix}V\\V\end{smallmatrix}$ is cumbersome; it might be easier to replace it by a more convenient symbol—say, turn the bottom one upside down thus: $\begin{smallmatrix}V\\ \Lambda\end{smallmatrix}$, or X, for 10. In this system, we get

47 ↔ XXXXVII,

which perhaps has a familiar ring. For larger numbers this is still inadequate, so we might wish for a symbol for five X's, say L, for ten X's, say C, etc. We now have the classical Roman numeral system. The subtractive feature of writing IV for IIII did not occur until a few centuries ago. (The reader should note that we are not claiming that Roman numerals were so discovered but are trying to plausibly compress thousands of years of numerical insights into a few paragraphs.) The system of Roman numerals is an additive system; i.e., each symbol stands for a certain value, and the string of symbols stands for the sum of all the values of the symbols involved. The system of Egyptian hieroglyphics was also of this type.

While numbers like 47 and other moderate numbers can be handled easily in this system and comparisons of two numbers are not difficult, arithmetic is not so easy. Also, large numbers get unwieldy; for example,

$$1968 \leftrightarrow \text{MDCCCCLXVIII}$$
$$1999 \leftrightarrow \text{MDCCCCLXXXXVIIII}.$$

2 EXERCISES

1 Can you devise addition and subtraction algorithms (methods) for Roman numerals?

2 Can you devise multiplication and division methods?

One method of avoiding this problem of great strings of symbols is to make more numerical symbols. Many cultures did this by making use of letters of the alphabet, e.g., the Greek, Hebrew, and Hindu-Brahmi systems. If we examine the Greek system, we see that they devised the following scheme [4, p. 53]:

α	β	γ	δ	ε	ϛ	ζ	η	θ	ι	κ	λ	μ	ν	ξ	ο	π	ϟ
1	2	3	4	5	6	7	8	9	10	20	30	40	50	60	70	80	90

ρ	σ	τ	υ	φ	χ	ψ	ω	ϡ	,α	,β	,γ	etc.
100	200	300	400	500	600	700	800	900	1,000	2,000	3,000	

M	$\overset{\beta}{M}$	$\overset{\gamma}{M}$	etc.
10,000	20,000	30,000	

Thus

$$47 \leftrightarrow \mu\zeta$$
$$1{,}999 \leftrightarrow {,}\alpha\varpi\varkappa\theta.$$

This has the advantage of brevity but makes arithmetic even harder, which perhaps helps explain why it did not replace the Roman system.

There were, however, other methods developed to help solve the great length of the names of big numbers in an additive system. One such method is using subtraction, i.e., writing 4 as IV, 9 as IX, 90 as XC, etc. This does make names a little shorter, but still, 1,888 is MDCCCLXXXVIII. Further, the added complications make arithmetic even harder.

There was another approach—the multiplicative approach adopted by other cultures, e.g., the Chinese-Japanese system. In order to maintain reasonably familiar symbols and to stay in a system which reads horizontally, we shall consider a hypothetical system of this type based on 5. In this system, a table of symbols for the first few numerals is devised. We shall use 1, 2, 3, 4 for their usual meaning. Further symbols standing for the base, its square, its cube, etc., are needed. We shall use the following:

$$5 \leftrightarrow h \text{ (hand)}$$
$$5^2 = 5 \cdot 5 \leftrightarrow q \text{ (a hand of hands)}$$
$$5^3 = 5 \cdot 5 \cdot 5 \leftrightarrow r \text{ (a hand of hands of hands)}$$
$$5^4 = 5 \cdot 5 \cdot 5 \cdot 5 \leftrightarrow s.$$

Since $47 = 1 \cdot 25 + 4 \cdot 5 + 2$, 47 in this system would be written as $1q4h2$; 1,999 would be $3s4q4h4$ since $1,999 = 3 \cdot 625 + 4 \cdot 25 + 4 \cdot 5 + 4$.

This method of notation is well on the way to the place notation we use tdoay. The Babylonians apparently were the first of several cultures which had an intermediate system, i.e., between the above multiplicative system and our present place-notational system. It is the one above without using the h, q, r, s, etc., or alternatively, it is our present system without the use of zero as a place marker. In this system, one had to know from context (using the familiar base 10 rather than the Babylonians' base 60) if 7 stood for 7, 70, 700, etc. Even harder to determine was whether things like 35 were 35, 350, 305, 3,005, etc.

The first known culture to achieve a fully developed place notation with zero was the Mayan civilization. About two thousand years ago, they developed a system of place notation based on 20. (Their system had one irregularity; for details, see [4, p. 69] or [17, Chap. 1].)

3 EXERCISES

1 Show that every natural number can be written as a sum of distinct powers of 2. How many ways can this be done?
2 Can every natural number be written as a sum of distinct powers of 3? What if repetitions are allowed, so that every power of 3 can be used twice if necessary?
3 How many other number systems (other than base-10) can you think of which are still used in certain words of our language or measures for certain trades?

Section 3 Place Notation for the Base b

In this section we shall develop a place-notation system to the base b, where b may be any positive integer exceeding 1.

4 Theorem

If n is any positive integer, then there is one and only one way to express n as a sum of powers of b such that no power of b occurs more than $b-1$ times. That is, we can find whole numbers k, a_0, a_1, ..., a_k such that $0 \leq a_0 < b-1$, $0 \leq a_1 \leq b-1$, ..., $0 \leq a_{k-1} \leq b-1$ and $1 \leq a_k \leq b-1$ and such that

$$n = a_k b^k + a_{k-1} b^{k-1} + \cdots + a_1 b + a_0 b^0.$$

Furthermore, there is exactly one way of doing this.

PROOF: We recall that by convention, $b^0 = 1$. Thus if $n \leq b-1$, we can write

$$n = n \cdot b^0 = a_0 b^0$$

with $a_0 = n$. Clearly this is the only way n can be expressed in the desired form. We proceed by induction to show that every natural number may be written in the desired form, the uniqueness will be handled later.

We suppose that every number less than n can be written uniquely in the desired form, and we show that n can be so written uniquely. Let $n - 1 = a_k b^k + a_{k-1} b^{k-1} + \cdots + a_1 b + a_0 b^0$.

If $a_0 \neq b-1$, then $n = (n-1) + 1 = a_k b^k + a_{k-1} b^{k-1} + \cdots + a_1 b + (a_0 + 1) b^0$ is a way of writing n in the desired form. If all the $a_i = b - 1$, then $n = (n-1) + 1 = (b-1)b^k + (b-1)b^{k-1} + \cdots + (b-1)b + (b-1) + 1 = b^{k+1}$ is a way of writing n in the desired form. If $a_0 = b - 1$ but not all the $a_i = b - 1$, pick j such that $a_0 = a_1 = \cdots = a_j = b - 1$, but $a_{j+1} \neq b - 1$. Thus

$$n = (n-1) + 1 = a_k b^k + \cdots + a_{j+1} b^{j+1} + (b-1) b^j + \cdots$$
$$+ (b-1)b + (b-1) + 1$$
$$= a_k b^k + \cdots + a_{j+1} b^{j+1} + b^{j+1}$$
$$= a_k b^k + \cdots + a_{j+2} b^{j+2} + (a_{j+1} + 1) b^{j+1}$$

Since $a_{j+1} \neq b - 1$ but $0 \leq a_{j+1} \leq b - 1$, we see that $0 \leq a_{j+1} + 1 \leq b - 1$, and n has been written in the desired form. Hence, in all cases, n can be written in the desired form.

We must now show that this representation is unique. Suppose not; then there is a smallest positive integer, say n, which can be written in two different ways.

Section 3 Place Notation for the Base b 397

Let
$$n = a_k b^k + a_{k-1} b^{k-1} + \cdots + a_1 b^1 + a_0 b^0$$
$$= c_l b^l + c_{l-1} b^{l-1} + \cdots + c_1 b^1 + c_0 b^0.$$

If $k = l$, then

$$n - b^k = (a_k - 1) b^k + a_{k-1} b^{k-1} + \cdots + a_1 b + a_0 b^0$$
$$= (c_k - 1) b^k + c_{k-1} b^{k-1} + \cdots + c_1 b + c_0 b^0,$$

which yields two representations of $n - b^k$, a contradiction since $n - b^k < n$. Thus $k \neq l$. We may suppose $k > l$. (Why?)

Consider $c_l b^l + c_{l-1} b^{l-1} + \cdots + c_1 b + c_0$. Since $c_i \leq b - 1$, we see that $n \leq (b - 1) b^l + (b - 1) b^{l-1} + \cdots + (b - 1) b + (b - 1)$. But $(b - 1) b^l + (b - 1) b^{l-1} + \cdots + (b - 1) + (b - 1) = (b^{l+1} - b^l) + (b^l - b^{l-1}) + \cdots + (b^2 - b) + (b - 1) = b^{l+1} - 1$ since each term, except the first and last, occurs once with a plus sign and once with a minus sign. Thus $n \leq b^{l+1} - 1$. On the other hand, $n = a_k b^k + \cdots + a_0 b^0 \geq b^k$. Since $k > l$, $n \geq b^{l+1}$. Putting these facts together, we get $n \geq b^{l+1} > b^{l+1} - 1 \geq n$, which is a contradiction. Thus the assumption of two distinct representations for n is contradictory and hence false.

5 Definition

If n is a positive integer, we say $a_k a_{k-1} \cdots a_1 a_0$ is *the base-b place-notational representation of n;* symbolically,

$$n = (a_k a_{k-1} \cdots a_1 a_0)_b$$

if and only if $0 \leq a_0 < b - 1$, $0 \leq a_1 < b - 1, \ldots, 0 \leq a_{k-1} \leq b - 1$, $1 \leq a_k \leq b - 1$, and $n = a_k b^k + a_{k-1} n^{k-1} + \cdots + a_1 b^1 + a_0$.

We shall write b in base-10 notation. Note that if $b = 10$, we normally don't write the parentheses or subscript 10. Thus $3^5 = (243)_{10} = 243$. On the other hand, $3^5 = (11110011)_2 = (100000)_3 = (183)_{12} = (\phi 3)_{16} = (\tau 3)_{20} = (43)_{60}$, where τ stands for 12 and ϕ for 15. In the future, when we write things to bases higher than 10, we shall use as required (up to $b - 1$) the new digits symbols listed below. This is necessary since $243 = (\tau 3)_{20} \neq (123)_{20} = 1 \cdot 20^2 + 2 \cdot 20 + 3 = 443$.

6 Notation

$10 \leftrightarrow t$ $15 \leftrightarrow \phi$
$11 \leftrightarrow e$ $16 \leftrightarrow s$
$12 \leftrightarrow \tau$ $17 \leftrightarrow \sigma$
$13 \leftrightarrow \theta$ $18 \leftrightarrow \eta$
$14 \leftrightarrow f$ $19 \leftrightarrow n$

7 EXERCISES

1. Prove that if $b \geq 2$ and n is any positive whole number, then two consecutive powers of b can be found so that

$$b^k \leq n < b^{k+1}$$

2. Find the expansion of 15 for bases 2, 3, 4, and 5.
3. Find the expansion of 23 for bases 2, 3, 4, 5, and 12.
4. Find the expansion of 37 for bases 2, 3, 12, 19, and 20.
5. Prove that if $b \geq 2$ and N is any integer, we can find numbers q and r (quotient and remainder) such that $N = qb + r$ and $0 \leq r < b$.

We still must consider the problem of how to find and write the number of a herd of sheep, a stack of dollars, or a gaggle of geese in a given base. The most frequently used method is simply to count the objects; however, in English we have a complete set of words for counting only in base 10. If the number is not too large, we have words for counting in base 12 (dozen, gross, etc.); while if it is small, we can count in base 20 (e.g., four score and seven). If we wish to count large numbers in base 12 or 20, we would need to make up new words. If we wished to count a natural number greater than the base, we would need new words in most bases.

If the objects to be counted are small and don't move by themselves, another method of determining their number in a given base, say b, is the following: We first put the objects into piles, each pile having b objects (divide by b). If there are some left over, we count that number (which is less than b); suppose it is a_0 (the remainder of the division). We next push the piles of b objects each into groups of b piles (divide the previous quotient by b). If there are some piles of b objects left over, we count the number of piles and call that number a_1 (the remainder of the second division). We continue grouping b of the groups together and counting how many are left over (divide the previous quotient by b and record the remainder as the next digit). We stop when we obtain fewer than b piles of a given size (when the quotient of the division becomes zero). When we have finished, $a_0, a_1, a_2, \ldots, a_k$ are the digits of the number for the base b; i.e., if N is the number of objects, then

$$N = (a_k a_{k-1} \cdots a_2 a_1 a_0)_b.$$

Section 3   Place Notation for the Base b   399

The division process proves the validity of the above method if it is followed through. Thus (see Problem 5 of Exercises 7),

$$N = q_1 b + a_0 \qquad 0 \leq a_0 < b$$

while $q_1 = q_2 b + a_1$, with $q_2 < q_1$ and $0 \leq a_1 < b$. So

$$N = (q_2 b + a_1)b + a_0 = q_2 b^2 + a_1 b + a_0$$

while $q_2 = q_3 b + a_2$, with $q_3 < q_2$ and $0 \leq a_2 < b$. So

$$N = q_3 b^3 + a_2 b^2 + a_1 b + a_0.$$

We continue until we get $q_k < b$, which must happen eventually since the q's are nonnegative and decreasing. Then we obtain

$$q_k = 0 \cdot b + a_k \qquad 0 \leq a_k = q_k < b$$

and thus

$$N = a_k b^k + a_{k-1} b^{k-1} + a_{k-2} b^{k-2} + \cdots + a_1 b + a_0.$$

The above method is often used in counting pennies.

A third method is the following: We find the greatest power of b, say b^k, which does not exceed the given number, say N. Thus $b^k \leq N < b^{k+1}$. We divide N by b^k and obtain $N = a_k b^k + r_1$, with $1 \leq a_k < b$ and $0 \leq r_1 < b^k$. We know that $a_k < b$ since $N < b^{k+1}$. We now divide r_1 by b^{k-1} and get $r_1 = a_{k-1} b^{k-1} + r_2$, with $0 \leq a_{k-1} < b$ and $0 \leq r_2 < b^{k-1}$. Thus $N = a_k b^k + a_{k-1} b^{k-1} + r_2$. We now repeat the procedure with r_2 and b^{k-2}, then with r_3 and b^{k-3}, etc. Eventually we get

$$N = a_k b^k + a_{k-1} b^{k-1} + \cdots + a_1 b^1 + a_0$$

where $1 \leq a_k < b$ and $0 \leq a_i < b$, $i = 0, 1, \ldots, k - 1$.

8 Examples

1 Write $N = (13{,}579)_{10}$ as a number in base 12.

METHOD 1: We could develop language—a dozen gross, for example, could be a "case"—and count, in base-12 language, 13,579 pencil marks. While theoretically this is the simplest method, we prefer the other two methods.

METHOD 2:   We compute (base 10) that

$$13{,}579 = 12 \cdot 1{,}131 + 7$$
$$1{,}131 = 12 \cdot 94 + 3$$
$$94 = 12 \cdot 7 + 10 = 12 \cdot 7 + t$$
$$7 = 0 \cdot 12 + 7$$

where we have written t for 10. We see that $(13{,}579)_{10} = 12[12(12 \cdot 7 + t) + 3] + 7$. Thus $(13{,}579)_{10} = (7t37)_{12}$.

METHOD 3:   The powers of 12 (written base 10) are 1, 12, 144, 1,728, 20,736 . . . . We calculate that

$$1{,}728 = 12^3 \leq N < 20{,}736 = 12^4$$
$$13{,}579 = 7 \cdot 12^3 + 1{,}483$$
$$1{,}483 = 10 \cdot 12^2 + 43$$
$$43 = 3 \cdot 12 + 7$$
$$7 = 7 \cdot 12^0 + 0$$

So $(13{,}579)_{10} = 7 \cdot 12^3 + (10 \cdot 12^2 + (3 \cdot 12 + 7))$. Thus $(13{,}579)_{10} = (7t37)_{12}$.

2   Write $N = (13{,}579)_{10}$ in base 3.

METHOD 2:   We compute that

$$13{,}579 = 3 \cdot 4{,}526 + 1$$
$$4{,}526 = 3 \cdot 1{,}508 + 2$$
$$1{,}508 = 3 \cdot 502 + 2$$
$$502 = 3 \cdot 167 + 1$$
$$167 = 3 \cdot 55 + 2$$
$$55 = 3 \cdot 16 + 1$$
$$18 = 3 \cdot 6 + 0$$
$$6 = 3 \cdot 2 + 0$$
$$2 = 3 \cdot 0 + 2.$$

Thus $(13{,}579)_{10} = (200121221)_3$.

METHOD 3:   The powers of 3 are 1, 3, 9, 27, 81, 243, 729, 2,187, 6,561, 19,683, ... We see that

$$3^9 = 6{,}561 \leq N \leq 19{,}683 = 3^{10}.$$

Section 3   Place Notation for the Base b

We compute
$$13{,}579 = 2 \cdot 6{,}561 + 457$$
$$457 = 0 \cdot 2{,}187 + 457$$
$$457 = 0 \cdot 729 + 457$$
$$457 = 1 \cdot 243 + 214$$
$$214 = 2 \cdot 81 + 52$$
$$52 = 1 \cdot 27 + 25$$
$$25 = 2 \cdot 9 + 7$$
$$7 = 2 \cdot 3 + 1$$
$$1 = 1 \cdot 1 + 0.$$

Thus $(13{,}579)_{10} = (200121221)_3$.

9   EXERCISES

1 Develop and prove an algorithm for addition of numbers in base-5 notation.

2 Develop and prove an algorithm for multiplication of numbers in base-3 notation.

3 How much memory work is needed to learn the usual multiplication and addition algorithms for the base b?

4 Perform the following:

$$\begin{array}{cc} (4t7)_{16} \\ +(e9)_{16} \\ \hline (\quad)_{16} \end{array} \qquad \begin{array}{cc} (t5e)_{12} \\ \times(398)_{12} \\ \hline (\quad)_{12} \end{array} \qquad (21)_4 \overline{)(3322113)_4}$$

5 In what base (or bases), if any, are each of the following operations valid? Prove your answers.

$$\begin{array}{r} 457 \\ +233 \\ \hline 712 \end{array} \qquad \begin{array}{r} 651 \\ +123 \\ \hline 774 \end{array} \qquad \begin{array}{r} 12211 \\ +22111 \\ \hline 112022 \end{array} \qquad \begin{array}{r} 321122 \\ +112211 \\ \hline 434033 \end{array}$$

6 In what base(s) are the following performed? Prove your answer.

$$\begin{array}{r} 231 \\ \times 133 \\ \hline 42323 \end{array} \qquad \begin{array}{r} 21 \\ \times 11 \\ \hline 231 \end{array} \qquad \begin{array}{r} 11011 \\ \times 111 \\ \hline 10111101 \end{array}$$

7 How many natural numbers have n or fewer digits when written in base-b notation?

Section 4 Some Properties of Natural Numbers Related to Notation

It is very easy to tell if a natural number written in base-10 notation is divisible by 10, namely, if it ends in 0; or if it is divisible by 5, namely, if it ends in 0 or 5; or if it is divisible by 2, namely, if it ends in 0, 2, 4, 6, or 8. Similarly, we can formulate tests for divisibility by the numbers which divide the base, for any base b.

10 Theorem

If d divides b and $d, b > 1$, then a natural number N is divisible by d if and only if d divides the last digit of the base-b representation of N.

PROOF: We shall need to recall the following lemma. (See Chapter 2, Exercises 36, Problem 6. Those readers who skipped Chapter 2 might wish to turn to it now and read paragraphs 36 to 41.)

11 Lemma

If $N = qd + r$, where q and r are integers, then N and r have the same remainder upon division by d.

In any case, the proof of the lemma is not difficult.

Let $N = (a_k a_{k-1} a_{k-2} \cdots a_1 a_0)_b$. Thus $N = (a_k a_{k-1} \cdots a_1)_b \cdot b + a_0$. Since d divides b, we may write $b = qd$ for some positive integer q. We obtain $N = (a_k a_{k-1} \cdots a_1)_b \cdot q \cdot d + a_0$. From the lemma we see that d divides N if and only if d divides a_0. The theorem is proved.

We recall (Chapter 2, paragraphs 35 to 41) that two numbers a and b are congruent for a modulus m, $a \equiv b \bmod m$, if and only if a and b have the same remainder upon division by m, or equivalently, m divides $a - b$. The previous theorem can be strengthened to:

12 Theorem

If d divides b and $N = (a_k a_{k-1} \cdots a_k a_0)_b$, then $N \equiv a_0 \bmod d$.

The proof of this theorem is left as an exercise.

13 EXERCISES

1. Prove Lemma 11.
2. Prove Theorem 12.
3. What is the remainder when 9 gross 7 dozen and 5 objects are divided into three piles?

Section 4   Some Properties of Natural Numbers   403

There are numbers other than the divisors of the base for which we obtain easy divisibility tests. For example, you may already know how to check arithmetic computation by casting out 9s (explained below) or test for divisibility by 3 in base 10. In general, we have:

14 Theorem

If d divides $b - 1$ and $N = (a_k \cdots a_1 a_0)_b$, then

$$N \equiv (a_k + a_{k-1} + \cdots + a_1 + a_1) \bmod d$$

15 Examples

1 $(53{,}796)_{10} \equiv 5 + 3 + 7 + 9 + 6 \equiv 30 \equiv 3 + 0 \equiv 3 \bmod 9$ or mod 3.

2 Casting out nines. This is a process for checking numerical computation based on the following propositions:

(a) If $a \equiv a'$ and $b \equiv b' \bmod 9$, then $a'b' \equiv ab$ and $a' + b' \equiv (a + b) \bmod 9$.

(b) Every number is congruent, mod 9, to the sum of its digits.

Consider the product $1{,}968 \times 5{,}376{,}941 = 10{,}581{,}819{,}888$ and the sum $1{,}968 + 5{,}376{,}941 = 5{,}378{,}909$. Now,

$$1{,}968 \equiv \dot{1} + 9 + 6 + \dot{8} \equiv 6 \bmod 9$$
$$5{,}376{,}941 \equiv \dot{5} + \ddot{3} + 7 + \ddot{6} + 9 + \dot{4} + 1 \equiv 8 \bmod 9$$
$$10{,}581{,}819{,}888 \equiv \dot{1} + 0 + 5 + \dot{8} + \dot{1} + \dot{8} + 9 + 8 +$$
$$8 + 8 \equiv 3 \bmod 9$$
$$5{,}378{,}909 \equiv 5 + 3 + 7 + 8 + 9 + 0 + 9 \equiv 5 \bmod 9$$

Furthermore, $6 \cdot 8 \equiv 3$ and $6 + 8 \equiv 5 \bmod 9$. If either of these last two congruences had been false, it would have indicated a numerical error in computation. The name "casting out nines" probably arose because in figuring the sum of the digits, all 9s and pairs of digits which add to 9 (marked with dots above) may be cast out (ignored) and in adding the other digits, anytime the sum exceeds 9 we may subtract 9 from it. (Why?)

3 $(156435)_7 \equiv (1 + 5 + 6 + 4 + 3 + 5) \bmod 6$. Here we have a practical choice to make: We can add in base 10, getting $(156435)_7 \equiv (24)_{10} \bmod 6$, and then actually divide 24 by 6 and get remainder zero, or we can add in base 7, $(156435)_7 \equiv (33)_7 \equiv 3 + 3 \equiv 6 \bmod 6$. In this example, the first choice is probably easier. The larger the number, the more likely it is

that the second method is better. Of course, your facility in working in other bases will also be a crucial factor.

4 $(9te\tau\theta 78e)_{15} \equiv 9 + t + e + \tau + \theta + 7 + 8 + e \equiv (56)_{15} \equiv 5 + 6 \equiv 11 \equiv 4 \bmod 7$.

5 This testing can be further simplified, if d is not too large, by replacing some or all of the digits by smaller numbers to which they are respectively congruent modulo d. Thus:

$$(53{,}796)_{10} \equiv 5 + 3 + 7 + 9 + 6 \equiv -4 + 3 - 2 + 0 - 3 \equiv -6 \equiv 3 \bmod 9$$

or $(53{,}796)_{10} \equiv -4 + 3 - 2 + 0 + 6 \equiv 3 \bmod 9$

$$(156435)_7 \equiv 1 + 5 + 6 + 4 + 3 + 5 \equiv 1 - 1 + 0 - 2 + 3 - 1 \equiv 0 \bmod 6$$

$$(9te\tau\theta 78e)_{15} \equiv 9 + t + e + \tau + \theta + 7 + 8 + e$$
$$\equiv 2 + 3 - 3 - 2 - 1 + 0 + 1 - 3$$
$$\equiv -3 \equiv 4 \bmod 7$$

or $(9te\tau\theta 78e)_{15} \equiv 2 + 3 - 3 - 2 - 1 + 9 + 1 + 4$
$$\equiv 4 \bmod 7$$

PROOF OF THEOREM 14: If d divides $b - 1$, then $b \equiv 1 \bmod d$. By the multiplicative properties of congruence, we have $b^2 \equiv 1 \cdot 1 \equiv 1 \bmod d$, $b^3 \equiv b^2 \cdot b \equiv 1 \cdot 1 \equiv 1 \bmod d$, $b^4 \equiv 1 \bmod d$, etc. By the definition of base notation, we have

$$N = a_k b^k + a_{k-1} n^{k-1} + \cdots + a_1 b + a_0.$$

Thus
$$N \equiv a_k \cdot 1 + a_{k-1} \cdot 1 + \cdots + a_1 \cdot 1 + a_0$$
$$\equiv (a_k + a_{k-1} + \cdots + a_1 + a_0) \bmod d.$$

16 EXERCISES

1 Devise a rule for testing divisibility by a number d which divides $b + 1$. Prove that your rule works. (Recall Chapter 2, Exercises 60, Problem 1.)

2 Do the same for divisors of $b^2 - 1$.

3 Likewise for divisors of $b^2 + 1$. (This is connected with Problem 1.)

4 What is the remainder upon dividing $(12{,}345{,}678{,}987{,}654{,}321)_{10}$ by 101?

5 What is the remainder upon dividing 11 gross 6 dozen and 5 by 13?

6 Explain why the casting out described in the last sentence of Examples 15 number 2 is a valid process.

Section 4   *Some Properties of Natural Numbers*   405

It is frequently important to know whether or not a whole number is a perfect square. For small numbers, say less than 500, it is not difficult for most people to tell; they can use their knowledge of the squares of the first few numbers and a little computation. If a number is large, say more than 10 digits, then it is rather difficult to determine if it is a square by straightforward computation and use of our experiences with small numbers. Since in base-10 notation, if $N = (a_k a_{k-1} \cdots a_1 a_0)_{10}$, then $N \equiv (a_1 a_0) \bmod 100$, we were able to show in Theorem 56 of Chapter 2 (as can be verified by looking at a table of squares of the first 100 numbers) that N cannot be a perfect square unless its last two digits are one of the pairs:

$$
\begin{array}{llllllll}
00 & 01 & 04 & 09 & 16 & 21 & 24 & 25 \\
29 & 36 & 41 & 44 & 49 & 56 & 61 & 64 \\
69 & 76 & 81 & 84 & 89 & 96.
\end{array}
$$

Since there are 22 pairs, this method eliminates, on the average, 78 out of every 100 numbers considered, without any computation. While this may seem very good, and indeed it is often quite helpful, let us consider the question of how many of the positive integers are squares. The first answer is that there are infinitely many squares, namely, 1, 4, 9, 25, 36, 49, ... This does not get at the real question of what fraction of the numbers are squares. Out of the first 100 numbers, 10 or $\frac{1}{10}$ of the numbers, are perfect squares; out of the first 10,000, there are 100, or $\frac{1}{100}$, perfect squares. In fact, out of the first N numbers, roughly $\sqrt{N}$ numbers are squares. Of course, $\sqrt{N}$ may not be an integer, in which case the number of squares will be the greatest whole number less than (or equal to) $\sqrt{N}$. We denote this whole number by $[\sqrt{N}]$.

This handy notation will be used again, and therefore we make the following definition.

17   Definition

If x is any real number, then $[x]$ denotes the greatest integer not exceeding x.

18   Examples

$[\sqrt{2}] = 1$, $[\pi] = 3$, $[100.73] = 100$, $[17] = 17$, $[-13] = -13$, $[-7.6] = -8$.

19 EXERCISES

1. Find $[\sqrt{7}]$, $[\sqrt{101}]$, $[\frac{8.5}{7}]$, $[-\frac{9.3}{5}]$.
2. Find $[\sqrt[3]{17}]$, $[-(4\pi/3)]$, $[\sqrt[5]{100}]$.
3. Show that if x is not an integer, then $[-x] \equiv -[x] - 1$.
4. Draw a graph of $y = [x]/x$ for $0 < x \leq 10$.
5. How large must N be, in terms of q, in order that the inequalities $1 \geq [x]/x \geq 1 - 1/q$ hold for all values of $x \geq N$?

We now prove the following theorem.

20 Theorem

Let q be a given positive integer. If $N > q^2$, then the ratio of the number of integers not exceeding N which are squares to the number of positive integers not exceeding N will be less than $1/q$.

PROOF: By our previous discussion, we know that the number of perfect squares less than or equal to N is $[\sqrt{N}]$. Thus the desired ratio is $[\sqrt{N}]/N$, but

$$\frac{[\sqrt{N}]}{N} \leq \frac{\sqrt{N}}{N} = \frac{1}{\sqrt{N}}.$$

If $N > q^2$, then $\sqrt{N} > q$, and thus $1/\sqrt{N} < 1/q$. Combining these inequalities, we obtain

$$\frac{[\sqrt{N}]}{N} < \frac{1}{q}.$$

Thus the ratio of squares to natural numbers not exceeding N is less than $1/q$, which proves the theorem.

21 Corollary

The ratio of squares to all positive integers can be made as close to zero as we like by considering a large enough initial segment of the positive integers.

This corollary is usually stated that the limit of the ratio of squares to positive integers is zero, or that the density of the squares is zero.

In view of Theorem 20, our test by looking at the last two digits in base-10 notation is not all that one could desire. Perhaps a more judicious choice of base would be helpful. We shall make an analysis for base 12.

The digits in the base-12 system are 0, 1, 2, 3, 4, 5, 6, 7, 8, 9, t, e. We shall prove the following theorem.

22 Theorem

A number $N = (a_k a_{k-1} \cdots a_1 a_0)_{12}$ cannot be a perfect square unless its last two digits are one of the following: 00, 01, 04, 09, 14, 21, 30, 41, 44, 54, 61, 69, 81, 84, 94, t1.

PROOF: We first note that for any number $n = (b_l b_{l-1} \cdots b_1 b_0)_{12}$,

$$n \equiv (b_1 b_0)_{12} \bmod 144.$$

Thus $n^2 \equiv (b_1 b_0)_{12}^2 \bmod 144$, so that the last two digits of n^2 are the same as the last two digits of $(b_1 b_0)_{12}^2$. Consequently, we need only consider the squares of one- and two-digit numbers, since $b_1 = 0$ is possible. In base-12 notation, there are 144 such numbers—0 through 143 inclusive.

We could simply square these numbers, but certain simplifications are possible.

First we notice that if $72 \leq m \leq 143$, then $(m - 72)^2 = m^2 - 144m + 72^2 \equiv m^2 \bmod 144$ and $0 \leq m - 72 < 72$. Thus we need only consider squares of numbers up to 72.

If m is a number less than 72, then

$$(72 - m)^2 \equiv 72^2 - 2 \cdot 72 + m^2 \equiv m^2 \bmod 144.$$

Thus $(72 - m)^2$ and m^2 have the same last digits in base-12 notation. Since for $37 \leq m \leq 72$, we have $0 \leq 72 - m \leq 36$, we need only consider the first 37 numbers, $0, 1, 2, \ldots, t, e, (10)_{12}, (11)_{12}, \ldots, (30)_{12}$. At this point, one can either make further reduction by considering the cases m even, $m \equiv 0 \bmod 3$, etc. Or we can simply square the above 37 numbers in base 12 and verify that all the squares are as stated in the theorem. Thus, where all the numbers in this chart are written base 12,

$$
\begin{array}{lll}
0^2 = 0 = 00 & 10^2 = 100 & 20^2 = 400 \\
1^2 = 1 = 01 & 11^2 = 121 & 21^2 = 441 \\
2^2 = 4 = 04 & 12^2 = 144 & 22^2 = 484 \\
3^2 = 9 = 09 & 13^2 = 169 & 23^2 = 509 \\
4^2 = 14 & 14^2 = 194 & 24^2 = 554 \\
5^2 = 21 & 15^2 = 201 & 25^2 = 5t1 \\
6^2 = 30 & 16^2 = 230 & 26^2 = 630 \\
7^2 = 41 & 17^2 = 261 & 27^2 = 681 \\
8^2 = 54 & 18^2 = 294 & 28^2 = 714 \\
9^2 = 69 & 19^2 = 309 & 29^2 = 769 \\
t^2 = 84 & 1t^2 = 344 & 2t^2 = 804 \\
e^2 = t1 & 1e^2 = 381 & 2e^2 = 861 \\
 & & 30^2 = 900
\end{array}
$$

Since every perfect square is congruent to one of these mod 12, the theorem is established.

We can see that base 10 allows 22 possibilities out of 100 while base 12 allows only 16 possibilities out of 144, or $11\frac{1}{9}$ percent. This is almost twice as good as the 22 percent of base 10.

This whole idea can be generalized to the study of what remainder types for various moduli can be perfect squares. Since in base-10 notation it is easy to deduce the remainder for certain moduli, e.g., 9, 10, 11, we could also look at the restrictions imposed by these moduli. By using combinations of these methods, one can frequently eliminate from consideration almost all numbers which are not perfect squares. Only a perfect square can be congruent to a perfect square for all moduli.

The student should note that questions concerning the last one, two, or three digits of a number in base-b notation can always be rephrased in terms of congruences modulo b, b^2, or b^3, respectively.

23 EXERCISES

1 What are the possible final two digits of a perfect square in base-8 notation? What are the possible values of the final digit? Verify your results.

2 What are the possibilities for the last digit of a perfect cube in base-10 notation? What about base 8? Comment.

3 What information can you deduce from $x^3 \equiv y^3$ mod 2? mod 3? mod 5?

4 What are the possible last two digits for fourth powers in base 8?

5 What are the possible last two digits for fourth powers in base 5? (See Fermat's theorem, Corollary 67, Chapter 2.)

*6 Prove that a number is a perfect square if and only if it is congruent to a square for every modulus. Hint: find a prime p so that p^{2n} doesn't divide N, but p^{2n-1} does divide N when N is not a perfect square.

We turn now to a different aspect of our notational system. Some numbers are distinctive simply because of the number system used. Consider these numbers: 1,000,000; 222,222; 777,777,555,555; and 123,123,123,123. They stand out as numbers of special form when written in base 10; however, in other bases, these numbers might be totally undistinguished.

The next several results will deal with the question which, crudely put, reads as follows: Do all the digits get used a fair share of the time in most numbers?

We need to define certain notions in order to make this question precise.

24 Definition

Given an increasing sequence of positive integers, for example, $a_1 < a_2 < a_3 < a_4 < \cdots$, consider the fraction whose numerator is the number of integers in the sequence which do not exceed n and whose denominator is the number n, we say that this *sequence has density d* if and only if for every positive integer q there can be found another positive integer Q such that the above fraction is between $d - 1/q$ and $d + 1/q$ whenever n is greater than Q.

This concept might also be defined as follows: Given $a_1 < a_2 < a_3 \leq \cdots$, let $C(n)$ be the number of terms of the sequences which are less than or equal to n. We say that the density of the sequence is d if and only if for every positive integer q there can be found another integer Q such that if $n > Q$, then $d - 1/q < C(n)/n < d + 1/q$.

Intuitively this says that the ratio $C(n)/n$ can be made as close to d as we like if we look at a long enough segment of the numbers; that is, $C(n)/n$ approaches d as a limiting value when n gets large.

25 Examples

1. Let $a_1 = 1$, $a_2 = 2$, $a_3 = 3$, and in general, $a_m = m$. In this case, $C(n) = n$; thus no matter what the value of q, we can take $Q = 0$ and $1 - 1/q < C(n)/n = 1 < 1 + 1/q$ will hold for all $n > Q$. Thus the sequence of all integers has density 1.

2. Let $a_1 = 2$, $a_2 = 4$, $a_3 = 6$, ..., $a_m = 2m$, ... Intuitively, since one-half the numbers are even, the density should be $\frac{1}{2}$. $C_n = [n/2]$; that is, the greatest integer is $n/2$. Thus $C(n)/n = [n/2]/n$. It follows that $C(n)/n \leq n/2 \cdot 1/n = \frac{1}{2}$. Also, since $[n/2] \geq (n-1)/2$, we see that $C(n)/n \geq (n-1)/2 \cdot 1/n = \frac{1}{2} - 1/2n$. Thus, for a given positive integer q, we may take $Q = q$. For, if $n > q$, then $\frac{1}{2} - 1/q < \frac{1}{2} - 1/2n \leq C(n)/n \leq \frac{1}{2} < \frac{1}{2} + 1/q$. Thus the sequence has density $\frac{1}{2}$.

26 EXERCISES

1 Show that the sequence 3, 6, 9, 12, . . . of numbers divisible by 3 has density $\frac{1}{3}$.

2 Show that for $m \geq 1$, the sequence of numbers divisible by m, that is, the sequence $m, 2m, 3m, 4m, \ldots$ has density $1/m$.

3 Show that the sequence 5, 15, 25, 35, . . . of numbers ending in 5 has density $\frac{1}{10}$.

4 Show that the sequence of all integers greater than 10, that is, 11, 12, 13, 14, . . . has density 1.

5 If k is any integer, show that the sequence of even integers which are greater than k has density $\frac{1}{2}$.

6 Show that the sequence of perfect squares, 1, 4, 9, 16, 25, . . . , has density 0 (see Theorem 20).

7 Show that the sequence of perfect cubes has density 0.

*8 Construct a sequence which has no density. (This is not the same as having a density which is 0.) Can you prove that it has no density?

We begin our discussion of the distribution of digits by discussing the density of the sequence of positive integers which have some digit equal to 9—for example, 905, 23,796, 111,159, and 43,102,918 are numbers in this sequence.

We shall use $C(n)$ to denote the number of integers in our sequence (i.e., with some digit equal to 9) among the first n positive integers 1, 2, . . . , n. Simple calculation reveals that $C(10) = 1$, $C(20) = 2$, $C(30) = 3$, $C(100) = 19$. Thus $C(10)/10 = 0.1$, $C(20)/20 = 0.1$, $C(30)/30 = 0.1$, $C(100)/100 = 0.19$. This might lead one to suspect that the density of this sequence (if it has a density) is between 0.1 and 0.2. In fact, this is not the case.

27 EXERCISES

1 Compute $C(7)$, $C(10)$, $C(60)$, and $C(67)$.

2 Compute $C(100)$, $C(300)$, $C(367)$.

3 Compute $C(1,000)$, $C(4,000)$, and $C(4,367)$.

4 Compute $C(10,000)$, $C(80,000)$, and $C(84,367)$.

5 What are $C(76,543)$ and $C(63,248)$?

6 Can you guess a formula for $C[(a_k a_{k-1} \cdots a_1 a_0)_{10}]$ if none of $a_0, a_1, a_2, \ldots, a_k$ is 9?

*7 Can you guess a formula for $C(N)$?

We now prove an intuitively surprising theorem.

28   Theorem
Let 9, 19, 29, . . . be the sequence of all positive integers which have some digit equal to 9 in their base-10 expansion. This sequence has density 1.

29   *Terminology*
If a sequence of positive integers with a given property (e.g., a digit equal to 9) has density 1, we say that almost all positive integers have the property (or almost every integer has the property). If the density is 0, we say that almost no integers have the property.

Thus Theorem 28 could be stated as:

30   Theorem
Almost all positive integers have some digit equal to 9 in their base-10 expansion.

We shall begin our proof of the theorem with some lemmas on the values of $C(n)$. It is not difficult to compute $C(10^k)$.

31   Lemma
If $k \geq 1$, $C(10^k) = 10^k - 9^k$.

PROOF: We first count the number of positive integers less than 10^k which have no 9 in their expansion. That is, we count the number of integers with k or fewer digits which have no digit equal to 9. There are nine choices —0, 1, 2, 3, 4, 5, 6, 7, 8—for the first digit; of course, if the first digit is 0, it is normally not written. There are the same nine choices for each of the k digits; thus there are 9^k choices all told. However, we should not count the choice of every digit equal to 0, since we are considering positive integers, but we should count 10^k which we have skipped. Hence we see that there are 9^k positive integers less than or equal to 10^k which have none of their digits equal to 9. Thus we conclude that $C(10^k) = 10^k - 9^k$, and the lemma is proved.

The next step toward a proof of Theorem 28 is to estimate $C(n)$ for arbitrary values of n. Although an exact expression for $C(n)$ in terms of the base-10 expansion of n is possible (Exercises 27, Problem 7), it will suffice for our purposes to obtain simpler upper and lower bounds.

32 Lemma

If $k \geq 1$ and $10^k > n > 10^{k-1}$, then $n \geq C(n) > n - 9^k$.

PROOF: The right-hand inequality is clear; we prove only the left-hand one. We first notice that $n - C(n)$ is the number of integers less than or equal to n which have no digit equal to 9 in their decimal expansion. Thus, if $n \leq m$, we see that $n - C(n) \leq m - C(m)$. In particular, since $n \leq 10^k$, we obtain

$$33 \qquad n - C(n) \leq 10^k - C(10^k).$$

From Formula 33 and Lemma 31, we get

$$C(n) \geq n - 10^k + C(10^k) = n - 10^k + 10^k - 9^k = n - 9^k,$$

which gives the desired inequality.

We need one final lemma.

34 Lemma

If t is a positive integer and $k \geq 9t$, then $(\frac{9}{10})^k < 1/t$.

PROOF: We recall the identity $(x - 1)(1 + x + x^2 + \cdots + x^{k-1}) = x^k - 1$, where k is any positive integer. Put $x = \frac{10}{9}$ and obtain

$$35 \qquad \tfrac{1}{9}[1 + \tfrac{10}{9} + (\tfrac{10}{9})^2 + \cdots + (\tfrac{10}{9})^{k-1}] = (\tfrac{10}{9})^k - 1.$$

Since $\frac{10}{9} > 1$, so are all the powers of $\frac{10}{9}$, i.e., $(\frac{10}{9})^2 > 1$, $(\frac{10}{9})^3 > 1$, etc. Thus $1 + \frac{10}{9} + (\frac{10}{9})^2 + \cdots + (\frac{10}{9})^{k-1} > 1 + 1 + \cdots + 1 = k$. Combining this inequality with Formula 35, we get $\frac{1}{9} \cdot k < (\frac{10}{9})^k - 1$. It follows that $k/9 < (\frac{10}{9})^k$. Or, taking reciprocals, we obtain $9/k > (\frac{9}{10})^k$. If $k \geq 9t$, we get the inequality

$$\left(\frac{9}{10}\right)^k < \frac{9}{k} \leq \frac{9}{9t} = \frac{1}{t}.$$

The lemma is established.

We are now in a position to prove Theorem 28 and thus Theorem 30.

PROOF: We must show that, given q, we can find a number Q so that if $n > Q$, the inequality $1 + 1/q > C(n)/n > 1 - 1/q$ holds.

We know from Lemma 32 that with k replaced by $k + 1$, $n \geq C(n) \geq n - 9^{k+1}$ if $10^{k+1} \geq n > 10^k$. Thus $1 \geq C(n)/n \geq 1 - 9^{k+1}/n$. Since $n > 10^k$, $9^{k+1}/n < 9^{k+1}/10^k = 9(\frac{9}{10})^k$. So we get

36 $$1 \geq \frac{C(n)}{n} \geq 1 - 9\left(\frac{9}{10}\right)^k.$$

Given q, we now pick $Q = 10^{81q}$. With this choice of Q, if $n > Q$, then $n > 10^{81q}$ and hence $k \geq 81q = 9(9q)$. From Lemma 34 we see that with $t = 9q$, we obtain $(\frac{9}{10})^k < 1/9q$. Thus we obtain from Formula 36

$$1 + \frac{1}{q} > 1 \geq \frac{C(n)}{n} \geq 1 - 9\left(\frac{9}{10}\right)^k > 1 - 9 \cdot \frac{1}{9q} = 1 - \frac{1}{q}.$$

The theorem is proved.

Much stronger and more surprising results are true; see, for example, Problem 5 of Exercises 37.

37 EXERCISES

1. Prove that $C(n) \leq n - 9^k$ for $10^k \leq n \leq 10^{k+1}$.
2. Prove that almost all integers have some digit equal to 8 in their base-10 representation.
3. Prove that almost all integers have some digit equal to zero in their base-10 expansion.
4. Prove that, for any fixed digit, say r, almost all integers have some digit equal to r.
*5. Prove that almost all integers have all 10 digits, 0, 1, 2, ... , 9, occurring in their base-10 expansion. [Hint: If D_r is the number of integers less than m which have no digit equal to r, then at least $m - (D_0 + D_1 + \cdots + D_9)$ integers have all 10 digits in their expansion.]
6. Prove that given a positive integer t, there is a positive integer t' such that for $k \geq t'$, $(\frac{1}{2})^k < 1/t$. (See Lemma 34.)
7. Prove that if $\alpha > 1$ and t is a positive integer, then there is a positive integer t' such that wherever $k > t'$ we know $\alpha^k < 1/t$.
8. Prove that almost every integer has some digit equal to 1 in its base-b expansion.

*9 Prove that if $0 \leq r < b$, then almost every integer has some digit equal to r in its base-b expansion.

*10 Prove that for $b > 1$, almost every integer has all the digits $0, 1, 2, \ldots, b-1$ occurring in its base-b expansion.

*11 Prove that almost every integer has all the digits $0, 1, 2, \ldots, 9$ occurring in its base-10 expansion and at the same time has all the digits $0, 1, 2, \ldots, t, e$ occurring in its base-12 expansion.

*12 Prove that almost every integer has the pair 27 occurring in its base-10 expansion.

It is possible to prove a strengthened version of Problem 11, namely: For any finite number of bases, almost all numbers have all possible digits in all of these bases.

It is also possible to show that almost all integers have each possible digit appearing roughly one-tenth of the time in their base-10 expansions.

Section 5 Fractions: First Comments

Historically, our present system, of Babylonian origin, for writing fractions as the ratio of two integers seems to have had one major rival. The rival was the Egyptian system, which dates back to perhaps 3400 B.C. (Smith [21]). We use the word "fraction" in the vulgar sense of one integer divided by another. The Egyptians had symbols for writing the reciprocals of integers, or *unit fractions*, as they are sometimes called—e.g. (in modern notation), $\frac{1}{2}, \frac{1}{3}, \frac{1}{4}, \frac{1}{20}, \frac{1}{507}$. Other fractions, with certain exceptions, were written as sums of these so-called unit fractions without repetitions. For example, in our present notation, the Egyptians might have written $\frac{1}{2} + \frac{1}{3}$ for $\frac{5}{6}$, $\frac{1}{2} + \frac{1}{4}$ for $\frac{3}{4}$, $\frac{1}{4} + \frac{1}{3}$ for $\frac{7}{12}$.

From a mathematical point of view, this method of writing fractions raises many interesting questions. Can every fraction be written this way? In how many ways can a given fraction be written? Can every fraction be written as a sum of two unit fractions? How can we add fractions in Egyptian notation? Multiply them?

Section 5   Fractions: First Comments

Leather scroll in the British Museum, containing simple relations between fractions. The scroll is said to have been found together with the Rhind papyrus (circa 1700 B. C.), and at first the contents were disappointing; every line gave a simple relation between fractions, such as $\overline{9} + \overline{18} = \overline{6}, \overline{5} + \overline{20} = \overline{4}$, etc. However, the scroll has proved its value because it supplies a key for understanding the first stages of calculation with fractions. (Courtesy of the Journal of Egyptian Archeology, Vol. 12 (1927), p. 232.)

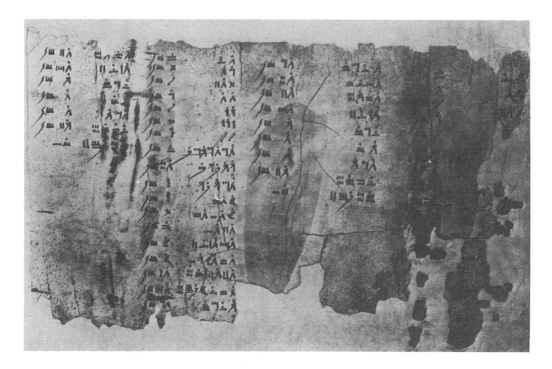

The Egyptian system seemed to be more dominant than our present system for some time; for example, Archimedes used this method of writing fractions. Because we are more familiar with the present system, it is convenient for us to study first our system, which seems to have come to Europe from the Hindus, and to return to the Egyptian system later when our understanding of the mathematical aspects of our system is better.

It is assumed that the reader can add, subtract, multiply, divide, reduce, and compare fractions when they are written in the usual way. It is assumed that the denominators of fractions are always positive integers.

38 EXERCISES

1 Are there other ways of writing $\frac{5}{6}$, $\frac{3}{4}$ as Egyptian fractions?

2 Write the following fractions in the Egyptian style (if possible): $\frac{5}{12}$, $\frac{9}{11}$, $\frac{17}{23}$.

3 Is there a method that enables one to write every fraction in the Egyptian system? Can you prove your answer correct?

Section 6 Farey Fractions

In 1816 a minerologist named Farey published a mathematical paper in which he discussed the properties of the sequence of all nonnegative reduced proper fractions with denominator not exceeding n arranged in increasing order. For every choice of $n = 1, 2, 3, \ldots$, we obtain such a sequence. These sequences have since been called the *Farey sequences of order n*, although he was not the first to consider them [7, Vol. 1]. We denote them by $\mathfrak{F}_n$.

Table 39 on the next page lists these sequences for $1 \leq n \leq 11$.

40 EXERCISES

1 Can you make any observations about these sequences? Can you prove any of your observations?

2 Prove that for $n \geq 2$, $\mathfrak{F}_n$ has an odd number of terms.

3 Write out $\mathfrak{F}_{12}$ and $\mathfrak{F}_{13}$ explicitly.

4 Show that if n is odd, the fractions next to $\frac{1}{2}$ in $\mathfrak{F}_n$ are $\frac{(n-1)/2}{n}$ and $\frac{(n+1)/2}{n}$. (What fractions with denominator $k < n$ are closest to $\frac{1}{2}$?)

5 Show that if n is even, the fractions next to $\frac{1}{2}$ in $\mathfrak{F}_n$ are the same as in $\mathfrak{F}_{n-1}$.

6 Compute the successive differences of the fractions in $\mathfrak{F}_5$. Do you note any pattern? Can similar remarks be made for $\mathfrak{F}_6$, $\mathfrak{F}_7$, $\mathfrak{F}_8$, $\mathfrak{F}_n$?

7 If p/q is in $\mathfrak{F}_n$, is $1 - p/q$ also in $\mathfrak{F}_n$?

Section 6 Farey Fractions

We begin by investigating which fractions can fall between other fractions.

41 Lemma

If a/b and c/d are two fractions, with $a/b < c/d$, and if h and k are any positive integers, then $a/b < (ha + kc)/(hb + kd) < c/d$.

39 Table

$\mathcal{F}_1$: $\frac{0}{1}$, $\frac{1}{1}$

$\mathcal{F}_2$: $\frac{0}{1}$, $\frac{1}{2}$, $\frac{1}{1}$

$\mathcal{F}_3$: $\frac{0}{1}$, $\frac{1}{3}$, $\frac{1}{2}$, $\frac{2}{3}$, $\frac{1}{1}$

$\mathcal{F}_4$: $\frac{0}{1}$, $\frac{1}{4}$, $\frac{1}{3}$, $\frac{1}{2}$, $\frac{2}{3}$, $\frac{3}{4}$, $\frac{1}{1}$

$\mathcal{F}_5$: $\frac{0}{1}$, $\frac{1}{5}$, $\frac{1}{4}$, $\frac{1}{3}$, $\frac{2}{5}$, $\frac{1}{2}$, $\frac{3}{5}$, $\frac{2}{3}$, $\frac{3}{4}$, $\frac{4}{5}$, $\frac{1}{1}$

$\mathcal{F}_6$: $\frac{0}{1}$, $\frac{1}{6}$, $\frac{1}{5}$, $\frac{1}{4}$, $\frac{1}{3}$, $\frac{2}{5}$, $\frac{1}{2}$, $\frac{3}{5}$, $\frac{2}{3}$, $\frac{3}{4}$, $\frac{4}{5}$, $\frac{5}{6}$, $\frac{1}{1}$

$\mathcal{F}_7$: $\frac{0}{1}$, $\frac{1}{7}$, $\frac{1}{6}$, $\frac{1}{5}$, $\frac{1}{4}$, $\frac{2}{7}$, $\frac{1}{3}$, $\frac{2}{5}$, $\frac{3}{7}$, $\frac{1}{2}$, $\frac{4}{7}$, $\frac{3}{5}$, $\frac{2}{3}$, $\frac{5}{7}$, $\frac{3}{4}$, $\frac{4}{5}$, $\frac{5}{6}$, $\frac{6}{7}$, $\frac{1}{1}$

$\mathcal{F}_8$: $\frac{0}{1}$, $\frac{1}{8}$, $\frac{1}{7}$, $\frac{1}{6}$, $\frac{1}{5}$, $\frac{1}{4}$, $\frac{2}{7}$, $\frac{1}{3}$, $\frac{3}{8}$, $\frac{2}{5}$, $\frac{3}{7}$, $\frac{1}{2}$, $\frac{4}{7}$, $\frac{3}{5}$, $\frac{5}{8}$, $\frac{2}{3}$, $\frac{5}{7}$, $\frac{3}{4}$, $\frac{4}{5}$, $\frac{5}{6}$, $\frac{6}{7}$, $\frac{7}{8}$, $\frac{1}{1}$

$\mathcal{F}_9$: $\frac{0}{1}$, $\frac{1}{9}$, $\frac{1}{8}$, $\frac{1}{7}$, $\frac{1}{6}$, $\frac{1}{5}$, $\frac{2}{9}$, $\frac{1}{4}$, $\frac{2}{7}$, $\frac{1}{3}$, $\frac{3}{8}$, $\frac{2}{5}$, $\frac{3}{7}$, $\frac{4}{9}$, $\frac{1}{2}$, $\frac{5}{9}$, $\frac{4}{7}$, $\frac{3}{5}$, $\frac{5}{8}$, $\frac{2}{3}$, $\frac{5}{7}$, $\frac{3}{4}$, $\frac{7}{9}$, $\frac{4}{5}$, $\frac{5}{6}$, $\frac{6}{7}$, $\frac{7}{8}$, $\frac{8}{9}$, $\frac{1}{1}$

$\mathcal{F}_{10}$: $\frac{0}{1}$, $\frac{1}{10}$, $\frac{1}{9}$, $\frac{1}{8}$, $\frac{1}{7}$, $\frac{1}{6}$, $\frac{1}{5}$, $\frac{2}{9}$, $\frac{1}{4}$, $\frac{2}{7}$, $\frac{3}{10}$, $\frac{1}{3}$, $\frac{3}{8}$, $\frac{2}{5}$, $\frac{3}{7}$, $\frac{4}{9}$, $\frac{1}{2}$, $\frac{5}{9}$, $\frac{4}{7}$, $\frac{3}{5}$, $\frac{5}{8}$, $\frac{2}{3}$, $\frac{7}{10}$, $\frac{5}{7}$, $\frac{3}{4}$, $\frac{7}{9}$, $\frac{4}{5}$, $\frac{5}{6}$, $\frac{6}{7}$, $\frac{7}{8}$, $\frac{8}{9}$, $\frac{9}{10}$, $\frac{1}{1}$

$\mathcal{F}_{11}$: $\frac{0}{1}$, $\frac{1}{11}$, $\frac{1}{10}$, $\frac{1}{9}$, $\frac{1}{8}$, $\frac{1}{7}$, $\frac{1}{6}$, $\frac{2}{11}$, $\frac{1}{5}$, $\frac{2}{9}$, $\frac{1}{4}$, $\frac{3}{11}$, $\frac{2}{7}$, $\frac{3}{10}$, $\frac{1}{3}$, $\frac{4}{11}$, $\frac{3}{8}$, $\frac{2}{5}$, $\frac{3}{7}$, $\frac{4}{9}$, $\frac{5}{11}$, $\frac{1}{2}$, $\frac{6}{11}$, $\frac{5}{9}$, $\frac{4}{7}$, $\frac{3}{5}$, $\frac{5}{8}$, $\frac{7}{11}$, $\frac{2}{3}$, $\frac{7}{10}$, $\frac{5}{7}$, $\frac{8}{11}$, $\frac{3}{4}$, $\frac{7}{9}$, $\frac{4}{5}$, $\frac{9}{11}$, $\frac{5}{6}$, $\frac{6}{7}$, $\frac{7}{8}$, $\frac{8}{9}$, $\frac{9}{10}$, $\frac{10}{11}$, $\frac{1}{1}$

PROOF: We recall that $a/b < c/d$ is equivalent to $ad < bc$. To show that $a/b < (ha + kc)/(hb + kd)$, we shall show that $a(hb + kd) < b(ha + kc)$. Multiplying both sides out, we obtain the equivalent inequality $ahb + akd < bha + bkc$. Since $ahb = bha$, this last inequality is equivalent to $akd < bkc$, which is true since $ad < bc$ and k is a positive integer.

Similarly, since $ad < bc$, we see that $had < hbc$. Thus $had + kcd < hbc + kcd$, which is equivalent to $(ha + kc)/(hb + kd) < c/d$. The proof of the lemma is complete.

42 Lemma

If $a/b < c/d$, $bc - ad = 1$, and $a/b < e/f < c/d$, then for some positive integers h and k, $e = ha + kc$ and $f = hb + kd$.

PROOF: We shall solve the equations

$$\text{43} \qquad ha + kc = e$$
$$\text{44} \qquad hb + kd = f$$

for the unknowns h and k and show that these solutions are positive integers. First we multiply Equation 44 by c and Equation 43 by $-d$ and add. Thus

$$\begin{aligned} hcb + kcd &= cf \\ -had - kcd &= -de \\ \hline h(bc - ad) &= cf - de. \end{aligned}$$

By hypothesis, $bc - ad = 1$, so that $h = cf - de$.

Similarly, we multiply Equation 43 by b and Equation 44 by $-a$ and obtain $k = eb - af$. Since a, b, c, d, e, and f are all integers, we see that h and k are also integers. Furthermore, since h is the numerator of the positive difference $c/d - e/f$, we see that h is positive. Similarly, k comes from the difference $e/f - a/b$, which is also positive. Thus h and k fulfill all the conditions, so the theorem is proved.

We are now in a position to prove two fundamental properties of Farey series.

45 Theorem

Let a/b and c/d be reduced fractions, with $a/b < c/d$.

1 If a/b and c/d are adjacent in $\mathfrak{F}_{n-1}$ but separated by e/f in $\mathfrak{F}_n$ for $n \geq 2$, then $e/f = (a + c)/(b + d)$ and, in fact, $e = a + c$ and $f = b + d = n$.
2 If a/b and c/d are adjacent in $\mathfrak{F}_n$, then $bc - ad = 1$.

If $n = 1$, we can interpret $\mathfrak{F}_{n-1} = \mathfrak{F}_0$ as the sequence consisting of no terms and statement 1 is true for $n = 1$ also.

PROOF: We note from Table 39 that the theorem is true for the first few values of n. We suppose that the theorem is true for $n \leq m$, and we shall show that it is true for $n = m + 1$. We prove statement 1 first.

Let a/b, c/d, e/f be as in the hypothesis of statement 1. Then a/b and c/d are adjacent in $\mathfrak{F}_m$, and by the induction hypothesis, $bc - ad = 1$. Thus the hypothesis of Lemma 41 is satisfied, so that all the fractions (reduced and unreduced) between a/b and c/d are precisely those of the form $(ha + kc)/hb + kd$ where h and k are positive integers. It is clear that the smallest denominator of any of the fractions between a/b and c/d occurs for $h = 1$ and $k = 1$, yielding the fraction $(a + c)/(b + d)$. Since we know that a/b and c/d are adjacent in $\mathfrak{F}_m$, $(a + c)/(b + d)$ is not in $\mathfrak{F}_m$; we deduce that $b + d \geq m + 1$. If h and k are not both 1, then $hb + kd > b + d \geq m + 1$. Thus there is at most one fraction $(a + c)/(b + d)$ with denominator $m + 1$ which separates a/b and c/d. We conclude that e/f is identical with $(a + c)/(b + d)$, that is, $e = a + c$ and $f = b + d$. Since f is in $\mathfrak{F}_{m+1}$, we know that $f \leq m + 1$; whence $f = m + 1$. Part 1 of the theorem is established.

We now prove statement 2 of the theorem.

Let $p/q < r/s$ be adjacent in $\mathfrak{F}_{m+1}$. If they are both in $\mathfrak{F}_m$, then $qr - ps = 1$ by the induction hypothesis. We give a proof for the case p/q is not in $\mathfrak{F}_m$. (The case r/s not in $\mathfrak{F}_m$ is handled similarly.) Then p/q separates two fractions of $\mathfrak{F}_k$, say a/b and c/d, where $a/b < c/d$. According to part 1, which we have already proved is valid for $n = m + 1$, p/q is the only fraction which separates a/b and c/d in $\mathfrak{F}_{m+1}$, and $p/q = (a + c)/(b + d)$. We conclude that $r/s = c/d$. We wish to show that $rq - ps = 1$, which is the same thing as $c(b + d) - d(a + c) = 1$. But $c(b + d) - d(a + c) = bc - ad$. This last expression is equal to 1 by the induction hypothesis since a/b and c/d are adjacent in $\mathfrak{F}_m$.

This completes the induction step, and the theorem follows.

46 Corollary

In a Farey series of order $n > 1$, any two adjacent Farey fractions have distinct denominators.

PROOF: It is clear from part 1 of the theorem that each time a higher denominator fraction e/f is introduced in going from $\mathfrak{F}_n$ to $\mathfrak{F}_{n+1}$, the adjacent fractions are a/b and c/d, which have lower denominators since $b + d = f$.

47 Corollary

If $p/q < r/s$ are adjacent Farey fractions, then $r/s = p/q + 1/(q \cdot s)$.

PROOF: See Problem 6, Exercises 49.

48 *Definition*

If a/b and c/d are any two fractions, we define the mediant of a/b and c/d to be $(a + c)/(b + d)$. We use the expression $a/b \# c/d$ to denote the mediant of a/b and c/d.

We can now look at $\#$ (mediation) as an operation similar to addition or multiplication which can be performed upon two fractions. The operation of mediation has many of the usual properties of the operation of addition. For example, if a/b, c/d, e/f are any three fractions,

$$\frac{a}{b} + \frac{c}{d} = \frac{c}{d} + \frac{a}{b} \quad \text{and} \quad \frac{a}{b} \# \frac{c}{d} = \frac{c}{d} \# \frac{a}{b}$$

$$\left(\frac{a}{b} + \frac{c}{d}\right) + \frac{e}{f} = \frac{a}{b} + \left(\frac{c}{d} + \frac{e}{f}\right) \quad \text{and} \quad \left(\frac{a}{b} \# \frac{c}{d}\right) \# \frac{e}{f} = \frac{a}{b} \# \left(\frac{c}{d} \# \frac{e}{f}\right)$$

$$\left(\frac{a}{b} + \frac{c}{d}\right)\frac{e}{f} = \frac{a}{b} \cdot \frac{e}{f} + \frac{c}{d} \cdot \frac{e}{f} \quad \text{and} \quad \left(\frac{a}{b} \# \frac{c}{d}\right) \cdot \frac{e}{f} = \frac{a}{b} \cdot \frac{e}{f} \# \frac{c}{d} \cdot \frac{e}{f}$$

However, there are significant differences.

For one thing, the definition of the mediant depends on the names of the fractions which are being operated upon, whereas the normal arithmetic operations don't depend on the names used. For example, $\frac{1}{2} + \frac{1}{3} = \frac{5}{6}$ and $\frac{4}{8} + \frac{1}{3} = \frac{12 + 8}{24} = \frac{20}{24} = \frac{5}{6}$, but $\frac{1}{2} \# \frac{1}{3} = \frac{2}{5}$ while $\frac{4}{8} \# \frac{1}{3} = \frac{5}{11} \neq \frac{2}{5}$. Thus $\frac{1}{2}$ and $\frac{4}{8}$ must be regarded as distinct for the operation of mediation.

The reader may want to amuse himself for a while by meditating on the differences between equals, on the difference between equality and identity, on when equal does not mean equal, etc.

49 EXERCISES

1 If $a/b < e/f < c/d$ are three adjacent Farey fractions, then $e/f = (a + c)/(b + d)$. However, find examples to show $e = a + c$ and $f = b + d$ do not necessarily hold.
2 If $a/b < c/d$ are two adjacent fractions in $\mathfrak{F}_n$, then $b > n/2$ or $d > n/2$.
3 If $a/b < c/d$ are two adjacent fractions in $\mathfrak{F}_n$, then $1/[n(n-1)] \leq c/d - a/b$.
4 If $a/b < c/d$ are two adjacent fractions in $\mathfrak{F}_n$, then $c/d - a/b \leq 1/n$.
5 Is equality possible in Problems 3 and 4? If so, when?
6 Prove Corollary 47. [Hint: Use Theorem 45, part (2).]

Section 7 Egyptian Fractions

We shall devote a considerable amount of time to Egyptian fractions, even though they are hardly central to modern number theory. Our reasons are partly historical—these problems arise from the oldest known mathematical manuscripts—but, more importantly, this theory has a large number of solved and unsolved problems easily accessible to the mathematical novice.

50 Definition

A fraction a/b is said to be written in *Egyptian form* if we have

$$\frac{a}{b} = \frac{1}{n_1} + \frac{1}{n_2} + \frac{1}{n_3} + \cdots + \frac{1}{n_k}$$

for distinct positive integers $n_1, n_2, n_3, \ldots, n_k$ for some $k \geq 1$.

51 Examples

$$\frac{2}{3} = \frac{1}{2} + \frac{1}{6}$$
$$\frac{3}{10} = \frac{1}{5} + \frac{1}{10} = \frac{1}{4} + \frac{1}{20}$$
$$\frac{3}{7} = \frac{1}{3} + \frac{1}{11} + \frac{1}{231} = \frac{1}{4} + \frac{1}{7} + \frac{1}{28}.$$

An observation made by the ancients which is very convenient in work with Egyptian fractions is the following.

52 Lemma

If n is a positive whole number, then $1/n = 1/(n+1) + 1/[n(n+1)]$.

PROOF: Add.

If follows immediately that if we have an Egyptian fraction, we can write it in many ways; e.g., $\dfrac{5}{6} = \dfrac{1}{2} + \dfrac{1}{3} = \dfrac{1}{2} + \dfrac{1}{4} + \dfrac{1}{12} = \dfrac{1}{2} + \dfrac{1}{4} + \dfrac{1}{13} + \dfrac{1}{12 \cdot 13} = \dfrac{1}{2} + \dfrac{1}{4} + \dfrac{1}{13} + \dfrac{1}{12 \cdot 13 + 1} + \dfrac{1}{12 \cdot 13(12 \cdot 13 + 1)}.$

In fact, we can prove:

53 Theorem

If a fraction can be written as an Egyptian fraction in one way, then it can be written as an Egyptian fraction in an infinity of ways.

PROOF: Suppose p/q can be written as an Egyptian fraction. Given any representation $p/q = 1/n_1 + 1/n_2 + \cdots + 1/n_k$, with $n_1 < n_2 < n_3 < \cdots < n_k$, we can by use of Lemma 52 get the representation $p/q = 1/n_1 + 1/n_2 + \cdots + 1/(n_k + 1) + 1/[n_k(n_k + 1)]$, which has length $k + 1$. We can now get a representation of length $k + 2$, etc. Thus it must have infinitely many distinct representations.

Stewart [23, Chap. 28] states that it is not known whether we can start with any fraction, say p/q, and proceed as follows (*the splitting method*) to obtain an Egyptian-fraction representation: First write

$$\frac{p}{q} = \overbrace{\frac{1}{q} + \frac{1}{q} + \cdots + \frac{1}{q}}^{p \text{ times}}$$

$$= \frac{1}{q} + \overbrace{\frac{1}{q+1} + \frac{1}{q(q+1)} + \cdots + \frac{1}{q+1} + \frac{1}{q(q+1)}}^{p-1 \text{ times}}$$

$$= \frac{1}{q} + \frac{1}{q+1} + \frac{1}{q(q+1)} + \overbrace{\frac{1}{q+1} + \cdots + \frac{1}{q+1}}^{p-2 \text{ times}}$$
$$+ \overbrace{\frac{1}{q(q+1)} + \cdots + \frac{1}{q(q+1)}}^{p-2 \text{ times}}$$

$$= \frac{1}{q} + \frac{1}{q+1} + \frac{1}{q(q+1)} + \overbrace{\left[\frac{1}{q+2} + \frac{1}{(q+1)(q+2)}\right]}^{p-2 \text{ times}}$$
$$+ \cdots + \left[\frac{1}{q+2} + \frac{1}{(q+1)(q+2)}\right]$$

$$+ \overbrace{\left\{\frac{1}{q(q+1)+1} + \frac{1}{q(q+1)[q(q+1)+1]}\right\} + \cdots}^{p-2 \text{ times}}$$
$$+ \left\{\frac{1}{q(q+1)+1} + \frac{1}{q(q+1)[q(q+1)+1]}\right\}$$

$= $ etc.

At first, the number of repetitions decreases, but after a while the same denominator might appear from two different sources and hence be repeated more times, and so on forever. No one has been able to prove that this splitting method must terminate, but no one has any examples when it doesn't. It appears that a similar problem on Egyptian fractions has recently been solved by Truman Botts [2 and 3], who showed that in similar circumstances one eventually gets all distinct fractions.

We work some examples.

54 Examples

1   $\frac{3}{7} = \frac{1}{7} + \frac{1}{7} + \frac{1}{7}$
$= \frac{1}{7} + \frac{1}{8} + \frac{1}{56} + \frac{1}{8} + \frac{1}{56}$
$= \frac{1}{7} + \frac{1}{8} + \frac{1}{56} + \frac{1}{9} + \frac{1}{72} + \frac{1}{57} + \frac{1}{3,192}$
$= \frac{1}{7} + \frac{1}{8} + \frac{1}{9} + \frac{1}{56} + \frac{1}{57} + \frac{1}{72} + \frac{1}{3,192}.$

Compare this with Examples 51. Of course, being clever is frequently faster than being, systematic. Thus if, in the second step above, we note that $\frac{1}{8} + \frac{1}{8} = \frac{1}{4}$ and $\frac{1}{56} + \frac{1}{56} = \frac{1}{28}$, we get a nicer expansion of $\frac{3}{7}$.

Although we are mainly interested in fractions between 0 and 1, we use the following example to indicate what may happen if the numerator is large. We take a small denominator so that the integers remain small.

2   $\frac{5}{2} = \frac{1}{2} + \frac{1}{2} + \frac{1}{2} + \frac{1}{2} + \frac{1}{2} = \overbrace{(\frac{1}{2})}^{\text{5 times}}$
$= \frac{1}{2} + \overbrace{(\frac{1}{3} + \frac{1}{6})}^{\text{4 times}}$
$= \frac{1}{2} + \frac{1}{3} + \frac{1}{6} + \overbrace{(\frac{1}{4} + \frac{1}{12} + \frac{1}{7} + \frac{1}{42})}^{\text{3 times}}.$

Applying our operation again, we obtain

$\frac{5}{2} = \frac{1}{2} + \frac{1}{3} + \frac{1}{4} + \frac{1}{6} + \frac{1}{7} + \frac{1}{12} + \frac{1}{42}$
$+ \left(\frac{1}{5} + \frac{1}{20} + \frac{1}{8} + \frac{1}{56} + \overbrace{\frac{1}{13} + \frac{1}{156}}^{\text{2 times}} + \frac{1}{43} + \frac{1}{1,806}\right)$

Rearranging the terms, we get

$$\frac{5}{2} = \frac{1}{2} + \frac{1}{3} + \frac{1}{4} + \frac{1}{5} + \frac{1}{6} + \frac{1}{7} + \frac{1}{8} + \frac{1}{12} + \frac{1}{13} + \frac{1}{20} + \frac{1}{42}$$
$$+ \frac{1}{43} + \frac{1}{56} + \frac{1}{156} + \frac{1}{1,806}$$
$$+ \left(\frac{1}{5} + \frac{1}{8} + \frac{1}{13} + \frac{1}{20} + \frac{1}{43} + \frac{1}{56} + \frac{1}{156} + \frac{1}{1,806}\right)$$

where all the terms in the parentheses are repeated terms.

If we apply the operation again, we find that $\frac{1}{5}$ splits into $\frac{1}{6}$ and $\frac{1}{30}$. But $\frac{1}{6}$ already appears, so splitting $\frac{1}{6}$, we get $\frac{1}{7}$ and $\frac{1}{42}$, both of which already appear. This illustrates the difficulty in attempting to prove that repetitions must disappear. It also shows that this method of writing Egyptian fractions is not as practicable as it might appear at first glance. Can you show that this method always works for fractions less than 1? Even that is unknown.

55 EXERCISES

1 Expand $\frac{5}{11}$ as an Egyptian fraction by the splitting method.
2 Expand $\frac{4}{13}$ as an Egyptian fraction by the splitting method.
3 Can you find nicer Egyptian-fraction expansions for $\frac{5}{11}$ or $\frac{4}{13}$?
4 Continue finding the expansion of $\frac{5}{2}$ by splitting. Will it terminate?
5 Try $\frac{8}{2}$ by the splitting method.
6 Show that if n divides $q + 1$, then n/q can be written as the sum of two unit fractions.
7 Show that if n divides $q + 1$, then $(n + 1)/q$ can be written as a sum of three unit fractions.

We shall develop two rather distinct methods of dealing with the problem of how to expand a given fraction as an Egyptian fraction. The first method makes use of our knowledge of the Farey series. We deal only with fractions between 0 and 1, although the theory can be extended.

Given a reduced fraction p/q, $0 < p/q < 1$, we consider the Farey series of order q, $\mathcal{F}_q$. The fraction p/q occurs in this series. Now preceding p/q will be some other fraction p_1/q_1. By Corollary 46, we see that $q_1 < q$, since no denominator of $\mathcal{F}_q$ is greater than q. We see from Theorem 45 that $0 \leq p_1 < p_1 + p' = p$ where p'/q' is the fraction adjacent to p/q from above. Now Corollary 47 tells us that $p/q = p_1/q_1 + 1/qq_1$. If $p_1 = 0$, then $p/q = 1/qq_1$, and we are done. If not, we can now repeat this process for the fraction p_1/q_1 in $\mathfrak{F}_{q_1}$ and obtain $p_1/q_1 = p_2/q_2 + 1/q_1q_2$, where $p_2 < p_1$ and $q_2 < q_1$. Thus $p/q = p_2/q_2 + 1/q_1q_2 + 1/qq_1$, where $0 \leq p_2 < p_1 < p$ and $1 \leq q_2 < q_1 < q$. We repeat this process over and over, and eventually we arrive at a stage where $p_k/q_k = p_{k+1}/q_{k+1} + 1/q_kq_{k+1}$, with $p_{k+1} = 0$. It follows that $q_{k+1} = 1$ since $0/1$ is the only Farey fraction with numerator 0. Thus $p_k/q_k = 1/q_k$. We obtain

$$\frac{q}{p} = \frac{1}{q_k} + \frac{1}{q_k q_{k-1}} + \cdots + \frac{1}{q_3 q_2} + \frac{1}{q_2 q_1}.$$

56 Examples

The reader should consult Table 39.

1 Let $p/q = \frac{5}{8}$. Examining $\mathfrak{F}_8$, we see $p_1/q_1 = \frac{3}{5}$. Examining $\mathfrak{F}_5$, $p_2/q_2 = \frac{1}{2}$. Examining $\mathfrak{F}_2$, $p_3/q_3 = \frac{0}{1}$. Thus $\frac{5}{8} = \frac{1}{2} + \frac{1}{10} + \frac{1}{40}$.
2 Turning to our old friend $p/q = \frac{3}{7}$, we see from $\mathfrak{F}_7$ that $p_1/q_1 = \frac{2}{5}$; from $\mathfrak{F}_5$, $p_2/q_2 = \frac{1}{3}$; from $\mathfrak{F}_3$, $p_3/q_3 = \frac{0}{1}$. Thus $\frac{3}{7} = \frac{1}{3} + \frac{1}{15} + \frac{1}{35}$. Compare this with Examples 51.

57 EXERCISES

1 Expand $\frac{4}{5}$ by both methods, splitting and Farey series.
2 Expand $\frac{5}{11}$ by Farey series (see Problems 1 and 3, Exercises 55).
3 Expand $\frac{4}{13}$ by Farey series (see Problems 2 and 3, Exercises 55).
4 Can you estimate beforehand how many steps there will be in expanding a fraction by this method? Can you make an upper estimate for the number of steps?
5 Can you estimate the size of the largest denominator involved?
6 Expand $\frac{3}{16}$ by the Farey-series method. Find a shorter expansion.

By being a little more careful and a little more formal in our treatment of the Farey-series method, we can gain more insight into the method.

+ 58 **Theorem**

Let p/q be a reduced fraction such that $0 < p/q < 1$. Then the Farey-series method yields an Egyptian-fraction expansion of p/q. Let $p/q = 1/n_1 + 1/n_2 + \cdots + 1/n_k$, $n_1 < n_2 < \cdots < n_k$, be the expansion obtained. Then $n_k \leq q(q-1)$ and $k \leq p$.

PROOF: The proof is by induction on the denominator q. For $q = 2$, the only fraction under consideration is $\frac{1}{2}$. Clearly, $\frac{1}{2} = \frac{1}{2}$ is an Egyptian expansion and satisfies all the claims of the theorem.

We make the induction hypothesis that the theorem is true for all fractions with denominator less than q. We consider the Farey series $\mathfrak{F}_q$. Adjacent to p/q on the left will be a fraction p_1/q_1, with $q_1 \neq q$. Also we know that $p_1 < p$ since $p = p_1 + p'$, where p'/q' is the adjacent fraction on the right. Further we know from Corollary 47 that $p/q = p_1/q_1 + 1/qq_1$. We know by the induction hypothesis that $p_1/q_1 = 1/n_1 + 1/n_2 + \cdots + 1/n_l$, where $l \leq p_1$ and $n_1 < n_2 < \cdots < n_l \leq q_1(q_1 - 1)$. Put $k = l + 1$ and $n_k = q \cdot q_1$. Since $q > q_1$, we see that $n_l < n_k$. Since $q_1 < q$, we see that $n_k \leq q(q-1)$. Also, since $l \leq p_1$ and $p_1 < p$, we conclude that $l + 1 = k \leq p$. The theorem follows by induction.

59 **EXERCISES**

1 Can you find a fraction p/q with $p > 5$ for which $k = p$? With $p > 10$? With $p > 100$?
2 Is there a largest value of p such that the expansion of p/q by the Farey-series method has $k = p$?
3 Can you find a fraction p/q with $q > 5$ for which $n_k = q(q-1)$? With $q > 10$? With $q > 100$?
4 Is there a largest value of q such that $n_k = q(q-1)$ in its Egyptian-fraction expansion by Farey series?
5 Can you find p/q such that both $k = p$ and $n_k = q(q-1)$? Can one or both of p and q be taken large?

There is one other method of writing Egyptian fractions which goes back at least to Fibonacci (Leonardo of Pisa) in his book *Liber Abaci* (1202 A.D.) and which was rediscovered independently by Sylvester [24]. It proceeds as follows: Given p/q, $0 < p/q < 1$, find a whole number n such that

60 $$\frac{1}{n} \leq \frac{p}{q} \text{ and } \frac{p}{q} < \frac{1}{n-1}.$$

The following is the obituary for Sylvester as published in the Illustrated London News, March 27, 1897.

"The world of science has lost one of the greatest of its living representatives and one of the most notable workers that it ever had in the sphere of pure mathematics by the death of James Joseph Sylvester, F.R.S., D.C.L., Savilian Professor of Geometry in the University of Oxford. Born eighty-three years ago, Professor Sylvester became Second Wrangler at Cambridge in the year of the Queen's accession to the throne, but in those days he was debarred from taking his degree by the fact that he was a Jew. He was called to the Bar, but abandoned practice to become Professor of Natural Philosophy at University College, London, and subsequently went out to Virginia to fill the mathematical chair of its University. Twenty years ago, after a period of work at the Woolwich Military Academy, he returned to America to become Professor of Mathematics in the newly founded Johns Hopkins University at Baltimore, but on the death of Henry Smith in 1883, he was made Savilian Professor at Oxford. He retired four years ago."
(Photograph and text by permission of the Granger Collection.)

These inequalities yield

61 $\qquad q \leq np$ and $np - p < q.$

Let $p_1/q_1 = p/q - 1/n = (np - q)/nq$. If $p_1/q_1 = 0$, we are done; otherwise, $p_1 = np - q$. From Formulas 61 we see that $0 < np - q < p$. Thus $1 \leq p_1 < p$. We now repeat this process, getting p_2/q_2, $p_3/q_3, \ldots, p_k/q_k$, where $p_k = 1$. This must happen since $1 \leq p_k < p_{k-1} < \cdots < p_1 < p$. Again, a formal inductive proof would yield:

62 Theorem

Let p/q be a fraction such that $0 < p/q < 1$. Then Fibonacci's method yields an Egyptian expansion for p/q. Let $p/q = 1/n_1 + 1/n_2 + \cdots + 1/n_k$, $n_1 < n_2 < \cdots < n_k$, be the expansion so obtained. Then:

1. $n_{i+1} > n_i(n_i - 1)$ for each $i = 1, 2, \ldots, k - 1$.
2. $k \leq p$.

PROOF: The proof is left as an exercise.

Compare this theorem with Theorem 58.

63 Examples

(Compare with Examples 56.)

1. Let $p/q = \frac{5}{8}$. We note that $1/n \leq p/q < 1/(n-1)$ is equivalent to $n - 1 < q/p \leq n$. So, to find $n - 1$, we divide p into q and throw away the remainder. Thus $\frac{1}{2} \leq \frac{5}{8} < \frac{1}{1}$. So $\frac{5}{8} = \frac{1}{2} + (\frac{5}{8} - \frac{1}{2}) = \frac{1}{2} + \frac{1}{8}$. The process stops.

2. Let $p/q = \frac{3}{7}$. Dividing 7 by 3, we obtain $n - 1 = 2$, that is, $\frac{1}{3} \leq \frac{3}{7} < \frac{1}{2}$. So $\frac{3}{7} = \frac{1}{3} + (\frac{3}{7} - \frac{1}{3}) = \frac{1}{3} + \frac{2}{21}$. We apply the process to $\frac{2}{21}$; $\frac{1}{11} \leq \frac{2}{21} < \frac{1}{10}$. Thus $\frac{3}{7} = \frac{1}{3} + \frac{1}{11} + (\frac{2}{21} - \frac{1}{11}) = \frac{1}{3} + \frac{1}{11} + \frac{1}{231}$.

If we compare this expansion of $\frac{3}{7}$ to the one obtained in Example 56, we find that this expansion has much larger denominators. The Fibonacci method finds the closest or best approximation at each step; however, in a global or overall sense, it often yields neither the shortest expansion nor the expansion with the smallest numbers. It happens frequently that if one uses a method which at each step yields the best result possible in the sense of that step but which is not influenced by previous steps or possible future steps, then the overall effect of the method is far from best.

64 EXERCISES

1. Prove Theorem 62.
$$\left[\text{Hint:} \quad \frac{1}{n_1} \leq \frac{p_1}{q_1} = \frac{n_0 p - q_0}{n_0 q_0} < \frac{1}{n_0} \cdot \frac{p}{q} < \frac{1}{n_0} \cdot \frac{1}{n_0 - 1}. \right]$$

2. Can you find a fraction p/q with $p \geq 5$ such that in the Fibonacci expansion of p/q, $k = p$?

3. Can you find a fraction p/q with $q \geq 5$ and such that $n_{i+1} = n_i(n_i - 1) + 1$ for each $i \leq k$. With $q \geq 10$? With $q \geq 100$?

4. Is there a largest value of q such that $n_{i+1} = n_i(n_i - 1) + 1$, $i = 1, 2, \ldots, k$?

5. Can you find a fraction with $k = p = 2$ and $n_2 = n_1(n_1 - 1) + 1$?

Peter Fillmore, while teaching Egyptian fractions to his freshman class, discovered that the answer is yes.

To the best of our knowledge, it is not known whether there exist fractions p/q with p large and $k = p$ in the Fibonacci expansion. It is possible to find fractions with k large and q large for which $n_{i+1} = n_i(n_i - 1) + 1$.

Section 7   *Egyptian Fractions*   429

Although no one knows for sure just how or why the Egyptians came upon their scheme of fractions, it does have some advantages in performing arithmetic operations. Thus, to add $\frac{3}{8} + \frac{5}{12}$ by our usual methods, we must first find the common denominator, convert each fraction into a fraction with that denominator, $\frac{9}{24} + \frac{10}{24}$, add the numerators and obtain $\frac{19}{24}$, and then reduce this if possible. If, on the other hand, we are given the same two fractions $\frac{1}{4} + \frac{1}{8}$ and $\frac{1}{4} + \frac{1}{6}$ to be added as Egyptian fractions, we write $\frac{1}{4} + \frac{1}{8} + \frac{1}{4} + \frac{1}{6}$. There are repetitions, and we must do something to remedy this. For example, we might use $\frac{1}{4} = \frac{1}{5} + \frac{1}{20}$ to get rid of one of the $\frac{1}{4}$ terms, or we might use $\frac{1}{4} + \frac{1}{4} = \frac{1}{2}$ to get rid of both $\frac{1}{4}$ terms. To multiply these two as Egyptian fractions, we write

$$(\tfrac{1}{4} + \tfrac{1}{8})(\tfrac{1}{4} + \tfrac{1}{6}) = \tfrac{1}{16} + \tfrac{1}{32} + \tfrac{1}{24} + \tfrac{1}{48}.$$

We see that the Egyptians had a workable system of dealing with fractions. In fact, as far as addition is concerned, it has some advantages over our present system. The main disadvantages are: First, it is difficult to compare Egyptian fractions; e.g., which is bigger, $\frac{1}{4} + \frac{1}{6}$ or $\frac{1}{3} + \frac{1}{8}$? Second, the expressions for some fractions get very complicated; try, for example, to write $\frac{21}{23}$ as an Egyptian fraction.

It is not known if there is a best or even a good method of finding a way to express a given fraction as an Egyptian fraction. The Fibonacci method has the disadvantage that the numerators get very large, the Farey-series method has the disadvantage that the procedure becomes very long, and the method of splitting and using the identity $1/n = 1/(n+1) + 1/[n(n+1)]$ has both the above disadvantages.

65   Examples

We shall expand $\frac{21}{23}$ by several methods.

1   *The Farey-series method.*   We first note that $\frac{21}{23}$ is between $\frac{1}{2}$ and $\frac{1}{1}$, so only the part of $\mathfrak{F}_{23}$ between $\frac{1}{2}$ and 1 is relevant. We take the mediant $\frac{2}{3}$. Since $\frac{21}{23}$ is between $\frac{2}{3}$ and $\frac{1}{1}$, we can ignore the other fractions of $\mathfrak{F}_{23}$ which occur between $\frac{1}{2}$ and $\frac{2}{3}$ (Why?). Taking the mediant of $\frac{2}{3}$ and $\frac{1}{1}$, we get $\frac{3}{4}$. Since $\frac{21}{23}$ is between $\frac{3}{4}$ and $\frac{1}{1}$, we can ignore the part of $\mathfrak{F}_{23}$ between $\frac{2}{3}$ and $\frac{3}{4}$. We continue this way, obtaining

$$\tfrac{0}{1} \quad \tfrac{1}{2} \quad \tfrac{2}{3} \quad \tfrac{3}{4} \quad \tfrac{4}{5} \quad \tfrac{5}{6} \quad \tfrac{6}{7} \quad \tfrac{7}{8} \quad \tfrac{8}{9} \quad \tfrac{9}{10} \quad \tfrac{10}{11} \quad \tfrac{1}{1},$$

where $\frac{21}{23}$ is between $\frac{10}{11}$ and $\frac{1}{1}$. We take the next mediant $\frac{11}{12}$ and find that it is bigger than $\frac{21}{23}$; hence the part of $\mathfrak{F}_{23}$ of interest is below $\frac{11}{12}$. We next take the mediant of $\frac{10}{11}$ and $\frac{11}{12}$ and get $\frac{21}{23}$, the desired fraction. Thus the relevant part of $\mathfrak{F}_{23}$ is

$$\tfrac{0}{1} \quad \tfrac{1}{2} \quad \tfrac{2}{3} \quad \tfrac{3}{4} \quad \tfrac{4}{5} \quad \tfrac{5}{6} \quad \tfrac{6}{7} \quad \tfrac{7}{8} \quad \tfrac{8}{9} \quad \tfrac{9}{10} \quad \tfrac{10}{11} \quad \tfrac{21}{23};$$

so

$$\tfrac{21}{23} = \tfrac{1}{2} + \tfrac{1}{6} + \tfrac{1}{12} + \tfrac{1}{20} + \tfrac{1}{30} + \tfrac{1}{42} + \tfrac{1}{56} + \tfrac{1}{72} + \tfrac{1}{90} + \tfrac{1}{110} + \tfrac{1}{253}.$$

The reader who does not understand how we shortened this method to avoid writing all of $\mathfrak{F}_{23}$ should do it the long way and observe how and why these are the only relevant fractions.

Fibonacci, or Leonardo of Pisa (c. 1180–1228), was an Italian merchant whose extensive travels around the Mediterranean brought him into contact with Arabic, Greek and Hebrew mathematical works, which he easily learned and augmented. His two main works are Liber Abaci *(1202) and* Practica Geometriae *(1220.) In the first, he introduced to European scholars the Arabic notation for numerals and algorithms of arithmetic, plus methods of calculating cube roots and square roots of integers. The second is a complete discussion of geometry and trigonometry as understood at that time. In a celebrated scientific tourney at the court of Emperor Frederic II of Hohenstaufen, he proved that $x^3 + 2x^2 + 10x = 20$ has no rational root nor any root which can be constructed with ruler and compass. (Courtesy of the Smith Collection, Columbia University.)*

2 *The Fibonacci method.* We divide 21 into 23 and get 1 plus a remainder. Thus the first term is $\frac{1}{2}$:

$$\tfrac{21}{23} = \tfrac{1}{2} + (\tfrac{21}{23} - \tfrac{1}{2}) = \tfrac{1}{2} + \tfrac{19}{46}.$$

We divide 19 into 46 and get 2 plus a remainder, so the next term is $\frac{1}{3}$:

$$\tfrac{21}{23} = \tfrac{1}{2} + \tfrac{1}{3} + (\tfrac{19}{46} - \tfrac{1}{3}) = \tfrac{1}{2} + \tfrac{1}{3} + \tfrac{11}{138}.$$

We divide 11 into 138 and get 12 plus a remainder, so the next term is $\frac{1}{13}$:

$$\frac{21}{23} = \frac{1}{2} + \frac{1}{3} + \frac{1}{13} + \left(\frac{11}{138} - \frac{1}{13}\right) = \frac{1}{2} + \frac{1}{3} + \frac{1}{13} + \frac{5}{1{,}794}.$$

Section 7   Egyptian Fractions

We divide 1,794 by 5 and obtain 358 plus a remainder, so the next term is $\frac{1}{359}$:

$$\frac{21}{23} = \frac{1}{2} + \frac{1}{3} + \frac{1}{13} + \frac{1}{359} + \left(\frac{5}{1,794} - \frac{1}{359}\right) = \frac{1}{2} + \frac{1}{3} + \frac{1}{11} + \frac{1}{359} + \frac{1}{1,794 \cdot 359}.$$

Since $1,794 \cdot 395 = 644,046$, we get the expansion

$$\frac{21}{23} = \frac{1}{2} + \frac{1}{3} + \frac{1}{11} + \frac{1}{359} + \frac{1}{644,046}.$$

3   *A modified Fibonacci method.*   We consider the step

$$\frac{21}{23} = \frac{1}{2} + \frac{1}{3} + \frac{1}{13} + \frac{5}{1,794}.$$

We notice that both 2 and 3 divide 1,794 and also $5 = 2 + 3$. Thus

$$\frac{5}{1,794} = \frac{2}{1,794} + \frac{3}{1,794} = \frac{1}{897} + \frac{1}{598}.$$

Thus $\frac{21}{23} = \frac{1}{2} + \frac{1}{3} + \frac{1}{13} + \frac{1}{598} + \frac{1}{897}$. This new expansion has considerably smaller denominators, but is still of length 5.

4   *A modified Farey-series method.*   Consider the relevant part of $\mathfrak{F}_{23}$:

$$\frac{0}{1} \quad \frac{1}{2} \quad \frac{2}{3} \quad \frac{3}{4} \quad \frac{4}{5} \quad \frac{5}{6} \quad \frac{6}{7} \quad \frac{7}{8} \quad \frac{8}{9} \quad \frac{9}{10} \quad \frac{10}{11} \quad \frac{21}{23} \quad \frac{11}{12} \quad \frac{1}{1}.$$

We notice that $\frac{5}{6} - \frac{1}{2} = \frac{1}{3}$ and $\frac{9}{10} - \frac{5}{6} = \frac{1}{15}$. Hence $\frac{9}{10} = \frac{1}{2} + \frac{1}{3} + \frac{1}{15}$. Thus

$$\frac{21}{23} = \frac{1}{11 \cdot 23} + \frac{10}{11} = \frac{1}{253} + \frac{1}{10 \cdot 11} + \frac{9}{10} = \frac{1}{253} + \frac{1}{110} + \frac{1}{15} + \frac{1}{3} + \frac{1}{2}.$$

So $\frac{21}{23} = \frac{1}{2} + \frac{1}{3} + \frac{1}{15} + \frac{1}{110} + \frac{1}{253}$. Being clever rather than systematic for the last two steps, we note that

$$\frac{21}{23} - \frac{9}{10} = \frac{3}{10 \cdot 23} = \frac{1}{10 \cdot 23} + \frac{2}{10 \cdot 23}.$$

Thus $\frac{21}{23} = \frac{1}{2} + \frac{1}{3} + \frac{1}{15} + \frac{1}{115} + \frac{1}{230}$.

It is impossible to express $\frac{21}{23}$ as an Egyptian fraction with fewer than five summands. To illustrate a technique by which one can show that $\frac{21}{23}$ cannot be expressed with four or fewer summands, we prove:

+ **66 Theorem**

It is not possible to express $\frac{21}{23}$ as an Egyptian fraction with fewer than four summands.

PROOF: It is impossible with one summand since $\frac{21}{23}$ is not a reciprocal of an integer. We next consider the case of two summands. If $\frac{21}{23} = 1/m + 1/n$ with $m < n$, then $1/m$ must be larger than $\frac{1}{2}(\frac{21}{23}) = \frac{21}{46}$. But $1/m > \frac{21}{46}$ implies that $m < \frac{46}{21}$, so that $m = 1$ or 2. The case $m = 1$ is clearly impossible, so the remaining case is $m = 2$. Thus, if it is possible to write $\frac{21}{23}$ as the sum of two unit fractions, one of them must be $\frac{1}{2}$. It follows that the other must be $\frac{21}{23} - \frac{1}{2} = \frac{19}{46}$. But there is no integer n such that $\frac{19}{46} = 1/n$. Thus it is impossible to write $\frac{21}{23}$ as a sum of two unit fractions.

We now consider the case of three summands $\frac{21}{23} = 1/m + 1/n + 1/q$, with $m < n < q$. We deduce that $1/m > \frac{1}{3}(\frac{21}{23})$, so that $m < \frac{69}{21}$. Thus $m = 1, 2,$ or 3. The case $m = 1$ is clearly impossible. Consider the case $m = 3$. In this case, $\frac{21}{23} - \frac{1}{3} = 1/n + 1/q$, or $\frac{40}{3 \cdot 23} = 1/n + 1/q$. Since $n < q$, we see that $1/n > \frac{1}{2}\left(\frac{40}{3 \cdot 23}\right)$, or $n < \frac{69}{20}$. Thus the possibilities are $n = 1, 2,$ or 3, but we must also have $n > m = 3$; since there is no value of n satisfying these conditions, $m = 3$ is impossible.

It remains to consider the case $m = 2$. In this case, we consider representations $\frac{21}{23} = \frac{1}{2} + 1/n + 1/q$, which is the same thing as $(\frac{21}{23} - \frac{1}{2}) = \frac{19}{46} = 1/n + 1/q$. If $\frac{19}{46}$ is to be represented as the sum of two terms, then $1/n > \frac{1}{2}(\frac{19}{46})$, or $n < \frac{92}{19}$. Also $n > m = 2$. Thus we need only consider $n = 3$ or 4. Now $\frac{19}{46} = \frac{1}{3} + 1/q$ is equivalent to $(\frac{19}{46} - \frac{1}{3}) = \frac{11}{138} = 1/q$, which is clearly impossible; while $\frac{19}{46} = \frac{1}{4} + 1/q$ is equivalent to $(\frac{19}{46} - \frac{1}{4}) = \frac{15}{92} = 1/q$, which is also impossible. Thus there is no way to represent $\frac{21}{23}$ with fewer than four factors, and the theorem is established.

By similar techniques (which, however, become somewhat lengthy) or by more complicated or more advanced techniques, one can prove that $\frac{21}{23}$ can be represented by no fewer than five summands.

Section 7   Egyptian Fractions

67   EXERCISES

1. Expand $\frac{2}{11}$ and $\frac{2}{17}$ by both the Farey-series method and Fibonacci's method.
2. Prove that for a fraction of the form $2/q$, the Fibonacci and Farey-series methods yield the same result.
3. Prove that if $q = q_1 \cdot q_2 \cdots q_l$, where each q_i is a prime, then each q_i must divide at least one of the n_j in any expansion $p/q = 1/n_1 + 1/n_2 + \cdots + 1/n_k$.
4. Prove that if one can find distinct positive integers A and B such that $A + B$ is divisible by p, A divides q, and B divides q, then $p/q = 1/n + 1/m$ is solvable for positive integers n and m.
5. Prove that if one can find distinct positive integers A, B, and C such that $A + B + C$ is divisible by p while all three of A, B, and C divide q, then $p/q = 1/l + 1/m + 1/n$ is solvable for distinct positive integers l, m, and n.
6. Expand $\frac{4}{15}$ by the method of Problem 4, the Farey-series method, and Fibonacci's method.
7. Expand $\frac{11}{138}$ by the method of Problem 5.
8. Use Problem 7, together with part 2 of Examples 65, to get a new, better expansion of $\frac{21}{23}$. (We think this is the best expansion, but we have been wrong on such guesses before.)
*9. Prove the converse of Problem 4.
*10. Prove that $\frac{21}{23}$ cannot be expanded in fewer than five terms.
*11. Show that if N is any positive integer, then $\frac{1}{2} + \frac{1}{3} + \frac{1}{4} + \frac{1}{5} + \frac{1}{6} + \cdots 1/n > N$ if n is large enough.
*12. Using Problem 11, show that a fraction $p/q > 1$ can also be written in Egyptian style.

The above exercises hint at the fact that neither the Farey-series nor the Fibonacci method yields the best expansions. We do not know if there are better *methods*. The disparity between the two methods increases with the size of the denominator.

68   Examples

1. By the Farey-series method, we obtain for $\frac{5}{121}$ that $\frac{1}{25} \frac{2}{49} \frac{3}{73} \frac{4}{97}$ $\frac{5}{121} \frac{1}{24}$ is the relevant part of $\mathfrak{F}_{121}$. Thus

$$\frac{5}{121} = \frac{1}{25} + \frac{1}{1{,}225} + \frac{1}{3{,}477} + \frac{1}{7{,}081} + \frac{1}{11{,}737}.$$

On the other hand, Fibonacci's method yields

$$\frac{5}{121} = \frac{1}{25} + \frac{1}{757} + \frac{1}{763,308} + \frac{1}{873,960,180,913} + \frac{1}{7,638,092,437,828,241,151,744}.$$

Hunting around yields

$$\frac{5}{121} = \frac{1}{25} + \frac{1}{759} + \frac{1}{208,725}.$$

There may be better ways to expand $\frac{5}{121}$, although using Problem 9 of Exercises 67, we can prove that they must have at least three terms.

2 For $\frac{23}{37}$ we find the relevant part of $\mathfrak{F}_{37}$:

Thus
$$\frac{1}{2} \quad \frac{3}{5} \quad \frac{8}{13} \quad \frac{13}{21} \quad \frac{18}{29} \quad \frac{23}{37} \quad \frac{5}{8} \quad \frac{2}{3} \quad \frac{1}{1}.$$

$$\frac{23}{37} = \frac{1}{2} + \frac{1}{10} + \frac{1}{65} + \frac{1}{273} + \frac{1}{609} + \frac{1}{1,073}.$$

Fibonacci's method yields

$$\frac{23}{37} = \frac{1}{2} + \frac{1}{9} + \frac{1}{96} + \frac{1}{1,056}.$$

However, if we consider the third step in the Fibonacci process, $\frac{23}{37} - \frac{1}{2} - \frac{1}{9} = \frac{7}{666}$, we can apply Problem 5 of Exercises 67 and obtain

$$\frac{7}{666} = \frac{6+1}{666} = \frac{1}{111} + \frac{1}{666}.$$

Thus
$$\frac{23}{37} = \frac{1}{2} + \frac{1}{9} + \frac{1}{111} + \frac{1}{666}.$$

It has been conjectured by the contemporary mathematicians Erdös and Strauss, and Sierpinski, respectively, that any fractions of the form $4/q$ and $5/q$, respectively, can be represented as the sum of three distinct unit fractions. However, at the time of this writing, neither of these problems has been solved, although some partial results are known. (See Stewart [23], Sec. 28.5].)

Section 8   *The Euclidean Algorithm*

Section 8   The Euclidean Algorithm

Frequently in working with numbers, both theoretically and practically, one wishes to find the greatest common divisor (e.g., when reducing fractions) or the least common multiple (e.g., when adding fractions) of two integers. An exposition of a simple method for finding them can be found in Euclid's seventh book. Although historical evidence indicates that it predates Euclid, this method is called the *Euclidean algorithm* (or algorism). The word "algorithm" derives from a mispronunciation of Al-Khowarizmi, the name of a famous Arabian mathematician (c. 825 A.D.). (See Smith [21, Vol. II, pp. 8–11] for an interesting account of the meanderings of this word or Ore [17 pp. 102–104].)

69   *Definition*

The *greatest common divisor* of two integers m and n, not both zero, is the largest positive integers d which divide both m and n. We shall sometimes denote it by g.c.d.(m, n).

70   *Definition*

The *least common multiple* of two nonzero integers m and n is the smallest positive integer l such that both m and n divide l. We shall sometimes use the notation l.c.m.(m, n).

If one of the integers m or n is zero, then since zero divides no positive integer, the least common multiple is not defined by Definition 70. Since this will cause a certain amount of inconvenience, by forcing us to prove that both m and n are positive whenever we write l.c.m.(m, n), we shall find it worthwhile to make a special convention in the case where one or both of m and n are zero.

71   *Conventions*

For any integer n, l.c.m.$(0, n)$ = l.c.m.$(n, 0)$ = 0.

72   Examples

1   g.c.d.$(10, 6) = 2$ g.c.d.$(3, 5) = 1$
    g.c.d.$(1, 7) = 1$ g.c.d.$(-2, 14) = 2$
    g.c.d.$(15, 10) = 5$ g.c.d.$(6, -10) = 2$
    g.c.d.$(21, 10) = 1$ g.c.d.$(0, 12) = 12$

2   l.c.m.$(10, 6) = 30$ l.c.m.$(3, 5) = 15$
    l.c.m.$(1, 7) = 7$ l.c.m.$(-2, 14) = 14$
    l.c.m.$(15, 10) = 30$ l.c.m.$(6, -10) = 30$
    l.c.m.$(21, 10) = 210$ l.c.m.$(0, 12) = 0$

73   The Euclidean Algorithm

Given two positive integers m and n, we proceed as follows: We divide m by n and write

$$m = q_0 n + r_1 \qquad 0 \leq r_1 < n.$$

Next we divide n by r_1 and get

$$n = q_1 r_1 + r_2 \qquad 0 \leq r_2 < r_1.$$

Next we divide r_1 by r_2 and get

$$r_1 = q_2 r_2 + r_3 \qquad 0 \leq r_3 < r_2.$$

We continue, getting

$$r_2 = q_3 r_3 + r_4 \qquad 0 \leq r_3 < r_4$$
$$\cdots \cdots \cdots \cdots$$
$$r_{k-2} = q_{k-1} r_{k-1} + r_k \qquad 0 \leq r_k < r_{k-1}$$
$$r_{k-1} = q_k r_k + 0.$$

We must eventually get zero remainder since the remainders are a decreasing sequence of nonnegative integers. The process stops when the zero remainder is obtained.

74   Theorem

The last positive remainder obtained in the Euclidean algorithm is the greatest common divisor of the original numbers. Symbolically, g.c.d.$(m, n) = r_k$.

PROOF: Let g.c.d.$(m, n) = d$. We shall prove that $d = r_k$.

We first prove that d divides r_k. From the definition of the g.c.d.(m, n), we see that d divides both m and n, and hence d divides $m - q_0 n = r_1$. Since d divides n and r_1, it divides $n - q_1 r_1 = r_2$. Since d divides r_1 and r_2, it divides $r_3 = r_1 - q_2 r_2$, etc. We follow the steps of the algorithm and finally obtain d divides $r_k = r_{k-2} - q_{k-1} r_{k-1}$.

Since d divides r_k and r_k is positive, it follows that $d \leq r_k$.

We next show that r_k divides both m and n. From the last line of the algorithm, it follows that r_k divides r_{k-1}. Thus r_k divides $r_{k-2} = q_{k-1} r_{k-1} + r_k$. We next see that r_k divides $r_{k-3} = q_{k-2} r_{k-2} + r_{k-1}$, etc. Continuing in this fashion, we obtain that r_k divides $n = q_1 r_1 + r_2$ and r_k divides $m = q_0 n + r_1$. Thus r_k divides both m and n.

By definition of the g.c.d.(m, n), d is the largest number which divides both m and n; hence $r_k \leq d$. Since we have shown above that $d \leq r_k$, it follows that $d = r_k$. The theorem is proved.

It should be noted that the two uses of the word "etc." in the proof of Theorem 74 are really a lazy man's shorthand for a proof by mathematical induction. This is frequently used because to put in all the details of induction would disrupt the continuity of thought and hide the real reasoning behind the proof.

75 Examples

1 Find g.c.d.$(1{,}353, 7{,}869)$.

$$7{,}869 = 5(1{,}353) + 1{,}104$$
$$1{,}353 = 1(1{,}104) + 249$$
$$1{,}104 = 4(249) + 108$$
$$249 = 2(108) + 33$$
$$108 = 3(33) + 9$$
$$33 = 3(9) + 6$$
$$9 = 1 \cdot 6 + 3$$
$$6 = 2 \cdot 3 + 0.$$

Thus g.c.d.$(1{,}353, 7{,}869) = 3$.

2 Find g.c.d.$(42{,}483, 40{,}460)$

$$42{,}483 = 1(40{,}460) + 2{,}023$$
$$40{,}460 = 20(2{,}023) + 0$$

Thus g.c.d.$(42{,}483, 40{,}460) = 2{,}023$.

76 EXERCISES

1 Compute g.c.d.$(7{,}361, 5{,}429)$.

2 Compute g.c.d.$(7{,}361, 40{,}460)$.

3 Reduce the fraction $\dfrac{5{,}124}{5{,}429}$.

4 Compute g.c.d.(m, n), l.c.m.(m, n), $m \cdot n$, and g.c.d.$(m, n) \cdot$ l.c.m.(m, n) for the following:

$$m = 21,\ n = 35$$
$$m = 6,\ n = 24$$
$$m = 5,\ n = 7$$
$$m = 32,\ n = 24.$$

Do you notice any pattern?

5 Write out the inductive proofs needed in the proof of Theorem 74.

6 Show that for m and n positive integers, g.c.d.$(m, n) =$ g.c.d.$(-m, n) =$ g.c.d.$(-m, n)$.

77 Theorem

If m and n are positive integers, then the product of m and n is equal to the product of their greatest common divisor and their least common multiple. That is, $m \cdot n =$ g.c.d.$(m, n) \cdot$ l.c.m.(m, n).

PROOF: Let g.c.d.$(m, n) = d$, l.c.m.$(m, n) = e$. Consider $mn/d = m(n/d) = n(m/d)$. Since both n/d and m/d are integers, mn/d is a common multiple of both m and n and hence $mn/d \geq e$. Thus $mn \geq ed$.

On the other hand, consider mn/e. Let $mn = se + r$, $0 \leq r < e$. Since both m and n divide e and mn, they must both divide the remainder $r = mn - se$. But e is the smallest *positive* integer which is divisible by both m and n. Hence the remainder must be zero, and mn/e is an integer.

Let $e = pm$ and $e = qn$, where we know p and q are positive integers since we know e is the l.c.m. of the positive integers m and n.

We notice that $mn/e = m/q = n/p =$ integer. But $m/q \cdot q = m$ and $n/p \cdot p = n$; thus mn/e divides both m and n. Since $d =$ g.c.d.(m, n), it follows that $mn/e \leq d$, or $mn \leq de$. Since we proved above that $mn \geq de$, we conclude that $mn = de$. The theorem is proved.

The efficiency of the Euclidean algorithm can be increased slightly if, instead of taking r_{i+1} to be the least nonnegative remainder on dividing r_{i-1} by r_i, we take it to be the absolutely least remainder, i.e., in the interval $-r_i/2 < r_{i+1} \leq r_i/2$. In this case, we may get $\pm d$ as the final term.

78 Example

(See Examples 75.) Find g.c.d.$(1{,}353, 7{,}869)$.

$$7{,}869 = 6(1{,}353) - 249$$
$$1{,}353 = 5(249) + 108$$
$$249 = 2(108) + 33$$
$$108 = 3(33) + 9$$
$$33 = 4 \cdot 9 - 3$$
$$9 = 3 \cdot 3.$$

We have reduced the number of steps by one-fourth. Of course, in practice one frequently sees what the g.c.d. is before the last step, since it must divide all the remainders.

79 EXERCISES

1 Find g.c.d.(7,361, 5,429) by the modified algorithm.
2 Find g.c.d.(7,361, 404,601) by the modified algorithm.
3 Prove that the modified algorithm works.
4 Prove Theorem 77 for the case when m or n is zero.
5 What modifications would be needed in Theorem 77 if m and n could take negative values as well? Can you prove it?

Kronecker (1823–1891) proved that no other modifications of the Euclidean algorithm can be more efficient than the modification given above. Lamé (1795–1870) proved that the number of steps involved in the original algorithm is at most five times the number of digits of the shorter number. (See Uspensky and Heaslet [25, Chap. III].)

Leopold Kronecker (1823–1891) was one of a triumvirate of great mathematicians at the University of Berlin in the late nineteenth century concentrating in the fields of number theory and algebra. He was akin to the Pythagoreans in his belief that everything of value could be reduced to the integers. "God created the integers, the rest is man's work." His writings were a strong impetus toward the philosophy of intuitionism in mathematics. (Courtesy of the Smith Collection, Columbia University.)

Gabriel Lamé (1795–1870) is chiefly remembered for his work in mathematical physics, particularly in heat and elasticity. He also made contributions to number theory. He showed that the number of steps of the Euclidean algorithm is less than five times the number of digits of the smaller number. Lamé proved that Fermat's last theorem (which states that if $n > 2$ then $x^n + y^n = z^n$ has no solution) is true for the case $n = 7$. (Courtesy of the Smith Collection, Columbia University.)

Section 9 Continued Fractions

The (original) Euclidean algorithm gives rise to yet another way of writing fractions, a way which is very important in making approximations.

Consider the algorithm applied to the numbers 5 and 23. For later convenience, we put in the apparently wasteful first step:

$$
\begin{aligned}
&1) \quad 5 = 0 \cdot 23 + 5 \\
&2) \quad 23 = 4 \cdot 5 + 3 \\
&3) \quad 5 = 1 \cdot 3 + 2 \\
&4) \quad 3 = 1 \cdot 2 + 1 \\
&5) \quad 2 = 2 \cdot 1.
\end{aligned}
$$

Dividing the first line by 23, we obtain

$$\frac{5}{23} = 0 + \frac{5}{23} = 0 + \frac{1}{\frac{23}{5}}.$$

Section 9   Continued Fractions

From the second line we obtain
$$\tfrac{23}{5} = 4 + \tfrac{3}{5}.$$
Thus
$$\frac{5}{23} = 0 + \frac{1}{\tfrac{23}{5}} = 0 + \frac{1}{4 + \tfrac{3}{5}} = 0 + \frac{1}{4 + \frac{1}{\tfrac{5}{3}}}.$$

From the third line we obtain $\tfrac{5}{3} = 1 + \tfrac{2}{3}$; thus
$$\frac{5}{23} = 0 + \frac{1}{4 + \frac{1}{1 + \tfrac{2}{3}}} = 0 + \frac{1}{4 + \frac{1}{1 + \frac{1}{\tfrac{3}{2}}}}.$$

While the fourth line yields $\tfrac{3}{2} = 1 + \tfrac{1}{2}$, so
$$\frac{5}{23} = 0 + \frac{1}{4 + \frac{1}{1 + \frac{1}{1 + \frac{1}{2}}}}.$$

Since the bottom line is now an integer, the process stops. We now have $\tfrac{5}{23}$ expressed as a compounded or continued fraction each part of which is like a unit fraction.

Let us examine the pieces of the continued fraction for $\tfrac{5}{23}$. The first term, 0, is the closest whole number less than $\tfrac{5}{23}$. The second stage, $0 + \tfrac{1}{4} = \tfrac{1}{4}$, is greater than $\tfrac{5}{23}$ and is the fraction of $\mathfrak{F}_4$ closest to $\tfrac{5}{23}$. Examining the third stage, we find $0 + \dfrac{1}{4 + \tfrac{1}{1}} = \dfrac{1}{5}$ is less than $\tfrac{5}{23}$ and is the fraction of $\mathfrak{F}_5$ which is closest to $\tfrac{5}{23}$. The next stage, $0 + \dfrac{1}{4 + \dfrac{1}{1 + \tfrac{1}{1}}} = \tfrac{2}{9}$, is greater than $\tfrac{5}{23}$ and is the fraction of $\mathfrak{F}_9$ closest to $\tfrac{5}{23}$. The final stage yields $\tfrac{5}{23}$ itself.

Thus if we wanted to approximate $\tfrac{5}{23}$ by a fraction with a smaller denominator, we could use $\tfrac{2}{9}$ for an approximation on the large side or $\tfrac{1}{5}$ for one on the small side. The errors would be $\tfrac{2}{9} - \tfrac{5}{23} = \tfrac{1}{207}$ and $\tfrac{5}{23} - \tfrac{1}{5} = \tfrac{2}{115}$, which are close in comparison with the size of the denominators involved in the approximation.

We now repeat the process for the fraction $\dfrac{7{,}869}{1{,}353}$. The Euclidean algorithm for this fraction was worked in part 1 of Examples 75.

We obtain

$$\frac{7,869}{1,353} = 5 + \cfrac{1}{\cfrac{1,353}{1,104}}$$

$$= 5 + \cfrac{1}{1 + \cfrac{1}{\cfrac{1,104}{249}}}$$

$$= 5 + \cfrac{1}{1 + \cfrac{1}{4 + \frac{249}{108}}}$$

$$= 5 + \cfrac{1}{1 + \cfrac{1}{4 + \cfrac{1}{2 + \cfrac{1}{\frac{108}{33}}}}}$$

$$= 5 + \cfrac{1}{1 + \cfrac{1}{4 + \cfrac{1}{2 + \cfrac{1}{3 + \cfrac{1}{3 + \cfrac{1}{1 + \frac{1}{2}}}}}}}.$$

The successive approximations obtained by chopping the continued fraction after an integer, are 5, $\frac{6}{1}$, $\frac{29}{5}$, $\frac{64}{11}$, $\frac{221}{38}$, $\frac{727}{125}$, $\frac{948}{163}$, $\frac{2,623}{451}$.

Computation will verify that the first, third, fifth, and seventh fractions are all too small but are the closest fractions in their respective Farey series, while the second, fourth, and sixth fractions are too large but are the closest fractions in their Farey series. Thus, for example, we know that $\frac{64}{11}$ is the best possible approximation with denominator less than or equal to 11 and that it is larger than $\frac{7,869}{1,353}$. In fact, the error is $\frac{1}{4,961} = 0.0020 \cdots$. If we consider $\frac{727}{125}$, the error is only $\frac{2}{56,375} = 0.0000354 \cdots$. For many purposes, $\frac{727}{125}$ would be more convenient than the less accurate 5.8159, which is $0.0000645 \cdots$ less than $\frac{7,869}{1,353}$. Furthermore, as the complexity of the original fraction and the desired accuracy

Section 9 Continued Fractions   443

increase, the number of decimal places needed increases much faster than (at least twice as fast as) the number of digits of the approximate fraction obtained from the continued fraction.

The main disadvantage of continued fractions, and it is a grave disadvantage, is that no one has figured out how to perform the arithmetic operations of addition, subtraction, and multiplication with them. Their main importance, which is considerable, has been in the theory and practice of approximation.

The theory of continued fractions is rich in many surprising and beautiful theorems. There are efficient ways to compute the approximating fractions (convergents) and many fascinating relationships between them. There is essentially only one way to write a given fraction as a continued fraction. The interested reader is referred to the excellent monograph by Khintchine [13]. Other good, but less complete, monographs are those by Moore and Olds [15, 16].

80 EXERCISES

1 Compute the continued fraction for $\frac{75}{101}$.

2 (a) Find the continued fractions of $\frac{7}{10}$, $\frac{8}{11}$, and $\frac{7}{16}$.
   (b) Notice how the Farey series build up around $\frac{7}{10}$, $\frac{8}{11}$, and $\frac{7}{16}$. Do you notice any connections? Is there something special about these numbers, or do the relationships hold in general?

3 Evaluate $1 + \cfrac{1}{1 + \cfrac{1}{2 + \cfrac{1}{3 + \cfrac{1}{4}}}}$.

4 Evaluate $\cfrac{1}{2 + \cfrac{1}{2 + \cfrac{1}{2 + \cfrac{1}{2}}}}$.

5 Consider $\dfrac{p}{q} = a_0 + \cfrac{1}{a_1 + \cfrac{1}{a_2 + \cfrac{1}{\ddots \cfrac{}{a_{n-1} + \cfrac{1}{a_n}}}}}$.

(a) If a_1 is increased, will p/q increase or decrease? Prove your answer.

(b) If a_2 is increased, will p/q increase or decrease? Prove your answer.

6 (a) What happens if a_1 is decreased in Problem 5?

(b) What happens if a_2 is decreased?

7 The approximating fractions (convergents) for $\frac{5}{23}$ are $0, \frac{1}{4}, \frac{1}{5}, \frac{2}{9}, \frac{5}{23}$. Compute the numerators of the successive differences.

8 Do the same for some other continued fractions. Look again at how the approximating fractions fit into the Farey series.

Section 10 Decimal Fractions

We are all familiar with decimal fractions and their use from our grammar school days. There are many things we know about them from experience; for example, some stop:

$$\frac{1}{4} = 0.25$$
$$\frac{1}{8} = 0.125$$
$$\frac{1}{25} = 0.04,$$

and some are repeating:

$$\frac{1}{3} = 0.3333 \cdots$$
$$\frac{2}{9} = 0.2222 \cdots$$
$$\frac{81}{111} = 0.729729729 \cdots .$$

Some repeating decimals are erratic for awhile and then repeat:

$$\frac{1}{6} = 0.166666 \cdots$$
$$\frac{13}{88} = 0.14772727272 \cdots$$
$$\frac{19}{112} = 0.1696428571428571428571 \cdots .$$

Others don't seem to repeat:

$$\frac{7}{29} = 0.2413793034482758620689655 17 \cdot \ \cdot$$
$$\frac{30}{61} = 0.4918032786885245901639344426 29508 \cdots$$
$$\frac{60}{97} = 0.6818556701030927835015 1546391 75257 \cdots .$$

We shall discuss each of these four types.

The first type are easy enough to understand. They are an abbreviated notation for fractions which can be written with denominator a power of 10. Thus

$$\frac{1}{4} = 0.25 = \frac{25}{100}$$
$$\frac{1}{8} = 0.125 = \frac{125}{1,000}$$
$$\frac{1}{25} = 0.04 = \frac{4}{100}.$$

All the properties of finite decimals can be derived from the fact that they are, in fact, ordinary fractions with a convenient choice of denominators. Thus a given fraction can be written as a finite decimal if and only if there is an equal fraction with denominator a power of 10. From this fact, it is easy to determine exactly which reduced fractions have finite decimal expansions.

81 Theorem

A reduced fraction p/q has a finite decimal expansion if and only if q divides a power of 10, that is, $q = 2^m \cdot 5^n$ for some nonnegative integers m and n.

PROOF: If p/q has a finite decimal expansion, then $p/q = N/10^k$ for some nonnegative integers N and k. But since p/q is reduced and $p/q = N/10^k$, we know N and 10^k have a common factor, say r, such that $N/r = p$ and $10^k/r = q$. But this implies that $10^k/q = r$. Thus q divides $10^k = 2^k \cdot 5^k$. It follows (Theorem 21, Chapter 2) that $q = 2^m 5^n$ with $0 \leq m \leq k, 0 \leq n \leq k$.

Conversely, if $q = 2^m \cdot 5^n$, then $q \cdot (2^n \cdot 5^m) = 2^{m+n} 5^{m+n} = 10^{n+m}$. Thus $p/q = (p \cdot 2^n \cdot 5^m)/10^{n+m}$, and p/q can be expressed as a finite decimal. The theorem is proved.

Instead of ending with 10^{n+m} as the denominator, we could end with 10^k, where k is any integer at least as large as the larger of m and n.

For reduced fractions with finite decimals, there is no difficulty in finding the fraction given the decimal or in finding the decimal given the fraction.

The notion of finite decimal fractions and an understanding of Theorem 81, at least in rudimentary form, appear to have come from the Babylonians, who used a number system based on 60 rather than 10.

The use of decimals was essentially lost to Europe for centuries and remained alive through Hindu and Arabian mathematics. Even so, its introduction into modern mathematical life did not come about until the sixteenth century. It first appears in 1530 in a calculation manual published by Christoff Rudolff in Augsburg. The first systematic treatment of the rules of arithmetic for decimals was written in 1585 by the Dutch mathematician Simon Stevin in a work entitled *De Thiende*, or *La Disme* in the French translation which came soon afterward. *La Disme* is subtitled "Teaching How All Computations That Are Met in Business May Be Performed by Integers Alone."

Simon Stevin (1549–1620) should be remembered for his versatility! He was a scholar, inventor, engineer, quartermaster general in the Dutch army, director of dikes and canals, and reorganizer of government bookkeeping; he published works on fortifications, mechanics, interest tables, hydrostatics, decimals, the metric system, languages, arithmetic, geometry, trigonometry. He introduced the decimal system and fractional exponents to Europe, and was the first to apply the Euclidean algorithm to polynomials in order to find their greatest common divisors. (Permission of the Granger Collection.)

Stevin was a versatile genius at both the theoretical and the practical and served his country, Holland, and the world in many capacities. For a brief sketch of his life, see Ore [17]. For other information on how decimals were introduced to our culture, see [4, pp. 147–148].

According to the translation of *La Disme* by Vera Sanford in D. E. Smith's book [22], the work begins as follows:

To astrologers, surveyors, measurers of tapestry, gaugers, stereometers in general, mintmasters and to all merchants Simon Stevin sends greeting:

A person who contrasts the small size of this book with your greatness, my most honorable sirs to whom it is dedicated, will think my idea absurd, especially if he imagines that the size of this volume bears the same ratio to human ignorance that its usefulness has to men of your outstanding ability; but in so doing he will have compared the extreme terms of the proportion which may not be done. Let him rather compare the third term with the fourth.

What is it that is here propounded? Some wonderful invention? Hardly that, but a thing so simple that it scarce deserves the name invention; for it is as if some stupid country lout chanced upon great treasure without using any skill in the finding. If anyone thinks that, in expounding the usefulness of decimal numbers, I am boasting of my cleverness in devising them, he shows without doubt that he has neither the judgment nor the intelligence to distinguish simple things from difficult, or else that he is jealous of a thing that is for the common good. However this may be, I shall not fail to mention the usefulness of these numbers, even in the face of this man's empty calumny. But, just as the mariner who has found by chance an unknown isle, may declare all its riches to the king, as, for instance, its having beautiful fruits, pleasant plains, precious minerals, etc., without its being imputed to him as conceit; so may I speak freely of the great usefulness of this invention, a usefulness greater than I think any of you anticipates, without constantly priding myself on my achievements.

Stevin also pushed for the metric system in his book.

As we stated earlier, we assume that the reader is familiar with the arithmetic of finite decimals, and so we return to more theoretical questions on the nature of decimals.

The infinite periodic decimals pose a more difficult problem, although many of the readers can probably juggle a decimal like $0.61616161\cdots$ and find a fraction with this infinite periodic expansion. Since the expression $0.616161\cdots$ does not really say that the 61 repeats but only says that we don't know what comes next, we underline the last appearance of the repeating digits to indicate exactly what the repeating pattern is. Thus $0.6161\underline{61}$, $\frac{1}{3} = 0.33\underline{3}$, $\frac{81}{111} = 0.729\underline{729}$, and $\frac{1}{6} = 0.166\underline{6}$.

Now we can write repeating decimals more precisely, and we can find repeating decimals from some fractions at least—in many junior high schools, students are taught how to convert a repeating decimal into a fraction. But what does the equality

$$\tfrac{1}{3} = 0.333\underline{3}$$

mean? In what sense does $\frac{1}{3}$ equal this unending row of 3s? More generally, what does

$$x = 0.10110111011110111110111110 \cdots$$

mean? Or is it meaningless, since the decimal digits neither terminate nor repeat? If meaningful, can we find a fractional value for x?

Before we can put ourselves on solid ground, it is necessary to define precisely the terms with which we are dealing.

82 Definition

A symbol of the form $N.a_1a_2a_3a_4 \cdots$ is called a *decimal* base 10 if and only if:

1 N is an integer in base-10 notation.
2 Each of the symbols a_1, a_2, a_3, ... is one of the numbers 0,1,2,3,4,5,6,7,8,9.

We next define what it means for a number to be equal to a decimal. Before reading the definition, the reader is urged to think about the problem and perhaps discuss it with his colleagues. In stating this definition, it is supposed that we are already familiar with finite decimals.

83 Definition

We say that a real number x is equal to the decimal $N.a_1a_2a_3 \cdots$ if and only if, for each positive integer n, the inequalities

84 $$N.a_1a_2a_3 \cdots a_n \leq x \leq N.a_1a_2a_3 \cdots a_n + \frac{1}{10^n}$$

hold.

85 Examples

1 $\frac{1}{3} = 0.3333\underline{}$ since $0.3 \leq \frac{1}{3} \leq 0.3 + \frac{1}{10} = 0.4$
$0.33 \leq \frac{1}{3} \leq 0.33 \leq \frac{1}{100} = 0.34$
$0.333 \leq \frac{1}{3} \leq 0.333 + \frac{1}{1,000} = 0.334.$

In general,

$$\underbrace{0.333 \cdots 3}_{n \text{ times}} \leq \tfrac{1}{3} \leq \underbrace{0.333 \cdots 3}_{n \text{ times}} + \frac{1}{10^n},$$

Section 10 Decimal Fractions 449

since $3(0.3333 \cdots 3) = 0.9999 \cdots 9 \leq 1$, while $1 \leq 3(0.33 \cdots 34) = 1.0000 \cdots 2$.

2 $\frac{1}{2} = 0.5\underline{0}$. Since $\frac{1}{2} = 0.5 \leq \frac{1}{2} \leq 0.5 + \frac{1}{10} = \frac{1}{2} + \frac{1}{10}$
$\frac{1}{2} = 0.50 \leq \frac{1}{2} \leq 0.50 + \frac{1}{100} = \frac{1}{2} + \frac{1}{100}$
$\frac{1}{2} = 0.500 \leq \frac{1}{2} \leq 0.500 + \frac{1}{1,000} = \frac{1}{2} + \frac{1}{1,000}$ etc.

86 EXERCISES

1 Verify $\frac{2}{3} = 0.66\underline{6}$.
2 Verify $\frac{2}{9} = 0.22\underline{2}$.
3 Verify $\frac{81}{111} = 0.7\underline{29}$.
4 Verify $\frac{1}{8} = 0.125\underline{0}$.
5 Verify $\frac{13}{88} = 0.147\underline{72}$.
6 Prove that if $0.a_1a_2a_3$ is a finite decimal, then $0.a_1a_2a_3 = 0.a_1a_2a_3\underline{0}$.
7 Prove that $N.a_1a_2a_3 \cdots a_k\underline{0} = N.a_1a_2 \cdots a_k$.
8 Verify $1 = 0.99\underline{9} = 1.00\underline{0}$.
9 Prove that if $a_k \neq 9$, then $0.a_1a_2 \cdots a_k\underline{9} = 0.a_1a_2 \cdots a_{k-1}b_k$, where $b_k = a_k + 1$.
10 Prove that if $x = 0.a_1a_2a_3\cdots$, then $10x = a_1.a_2a_3\cdots$.
11 Prove that if $x = 0.a_1a_2a_3\cdots$, then $10^k x = a_1a_2a_3 \cdots a_k.a_{k+1}a_{k+2}\cdots$.
12 Prove that if $10x = a_1.a_2a_3\cdots$, then $x = 0.a_1a_2a_3\cdots$.
13 Prove that if $10^k x = a_1a_2a_3 \cdots a_k.a_{k+1}a_{k+2}\cdots$, then $x = 0.a_1a_2a_3\cdots$.
14 Define "decimals" for the base 2. Define equality for decimals base 2. Find the expansion of $\frac{1}{3}$ for decimals base 2.
15 Prove the analogs of Problems 9, 11, and 13 for base 2.
16 Define "decimals" and equality of decimals for base $b > 1$.
17 Prove analogs for Problems 9, 11, and 13 for decimals base $b > 1$.
18 What fraction will have finite decimals in base b, $b > 1$?

We now ask some questions which we shall try to answer in the forthcoming discussion.

Does every number have at least one decimal? How many decimals can a given number have (Problem 8, Exercises 86)?

Does every decimal correspond to some number? How many?

Do different numbers correspond to different decimals? (We know that different decimals may correspond to the same number.)

Can we obtain a method for finding the decimal for a given number?

Let us begin by an examination of fractions. We were all taught, very early in life, that to get the decimal for a fraction, we divide the numerator by the denominator, adding zeros to the numerator as needed to continue until we have computed as many decimal places as we like. Let us examine this method when applied to $\frac{7}{29}$; the first 28 places of its decimal was given at the beginning of the chapter. On the left is the usual division algorithm; on the right are the steps involved in performing this algorithm, each written on a separate line.

$$
\begin{array}{r}
0.24137 \\
29 \overline{)7.00000} \\
0 \\
\overline{70} \\
58 \\
\overline{120} \\
116 \\
\overline{40} \\
29 \\
\overline{110} \\
87 \\
\overline{230} \\
203 \\
\overline{27}
\end{array}
$$

$7 = 0 \cdot 29 + 7$

$70 = 2 \cdot 29 + 12$

$120 = 4 \cdot 29 + 4$

$40 = 1 \cdot 29 + 11$

$110 = 3 \cdot 29 + 23$

$230 = 7 \cdot 29 + 27.$

If we divide each equation on the right by 29, we obtain the equations

$$\frac{7}{29} = 0 + \frac{7}{29}$$
$$10 \cdot \frac{7}{29} = 2 + \frac{12}{29}$$
$$10 \cdot \frac{12}{29} = 4 + \frac{4}{29}$$
$$10 \cdot \frac{4}{29} = 1 + \frac{11}{29}$$
$$10 \cdot \frac{11}{29} = 3 + \frac{23}{29}$$
$$10 \cdot \frac{23}{29} = 7 + \frac{27}{29}.$$

In each equation, the right-hand side is written as a whole number plus a fractional part, with the fractional part nonnegative

Section 10   Decimal Fractions   451

and less than 1. We multiply the fractional part by 10 (bring down a zero, in long division), use it as the left-hand side of the next row, and continue the process, obtaining the digits of the decimal expansion.

Using this as a model, we define the *decimal algorithm*.

87   The Decimal Algorithm

Let x be any positive number. Find N and r_0 such that $x = N + r_0$ where N is an integer and $0 \leq r_0 < 1$. Repeat for $10r_0$, etc.:

$10r_0 = a_1 + r_1$ where a_1 is an integer and $0 \leq r_1 < 1$
$10r_1 = a_2 + r_2$ where a_2 is an integer and $0 \leq r_2 < 1$
$10r_2 = a_3 + r_3$ where a_3 is an integer and $0 \leq r_3 < 1$
$10r_3 = a_4 + r_4$ where a_4 is an integer and $0 \leq r_4 < 1$.

In general, $N = [x]$, where $[x]$ means the largest integer less than or equal to x and $r_1 = x - [x]$. Given r_k, we get r_{k+1} and a_{k+1} by the formulas $a_{k+1} = [10r_k]$ and $r_{k+1} = 10r_k - [10r_k]$.

88   Theorem

If N, a_1, a_2, a_3, ... are obtained from x by the decimal algorithm, then $0 \leq a_i < 9$ and $x = N.a_1a_2a_3\cdots$.

PROOF: Since each $r_i < 1$, we know $10r_i < 10$. Thus $a_i < 10$. On the other hand, since $10r_1 \geq 0$, we see that a_i is a nonnegative integer; thus $0 \leq a_i \leq 9$.

To show the desired equality, we must show that for each k,

$$N.a_1a_2 \cdots a_k \leq x < N.a_1a_2 \cdots a_k + \frac{1}{10^k}.$$

First we note that

$$x = N + r_0 \text{ and } r_0 = \frac{a_1}{10} + \frac{r_1}{10} \qquad 0 \leq r_1 < 1.$$

Thus $x = N + a_1/10 + r_1/10$. But $N + a_1/10 = N.a_1$; thus

$$N.a_1 \leq x = N.a_1 + \frac{r_1}{10} \leq N.a_1 + \frac{1}{10} \qquad \text{since } 0 \leq r_1 < 1.$$

Since $x = N + a_1/10 + r_1/10$ and $r_1 = a_2/10 + r_2/10$, we get $x = N + a_1/10 + a_2/100 + r_2/100 = N.a_1a_2 + r_2/100$. Thus $N.a_1a_2 \leq x = N.a_1a_2 + r_2/100 < N.a_1a_2 + 1/100$ since $0 = r_2 < 1$.

We make the induction hypothesis that

89
$$x = N.a_1a_2 \cdots a_n + \frac{r_n}{10_n}$$

and we show that

90
$$x = N.a_1a_2 \cdots a_{n+1} + \frac{r_{n+1}}{10^{n+1}}.$$

From the $(n + 1)$st line of the algorithm, we have $10r_n = a_{n+1} + r_{n+1}$, or $r_n = a_{n+1}/10 + r_{n+1}/10$. Substituting for r_n in Formula 89, we get

$$x = N.a_1a_2 \cdots a_n + \frac{a_{n+1}}{10^{n+1}} + \frac{r_{n+1}}{10^{n+1}}.$$

But $N.a_1a_2 \cdots a_n + a_{n+1}/10^{n+1} = N.a_1a_2 \cdots a_{n+1}$, so that Formula 90 is valid. It follows by induction that Formula 89 is valid for every value of n.

We see immediately from Formula 89 that

$$N.a_1a_2 \cdots a_n \le x = N.a_1a_2 \cdots a_n + \frac{r_{n+1}}{10^n} \le N.a_1a_2 \cdots a_n + \frac{1}{10^n}$$

since $0 \le r_{n+1} < 1$. The theorem is proved.

We shall need Formula 89 again in the proof of Theorem 113. Historically, this algorithm appears to have arisen from the observation that $x \approx [100x]/100$, where $\approx$ denotes approximate equality and $[100x]$ denotes the greatest integer in $100x$. For example, if $x = \sqrt{2}$, then we get

$$\sqrt{2} \approx \frac{[100\sqrt{2}]}{100} = \frac{[\sqrt{20,000}]}{100} = \frac{141}{100}$$

since $141^2 < 20,000 < 142^2$, and if $x = \frac{1}{3}$, we get

$$\frac{1}{3} \approx \frac{[100/3]}{100} = \frac{33}{100}.$$

(See [4, p. 147].)

91 EXERCISES

1 How big can $x - [100x]/100$ be? What about $x - [1,000x]/1,000$?

2 How big can $x - [10^k x]/10^k$ be?

3 Prove that $[100x]/100 = 0.a_1a_2$, where a_1 and a_2 are the first two decimal digits of x.

4 Prove that $[10^k x]/10^k = 0.a_1 a_2 a_3 \cdots a_k$, where $a_1, a_2, \ldots, a_k$ are the first k decimal digits in the decimal expansion of x.

We now know that each number can be written as a decimal, that the decimal algorithm is equivalent to dividing the numerator by the denominator for fractions, and by examining the algorithm, that there is only one decimal expansion which comes from the algorithm since at each step the choices are uniquely determined. We don't yet know when or how often there are other decimal expansions for a given number not coming from the decimal algorithm.

We could avoid some problems by changing our definition of decimal to allow as decimals only those expansions which come from the algorithm. If this approach were taken, then clearly every number would correspond to one and only one decimal and conversely. This approach might seem like a simplification, but certain difficulties arise.

For example, $3 \cdot \frac{1}{3} = 1$, but $3(0.333\underline{3}) = 0.999\underline{9}$, and $0.999\underline{9}$ would not be a decimal in the procedure of the above paragraph. (Can you prove it would not be?) Furthermore, if we wrote down a string of symbols, such as $0.01001000100001000001 \cdots$, we would often be unable to tell if it was permissible or not unless we had made a thorough study of which decimals can come from numbers. To do that, it would be necessary to make a thorough study of the number system, and we should be led very far afield.

Thus we continue our present approach, defining decimals and equality of decimals as in Definitions 82 and 83. We shall need to make an additional assumption about our number system (in any axiomatic approach, we would have to make a similar assumption).

92 The Completeness Axiom

Given any decimal, $N.a_1 a_2 a_3 \cdots$, there is some real number x such that $x = N.a_1 a_2 a_3 \cdots$.

This assumption ensures that all the numbers are there, or that the system is complete.

93 Theorem

Let $0.a_1 a_2 a_3 \cdots = 0.b_1 b_2 \cdots$; then either:

1 $a_j = b_j$ for each positive integer j.

2   There is a positive integer k such that $a_j = b_j$ for $1 \leq j < k$, $a_k = b_k + 1$, and $a_j = 0$ and $b_j = 9$ for $j > k$.
3   There is a positive integer k such that $a_j = b_j$ for $1 \leq j < k$, $b_k = a_k + 1$, and $b_j = 0$ and $a_j = 9$ for $j > k$.

This theorem says that the only possible ambiguities of decimals which can occur is the one discussed in Problem 9, Exercises 86. We also note that the only difference between statements 2 and 3 of the theorem is that the a's and the b's are interchanged.

PROOF: Let $x = 0.a_1a_2a_3 \cdots = 0.b_1b_2b_3 \cdots$. If $a_j = b_j$ for each j, then statement 1 is true and there is nothing to prove. If not, let k be the first number for which $a_k \neq b_k$. There are two possibilities: $a_k > b_k$ or $a_k < b_k$. We argue the case when $a_k > b_k$, and we shall show that statement 2 holds. If $a_k < b_k$, then statement 3 holds by the exact same argument with the a's and b's interchanged.

By hypothesis, $x = 0.a_1a_2a_3 \cdots$. From the definition of equality of decimals,

94         $$0.a_1a_2 \cdots a_n \leq x \leq 0.a_1a_2 \cdots a_n + \frac{1}{10^n}.$$

Similarly,

95         $$0.b_1b_2 \cdots b_n \leq x \leq 0.b_1b_2 \cdots b_n + \frac{1}{10^n}.$$

From the left-hand inequality of Formula 94 and the right-hand inequality of Formula 95, with $n = k$, we deduce

$$0.a_1a_2 \cdots a_k \leq 0.b_1b_2 \cdots b_k + \frac{1}{10^k}.$$

From the meaning of finite decimal fractions and the fact that $a_1 = b_1$ and $a_2 = b_2, \ldots, a_{k-1} = b_{k-1}$, we infer that $a_k \leq b_k + 1$. But we are in the case $b_k < a_k$, whence $b_k + 1 \leq a_k$; it follows that $a_k = b_k + 1$.

We next show that for $j > k$, $b_j = 9$. We suppose that for some $n > k$, $b_n < 9$, and we show that this leads to a contradiction. From Formulas 94 and 95, we deduce

$$0.a_1a_2a_3 \cdots a_{k-1}a_k \cdots a_n \leq 0.b_1b_2b_3 \cdots b_{k-1}b_k \cdots b_n + \frac{1}{10^n}.$$

Since $a_j = b_j$ for $j < k$ and since $b_n + 1 \leq 9$, we can rewrite the above inequality as

96      $$0.a_1a_2a_3 \cdots a_{k-1}a_k \cdots a_n \leq 0.a_1a_2a_3 \cdots a_{k-1}b_k \cdots b_{n-1}c_n$$

where $c_n = b_n + 1$.

Section 10   Decimal Fractions   455

On the other hand, given two finite decimals we find the larger by determining which is larger at the first decimal place at which they disagree. Thus, since $a_k > b_k$,

97 $0.a_1a_2a_3 \cdots a_{k-1}a_k \cdots a_n > 0.a_1a_2 \cdots a_{k-1}b_k \cdots b_{n-1}c_n.$

Formulas 96 and 97 contradict one another, and thus our assumption that for some $n > k$, $b_n < 9$ must be a false assumption. Therefore, for all $n > k$, $b_n = 9$.

We must still prove that $a_j = 0$ for $j > k$. Since $b_j = 9$ for $j > k$ and $a_k = b_k + 1$, we see that for $n > k$,

98 $0.b_1b_2 \cdots b_k \cdots b_n + \dfrac{1}{10^n} = 0.b_1b_2 \cdots b_{k-1}a_k$

$$= 0.a_1a_2 \cdots a_k.$$

From Formulas 94, 95, and 98, we get

$$0.a_1a_2 \cdots a_k \cdots a_n \leq 0.a_1a_2 \cdots a_k.$$

This can only happen if $a_{k+1} = a_{k+2} = \cdots = a_n = 0$. Thus statement 2 holds.

The same argument with the a's and b's interchanged shows that if $b_k > a_k$, then statement 3 holds. The theorem is proved.

99   Corollary
A nonrepeating decimal can never equal a repeating decimal.

100   Corollary
Except for nonzero fractions which have a finite decimal expansion, every number has a unique decimal expansion.

101   Corollary
Except for fractions which can be written in the form $p/2^m 5^n$, $p \neq 0$, every number has a unique decimal expansion.

We now return to fractions and decimals which are repeating from some point on.

102   Theorem
Every decimal which eventually repeats is equal to a fraction. When reduced, the denominator of this fraction divides $10^k \cdot (10^j - 1)$, where k is the number of terms before the repeating pattern begins and j is the period length.

PROOF: Let $0.a_1a_2a_3 \cdots a_ka_{k+1} \cdots a_{k+j}$ be a decimal which eventually repeats. We know it is equal to some number, say x.

We first handle the case of a pure repeating decimal, that is, $k = 0$. Thus
$$x = 0.\underline{a_1a_2a_3 \cdots a_j}.$$

It follows that (Problem 11, Exercises 86)
$$10^j x = a_1a_2a_3 \cdots a_j \cdot \underline{a_1a_2a_3 \cdots a_j}$$
$$= a_1a_2a_3 \cdots a_j + 0.\underline{a_1a_2a_3 \cdots a_j}$$
$$= a_1a_2a_3 \cdots a_j + x.$$
Thus
$$(10^j - 1)x = a_1a_2a_3 \cdots a_j,$$
so
$$x = \frac{a_1a_2a_3 \cdots a_j}{10^j - 1}$$

which proves that x is a fraction. Since the denominator is $10^j - 1$ in this form, the reduced denominator certainly divides $10^j - 1$.

If $k \geq 1$, then we begin with
$$x = 0.a_1a_2 \cdots a_k\underline{a_{k+1} \cdots a_{k+j}}.$$
It follows
$$10^k x = a_1a_2 \cdots a_k + 0.\underline{a_{k+1} \cdots a_{k+j}}.$$

The second term on the right-hand side is a pure repeating decimal and hence a fraction by the previous paragraph. Thus, for proper choice of integers p, q, we get
$$10^j x = a_1a_2 \cdots a_k + \frac{p}{q},$$

where q divides $10^j - 1$.

Solving for x, we obtain

103
$$x = \frac{q \cdot (a_1a_2 \cdots a_k) + p}{10^k \cdot q}.$$

Thus x is a fraction whose denominator is a divisor of $10^k(10^j - 1)$, and the theorem is proved.

The method of proving the theorem enables us to find the fractional form of a number given to us by an eventually repeating decimal. This theorem and Formula 103 were first proved by J. H. Lambert in 1758. G. W. von Leibniz stated, in 1677, that $1/n$ is periodic when expanded as a decimal and that the period length divides $n - 1$.

Section 10   Decimal Fractions

Gottfried Wilhelm von Leibniz (1646-1716) lived in Leipzig and studied law at the University. In 1672, on a political mission to Paris, he met G. Huygens, who led him to mathematics. Much of Leibniz' work is slighted in English-speaking countries because of the controversy between British and Continental mathematicians as to whether Newton or Leibniz has priority for inventing calculus. In 1675 Leibniz introduced the modern symbols for calculus and also the use of $=$ for proportion, $\sim$ for similar, and $\simeq$ for congruence. According to P. E. B. Jourdain, "Leibniz attributed all his mathematical discoveries to his improvements in notation." (Courtesy of the Smith Collection, Columbia University.)

104 Examples

1 We find the fraction equal to $0.1\overline{59}$. Let $x = 0.1\overline{59}$, then $1{,}000x = 159.\overline{159}$. Thus $1{,}000x = 159 + x$. Solving for x, we obtain $x = \frac{159}{999}$. Since $1 + 5 + 9$ is divisible by 3, so is 159 (Theorem 14), while 999 is obviously divisible by 3. Thus $x = \frac{53}{333}$.

2 We find the fraction for $0.23\overline{5767}$. Let $x = 0.23\overline{567}$; then $1{,}000x = 235 + 0.\overline{67}$. Let $y = 0.\overline{67}$; then $100y = 67 + y$, so $y = \frac{67}{99}$. Thus $1{,}000x = 235 + \frac{67}{99}$, so $x = \frac{99 \cdot 235 + 67}{1{,}000 \cdot 99} = \frac{23{,}265 + 67}{1{,}000 \cdot 99} = \frac{23{,}332}{1{,}000 \cdot 99} = \frac{5{,}833}{250 \cdot 99} = \frac{5{,}833}{24{,}750}$.

105 EXERCISES

1 Find a fraction equal to (a) $0.010\overline{101}$, (b) $0.\overline{456456}$, (c) $0.0123\overline{434}$, (d) $0.876\overline{5123}$.

2 Let $x = 0.a_1 a_2 \cdots a_k \overline{a_{k+1} \cdots a_{k+j}}$. Show that

$$x = \frac{(10^j - 1)(a_1 \cdots a_k)_{10} + (a_{k+1} \cdots a_{k+j})_{10}}{10^k \cdot (10^j - 1)}.$$

3 Is Leibniz' conjecture about the period length of $1/n$ correct?

The following theorem first appeared in John Wallis' book, *Treatise of Algebra Both Historical and Practical*, written in 1676 and published in 1685 [26].

106 Theorem

If p/q is a reduced fraction, then its decimal is periodic. The length of the period is at most $q - 1$.

PROOF: Consider the decimal algorithm:

$$\frac{p}{q} = N + r_0$$
$$10r_0 = a_1 + r_1$$
$$10r_1 = a_2 + r_2$$
$$10r_2 = a_3 + r_3$$
$$\vdots$$

We notice first that each step in the algorithm is completely determined by the remainder in the previous steps. Also we note that $r_0 = (p/q) - N = p_0/q$, since N is an integer. Also, since $0 \leq r_0 < 1$, we see that $0 \leq p_0 < q$. Similarly, since $r_0 = p_0/q$, $r_1 = p_1/q$, with $0 \leq p_1 < q$, etc. In general, $r_i = p_i/q$, with $0 \leq p_i < q$. If any $p_i = 0$, then all the succeeding a_i and p_i are zero, and the decimal repeat zeros (is finite).

If no p_i is zero, then $0 < p_i < q$, or $1 \leq p_i \leq q - 1$. Since there are only $q - 1$ choices for p_i, some two of the numbers $p_0, p_1, p_2, \ldots, p_{q-1}$ must be the same, say $p_k = p_{k+j}$ where $0 \leq k < k + j \leq q - 1$. Thus $r_k = p_k/q = p_{k+j}/q = r_{k+j}$. But we saw that the value of r_k completely determines the succeeding steps of the algorithm $a_k = a_{k+j}$ and $r_{k+1} = r_{k+j+1}$, $a_{k+2} = a_{k+j+2}$ and $r_{k+2} = r_{k+j+2}, \ldots, a_{k+j} = a_{k+2j}$ and $r_{k+j} = r_{k+2j}$.

Thus the terms from a_k to a_{k+j-1} are identical with the terms from a_{k+j} to a_{k+2j-1}. Similarly, since $r_{k+j} = r_{k+2j}$, the terms between a_{k+2j} and a_{k+2j-1} are identical with the terms from a_k to a_{k+j-1}, etc., forever. Thus the decimal is repeating with period length j. Since $0 \leq k < k + j \leq q - 1$, it follows that $j = k + j - k \leq q - 1$. The theorem is proved.

107 Corollary

A number is a fraction if and only if its decimal is periodic.

PROOF: Immediate from the last two theorems.

108 Corollary

There are numbers which are not fractions.

PROOF: $0.10110111011110 \cdots$ is not periodic—hence it is not a fraction.

Section 10 Decimal Fractions

109 *Definition*

A number which can be written as a fraction (a ratio of integers) is called *rational*. Other numbers are called *irrational*.

Given a periodic decimal, we can make a nonperiodic decimal by inserting one zero after the first period, two zeros after the second, three zeros after the third, etc. Since there are many other nonperiodic decimals, this shows that in some sense there are at least as many irrationals as rationals. In fact, there are many more.

110 Examples

1 The decimal of $\frac{5}{7}$ has period length of at most 6 according to the theorem. In fact, $\frac{5}{7} = 0.\underline{714285}$, which is pure periodic of maximum length.
2 Recall the examples at the beginning of the section: $\frac{1}{3} = 0.\underline{3}$, $\frac{2}{9} = 0.\underline{2}$, $\frac{81}{111} = 0.\underline{829}$ are all pure periodic, but they are not of maximal length as permitted by the theorem.
3 Again from the examples at the beginning of the section, $\frac{1}{6} = 0.1\underline{6}$, $\frac{13}{88} = 0.147\underline{72}$, and $\frac{19}{112} = 0.1696428\underline{571}$. None is pure periodic, and none is of maximal length.
4 Look at the expansion of $\frac{7}{29}$ given at the beginning. It is periodic with a period length of at most 28. In fact, it is pure periodic, and the digits given in the expansion comprise precisely one period.

111 Lemma

If g.c.d.$(q, 10) = 1$ and p/q is a reduced fraction, then so are the remainders $r_0 = p_0/q$, $r_1 = p_1/q$, etc., which arise in the decimal algorithm.

The proof is left as an exercise to the reader.

112 EXERCISES

1 Prove Lemma 111.
2 Use Problem 1 to show that the length of the period in the decimal expansion of p/q is at most $\phi(q)$, where $\phi(q)$ is the number of integers n with $1 \leq n \leq q$ and g.c.d.$(n, q) = 1$. (See Chapter 2, Problem 6 of Exercises 74.)

3 Find some numbers for which $\phi(q) = q - 1$.
4 Find some numbers for which $\phi(q) < q - 1$.
5 Find some fractions p/q for which $\phi(q) = q - 1$ and some for which $\phi(q) < q - 1$ and which have decimal periods of length $\phi(q)$.
6 Find some fractions for which $\phi(q) = q - 1$ and some for which $\phi(q) < q - 1$ and which have decimal periods of length less than $\phi(q)$. Do you notice any connection between $\phi(q)$ and the period lengths?
7 Find an example to disprove Leibniz' claim that $1/n$ has period length a divisor of $n - 1$.
8 Prove J. H. Lambert's theorem: If $1/n$ has period length $n - 1$, then n is prime.
9 Give induction proofs for the "etc." statements in the proof of Theorem 106.

Is it possible to tell, before expanding a fraction into its decimal form, how long its periodic will be and whether or not it will be pure periodic? The first known steps in this direction are found in John Wallis' book [26], which gives some rules for determining period lengths. The complete solution of this problem was worked out by K. F. Gauss [9] in 1801. The solution is given by the following theorem.

113 Theorem

A reduced fraction r/s will have period length l after e nonperiodic terms where l and e are determined as follows:

1 Write $s = 2^m 5^n q$ where neither 2 nor 5 divide q, that is, g.c.d.$(q, 10) = 1$. Then e equals the larger of m and n.
2 If $q = 1$, let $l = 1$; otherwise, let l be the smallest number such that q divides $10^l - 1$.

Before proving Theorem 113, it is convenient to prove several lemmas. A different proof is given by Problems 3 to 6, Exercises 123.
    We begin by defining some notation which will be helpful for the remainder of this section. Let p/q be a reduced fraction with $0 < p < q$; then $a_1, a_2, a_3, \ldots$ and $p_0, p_1, p_2, p_3, \ldots$ are defined

Section 10 Decimal Fractions

by the decimal algorithm as follows:

$$a_0 = 0, \ p_0 = p \qquad 0 \leq p_0 < q$$

114

(1) $\quad \dfrac{10 p_0}{q} = a_1 + \dfrac{p_1}{q} \qquad 0 \leq p_1 < q$

(2) $\quad \dfrac{10 p_1}{q} = a_2 + \dfrac{p_2}{q} \qquad 0 \leq p_2 < q$

(3) $\quad \dfrac{10 p_2}{q} = a_3 + \dfrac{p_3}{q} \qquad 0 \leq p_3 < q$

$\qquad \cdots \cdots \cdots \cdots \cdots$

(k) $\quad \dfrac{10 p_{k-1}}{q} = a_k + \dfrac{p_k}{q} \qquad 0 \leq p_k < q$

$\qquad \cdots \cdots \cdots \cdots \cdots$

Thus

115 $\qquad \qquad \dfrac{p}{q} = 0.a_1 a_2 a_3 a_4 \cdots .$

116 *Notation*

For p/q with decimal expansion given by Formula 115, we let

$$A_m = a_1 \cdot 10^{m-1} + a_2 10^{m-2} + a_3 \cdot 10^{m-3} + \cdots + a_{m-1} \cdot 10 + a_m$$
$$= (a_1 a_2 a_3 \cdots a_m)_{10}$$
$$A_{n,m} = a_{n+1} 10^{m-1} + a_{n+2} 10^{m-2} + \cdots + a_{m+n-1} 10 + a_{m+n}$$
$$= (a_{n+1} a_{n+2} \cdots a_{n+m})_{10}.$$

Clearly, $A_m = A_{0,m}$.

117 Examples

1 If $p/q = \tfrac{1}{7} = 0.\underline{142857}$, then $A_3 = 142$, $A_5 = 14{,}285$, $A_{2,3} = 285$, $A_{2,5} = 28{,}571$, $A_{6,3} = A_3 = 142$, and in general, $A_{6,m} = A_m$.

2 If $p/q = \tfrac{1}{19} = 0.\underline{052631578947368421}$, then $A_7 = 526{,}315$, $A_{2,12} = 263{,}157{,}894{,}736$, and $A_{4,20} = 31{,}578{,}947{,}368{,}421{,}052{,}631$.

118 Lemma

If p/q is a reduced fraction, then $10^j p_k = p_{k+j} + q A_{k,j}$.

We apply Formula 89 with $x = p/q$. Then $N = 0$ since $0 < p/q < 1$ and $r_i = p_i/q$. Thus

$$\dfrac{p_0}{q} = 0.a_1 a_2 \cdots a_j + \dfrac{p_j}{q \cdot 10^j}.$$

Multiplying by $q \cdot 10^j$, we get

$$10^j p_0 = (a_1 a_2 \cdots a_j)_{10} q + p_j.$$

which proves the lemma for the case $k = 0$.

Since the expansion p_k/q is simply given by the algorithm for p_0/q by ignoring the first k steps, we get

$$\frac{p_k}{q} = 0.a_{k+1}a_{k+2}a_{k+3} \cdots = 0.b_1 b_2 b_3 \cdots .$$

Thus, by the above paragraph,

$$10^j p_k = (b_1 \cdots b_j)q + p_{k+j} = q A_{k,j} + p_{k+j}.$$

119 EXERCISE

Prove Lemma 118 directly without using Formula 89.

120 Lemma

If g.c.d.$(10, q) = 1$ and p/q is reduced, then $p_k = p_{k+l}$ if and only if q divides $10^l - 1$.

PROOF: Suppose q divides $10^l - 1$. Then, in congruence notation, $10^l \equiv 1$ mod q. From Lemma 118 we get $10^l p_k = p_{k+l} + q A_{k,l}$. Since $10^l \equiv 1$ mod q and $q \equiv 0$ mod q, we get

$$1 \cdot p_k \equiv 10^l p_k = p_{k+l} + q A_{k,l} \equiv p_{k+l} + 0 \cdot A_{k,l} \equiv p_{k+l} \text{ mod } q.$$

Thus $p_k \equiv p_{k+l}$ mod q. But $0 \leq p_k < q$ and $0 \leq p_{k+l} < q$; thus $p_k = p_{k+l}$.

Suppose now that $p_k = p_{k+l}$. From Lemma 118 we get $10^l p_k = p_{k+l} + q A_{k,l}$.

Since $p_k = p_{k+l}$, we get

$$10^l p_k = p_k + q A_{k,l} \equiv p_k \text{ mod } q.$$

Thus $10^l p_k \equiv 1 \cdot p_k$ mod q. Since g.c.d.$(10, q) = 1$, it follows from Lemma 111 that g.c.d.$(p_k, q) = 1$. We deduce (Chapter 2, Theorem 63) that $10^l \equiv 1$ mod q. The proof is completed.

121 Corollary

If g.c.d.$(10, q) = 1$, there is a number $l \leq q - 1$ such that $10^l \equiv 1$ mod q.

PROOF: By Theorem 106, the first repetition of remainders begins before $q - 1$ steps. Thus, for some choices of k and l with $k + l \leq q - 1$, $p_k = p_{k+l}$, and so $10^l \equiv 1$ mod q by Lemma 120.

Section 10   Decimal Fractions

If we had developed base-b ($b > 1$) decimals, we would have obtained Corollary 121, with 10 replaced by b. A little more effort would prove that if q is prime, l divides $q - 1$. This would lead to another proof of Fermat's theorem (Chapter 2, Corollary 67): If q is prime and g.c.d.$(q, b) = 1$, then $b^{q-1} \equiv 1 \mod q$.

If q is not necessarily prime, we know that $l \leq \phi(q)$ (Problem 2, Exercises 112). Again, some additional effort would show that l divides $\phi(q)$. In this way, we could prove Euler's generalization of Fermat's theorem (Chapter 2, Problem 6 of Exercises 74): If g.c.d.$(b, q) = 1$, then $b^{\phi(q)} \equiv 1 \mod q$.

We are now in a position to prove Theorem 113.

PROOF OF THEOREM 113: Let q, r, s, e, l, m, and n be as in the statement of the theorem.

Consider the expansion of a reduced fraction p/q, $0 < p < q$. This fraction will satisfy $p_i = p_{i+l}$ for every value of i, $i = 0, 1, 2, 3, \ldots$ by Lemma 120, and $p_i \neq p_{i+j}$ for $j < l$. But if $p_i = p_{i+l}$, then

$$a_{i+1} = [10(p_i/q)] = [10p_{i+l}/q] = a_{i+l+1} \text{ for } i = 0, 1, 2, \ldots.$$

Thus every l terms, the decimal digits repeat; however, they might repeat every $l/2$ terms or every $l/3$ terms, etc. Of course, we know, by Lemma 120, that the p_i's don't repeat before l steps. We now show that the repetition after l steps is the first possible repetition and thus that the period length is exactly l.

Suppose $a_i = a_{i+j}$ for $i = 1, 2, 3, \ldots$. By Theorem 102 we know that q divides $10^t \cdot 10^j - 1$. Since g.c.d.$(q, 10) = 1$, we deduce that q divides $10^j - 1$. But by definition of l, we then know that $l \leq j$, so the period length is exactly l.

We consider now the expansion of $r/s = r/2^m 5^n q = 0.b_1 b_2 b_3 \cdots$. Multiplying by 10^e, we get

$$\frac{10^e \cdot r}{s} = \frac{2^{e-m} 5^{e-n} r}{q} = b_1 b_2 \cdots b_e . b_{e+1} b_{e+2} \cdots$$

Let

$$\frac{p}{q} = \frac{2^{e-m} 5^{e-q} r}{q} - (b_1 b_2 \cdots b_e)_{10} = 0.b_{e+1} b_{e+2} \cdots.$$

Since g.c.d.$(10, q) = $ g.c.d.$(r, q) = 1$, we see that p/q is reduced. If $p = 0$ or $q = 1$, then the decimal is either all zeros or all 9s; otherwise, by the previous paragraph, we know the expansion of p/q is periodic of length exactly l. Thus the expansion of r/s has period length l and there are at most e terms, $b_1, b_2, \ldots, b_e$, before the periodicity starts. We must yet show that the last digits of $b_1, b_2, \ldots, b_e$ cannot be such that the periodicity begins sooner. Suppose the periodicity begins after e' digits, with $e' \leq e$. Then by Theorem 102, s divides $10^{e'} \cdot (10^l - 1)$. But $s = 2^m 5^n \cdot q$, and since neither 2 nor 5 divides $10^l - 1$, we deduce that 2^m and 5^n divide $10^{e'}$. This implies that $e' \geq m$ and $e' \geq n$. Since e is the larger of m and n, it follows that $e' \geq e$. However, we already know that $e' \leq e$; whence $e' = e$. The theorem is proved.

122 Examples

1. The decimal for $\frac{7}{222}$ will have one nonperiodic term and period length 3 since $222 = 2 \cdot 111$ and $10^3 = 9 \cdot 111 + 1$.
2. The decimal for $\frac{1}{7}$ is pure periodic and has period length 6 since

$$10 \equiv 3 \bmod 7$$
$$10^2 \equiv 10 \cdot 10 \equiv 9 \equiv 2 \bmod 7$$
$$10^3 \equiv 10^2 \cdot 10 \equiv 6 \bmod 7$$
$$10^4 \equiv 10^3 \cdot 10 \equiv 18 \equiv 4 \bmod 7$$
$$10^5 \equiv 10^4 \cdot 10 \equiv 12 \equiv 5 \bmod 7$$
$$10^6 \equiv 10^5 \cdot 10 \equiv 15 \equiv 1 \bmod 7.$$

Thus 10^6 has remainder 1 and is the first power of 10 to do so.

Notice that the characteristic of the decimal expansion depends only on q and not on p so long as p/q is reduced.

123 EXERCISES

1. State, without computing the decimal expansion, the lengths of the periods and the number of initial nonperiodic terms for:
   (a) $\frac{1}{16}$ (b) $\frac{5}{18}$ (c) $\frac{7}{505}$ (d) $\frac{11}{350}$
2. Find a fraction with period length j and e nonperiodic terms for:

   (a) $j = 3, e = 4$ (b) $j = 2, e = 7$
   (c) $j = 4, e = 2$ (d) $j = 6, e = 0$

3. If $q = 2^m 5^n$, then Theorem 113 is true for p/q, g.c.d.$(p, q) = 1$.
4. Prove that if $p = a_1 10^{j-1} + a_2 10^{j-2} + \cdots + a_j 10^0$, then $p/(10^{j-1} - 1) = 0.a_1 a_2 \cdots a_j$. (Of course, if $a_1 = a_2 = \cdots = a_j = 1$, then this decimal has period less than j.)
5. Prove that if $p/q = 0.a_1 a_2 \cdots a_j$, then q divides $10^j - 1$.
6. Combine Problems 4 and 5 to prove Theorem 113 when g.c.d.$(q, 10) = 1$.

Section 10 Decimal Fractions 465

7 Prove Theorem 113 using Problems 3 to 6 above.
8 Compute the decimals of $\frac{1}{7}, \frac{2}{7}, \frac{3}{7}, \ldots, \frac{6}{7}$. Make some observations.
9 Compute $\frac{1}{17}, \frac{2}{17}, \frac{3}{17}$, and $\frac{12}{17}$. Look at the computation in Problem 8. Do you have some conjectures? You should, but more experimental evidence would be helpful. If you have a calculator or access to a computer, try experimenting.
10 Define the "decimal" algorithm for the base $b > 1$.
11 Find the expansions for the following fractions for the base indicated.

(a) $\frac{1}{3}, \frac{5}{8}, \frac{3}{7}$ for base 2.
(b) $\frac{1}{2}, \frac{5}{81}, \frac{4}{39}$ for base 3.
(c) $\frac{1}{16}, \frac{5}{33}, \frac{7}{19}$ for base 12.

12 State the analog of Theorem 113 for the base $b > 1$.
*13 Prove the analog of Theorem 113 for base $b > 1$.
*14 Let l be the least positive integer such that q divides $10^l - 1$. Prove that if q divides $10^j - 1$, then l divides j.
15 Prove Corollary 121 with "$l \leq q - 1$" replaced by "$l \leq \phi(q)$."

Problems 8 and 9 above should yield some idea of the usefulness of tables of decimal expansions of fractions. The first good tables seem to have appeared at the end of the eighteenth century and the beginning of the nineteenth, through the efforts of J. Bernoulli, K. F. Gauss, H. Goodwyn, J. C. Burckhardt, and others. From these tables, Bernoulli and Goodwyn noted that the periods of all fractions with certain kinds of denominator have the same digits in the same order but with different starting places. The complete theory of how the periods of fractions with the same denominator fit together was given and proved by Gauss [9]. Goodwyn further noted in 1802 that if $1/q$, for prime q, had an even period length, say $l = 2t$, then $a_j + a_{j+t} = 9$ for all j. E. T. Poselger in 1827 showed that for prime q, if $1/q$ has period length $l = 2t$, then $p_j + p_{j+t} = q$, that is, $r_j + r_{j+t} = 1$. We prove first the following theorem of J. Bernoulli, 1771.

124 Theorem

If $1/q = 0.a_1a_2 \cdots a_l$ has period length $l = q - 1$, and $1 \leq m < q$ then $m/q = 0.\overline{a_{j+1} \cdots a_l a_1 a_2 \cdots a_{j-1} a_j}$ for the proper choice of j. In fact, j is chosen so that $10^j \equiv m \bmod q$.

PROOF: Let the decimal algorithm for $1/q$ be given by Formulas 114 with $p = p_0 = 1$.

We have already seen that the decimal expansion from any point on is completely determined by the remainder p_k/q of the previous step. Since the decimal doesn't repeat until the qth step, none of the remainders can repeat until the $(q-1)$st step. Thus $p_0, p_1, p_2, \ldots, p_{q-2}$ are distinct. Since $0 < p_k < q$, each of $1, 2, 3, \ldots, q-1$ occurs exactly once among the p_i's. Choose j so that $p_j = m$. From Lemma 118 with $k = 0$, we get

125 $$10^j p_0 = p_j + qA_j.$$

Since $p_0 = 1$, $p_j = m$, and $q \equiv 0 \bmod q$, we get $10^j \equiv m \bmod q$. Dividing Formula 125 by q, we get $10^j(1/q) = (m/q) + (a_1 a_1 \cdots a_j)_{10}$. But $10^j(1/q) = a_1 a_2 \cdots a_j \cdot \underline{a_{j+1} a_{j+2} \cdots a_{j+q-1}}$. Thus

$$m/q = a_1 a_2 \cdots a_j \cdot \underline{a_{j+1} a_{j+2} \cdots a_{j+q-1}} - (a_1 a_2 \cdots a_j)_{10}.$$

Subtracting, we get

$$\frac{m}{q} = 0.\underline{a_{j+1} a_{j+2} \cdots a_{q-1} a_1 a_2 \cdots a_j},$$

which proves the theorem.

If $1/q$ has period length $\phi(q)$, a similar argument holds for the reduced fraction p/q even if q is not prime. If the period length of $1/q$ is l, a proper divisor of $\phi(k)$, the theorem and proof are of the same general nature but are more complicated. For details, see Gauss [9].

126 Examples

1. According to part 2 of Examples 122, we have $10^0 \equiv 1$, $10^2 \equiv 2$, $10^1 \equiv 3$, $10^4 \equiv 4$, $10^5 \equiv 5$, $10^3 \equiv 6 \bmod 7$. Thus

    $$\frac{1}{7} = 0.\overline{142857}$$
    $$\frac{2}{7} = 0.\overline{285714}$$
    $$\frac{3}{7} = 0.\overline{428571}$$
    $$\frac{4}{7} = 0.\overline{571428}$$
    $$\frac{5}{7} = 0.\overline{714285}$$
    $$\frac{6}{7} = 0.\overline{857142}$$

2. $\frac{1}{19} = 0.\overline{052631578947368421}$. Since $10^2 = 100 \equiv 5 \bmod 19$, we see that

    $$\frac{5}{19} = 0.\overline{263157894736842105}$$

On the other hand, by computation the first two digits of $\frac{11}{19}$ are 57. Thus

$$\frac{11}{19} = 0.\underline{578947368421052631}$$

and $10^6 \equiv 11 \bmod 19$. Since $\frac{18}{19} > \frac{9}{10}$, we know that the expansion of $\frac{18}{19}$ must begin $0.9\cdots$. Thus

$$\frac{18}{19} = 0.\underline{947368421052631578}$$

and $10^9 \equiv 18 \bmod 19$.

127 EXERCISES

1. Compute $\frac{15}{17}$. (See Problem 9 of Exercises 123.)
2. Solve $10^j \equiv 8 \bmod 17$ for j.
3. Prove Theorem 124 with hypothesis g.c.d.$(m, q) = 1$ and $l = \phi(q)$ rather than $l = q - 1$.

We see from the last examples and exercises that we can use decimal expansions to gain information about divisibility and congruences.

The following two theorems present some interesting and little-known facts, which perhaps some of you have already observed. They were noted first by Goodwyn in 1802.

128 Theorem

If q is a prime and $1/q = 0.a_1 a_2 \cdots a_l$ has an even period length $l = 2t$, then $p_j + p_{j+t} = q$ and $a_j + a_{j+t} = 9$.

The part of this theorem dealing with the a_i's is frequently called *Midy's theorem* (see [14]) after E. Midy, who in 1836 published an article discussing a stronger theorem which did not require the primality of q.

In order to prove Theorem 128, it is convenient to prove first a theorem which is stronger in some ways and weaker in others.

129 Theorem

If $1/q = 0.\overline{a_1 a_2 \cdots a_l}$ has period length $2t = l$ in its decimal expansion and $10^t \equiv -1 \bmod q$, then $p_k + p_{k+t} = q$ and $a_k + a_{k+t} = 9$.

PROOF: From Lemma 118, we deduce $10^t p_k \equiv p_{k+t}$ mod q. Thus $p_k + p_{k+t} \equiv p_k + 10^t p_k$ mod q. But by hypothesis, $10^t \equiv -1$; therefore,

$$p_k + p_{k+t} \equiv p_k + (-1)p_k \equiv 0 \text{ mod } q.$$

Thus q divides $p_k + p_{k+t}$. However, $0 < p_k + p_{k+t} \leq (q-1) + (q-1) < 2q$. It follows that $p_k + p_{k+t} = q$. This proves the first claim of the theorem.

From the above paragraph with k replaced by $k-1$, we conclude that $10p_{k-1} + 10p_{k-1+t} = 10q$, which may be rewritten

130 $$\frac{10p_{k-1}}{q} + \frac{10p_{k-1+t}}{q} = 10.$$

By definition, $a_k = [10p_{k-1}/q]$ and $a_{k+t} = [10p_{k-1+t}/q]$.

We note that $a_k \neq 10p_{k-1}/q$ since this equality would imply $p_k = 0$, which is impossible. Thus

131 $$a_k < \frac{10p_{k-1}}{q} < a_k + 1.$$

From Formulas 130 and 131, we deduce

132 $$10 - (a_k + 1) < \frac{10p_{k-1+t}}{q} < 10 - a_k.$$

Since $a_{k+t} = [10p_{k-1+t}/q]$, Formula 132 tells us that $a_{k+t} = 10 - (a_k + 1)$. It follows that $a_k + a_{k+t} = a_k + 10 - (a_k + 1) = 9$. This is precisely the second claim of the theorem.

To prove Theorem 128, it would be sufficient to show that under the hypothesis of Theorem 128, $10^l \equiv -1$ mod q, for then Theorem 129 would yield the desired result. Since we know that $10^l = 10^{2t} \equiv 1$ mod q and l is the least such number, we see that $10^t \not\equiv 1$ mod q but $(10^t)^2 \equiv 1$ mod q. If this were ordinary rather than congruence algebra, we would know that $x^2 = 1$ and $x \neq 1$ imply $x = -1$. We now show that if q is a prime, this same conclusion can be drawn for congruences modulo q. We do this by showing that a number can have at most two square roots modulo q.

133 Lemma

If q is a prime and $x^2 \equiv y^2$ mod q, then $x \equiv y$ or $x \equiv -y$ mod q.

PROOF: Let $x^2 \equiv y^2$ mod q; then $x^2 - y^2 \equiv 0$ mod q. Factoring $x^2 - y^2$, we get $(x+y)(x-y) \equiv 0$ mod q. This says that q divides the product of $(x+y)(x-y)$, and from this we deduce that either q divides $x+y$ or q divides $x-y$. (See Chapter 2, Theorems 16 and 40.) It follows that $x+y \equiv 0$ or $x-y \equiv 0$ mod q; whence $x \equiv -y$ or $x \equiv y$ mod q.

134 Corollary
If q is a prime and $x^2 \equiv 1 \bmod q$, then $x \equiv \pm 1 \bmod q$.

PROOF: Take $y = 1$ in Lemma 133.

Theorem 128 follows from Theorem 129 and Corollary 134. For a short discussion of which primes satisfy the hypothesis of Theorem 128 and of some other numbers which satisfy the hypothesis of Theorem 129, see W. G. Leavitt [14].

Wacław Sierpiński (1882–1969), a well-known contemporary mathematician, is the first Polish scholar since Madame Curie to be elected a foreign member of the French Academy of Sciences. He is known for his works in many branches of mathematics. We see him here lecturing to his seminar on elementary number theory at the Mathematical Institute of the Polish Academy of Sciences in 1961. He is speaking about continued fractions; the first several entries of the continued fraction expansion of π can be seen on the blackboard. "Incidentally," wrote H. S. M. Coxeter to the authors, "it is amusing to note that Sierpiński missed out the 1 between 15 and 292. (Even the greatest have their weak moments.)" (Photograph by W. Prażuch.)

135 EXERCISES

1. Find some composite numbers q such that $x^2 \equiv 1 \bmod q$ if and only if $x \equiv \pm 1 \bmod q$. If $1/q$ has an even period length, what conclusions can you draw?
2. Find some numbers q such that for some $x \not\equiv \pm 1 \bmod q$, $x^2 \equiv 1 \bmod q$.
3. Find the expansion of $\frac{1}{101}$ given that $11 \cdot 101 \cdot 9 = 9{,}999$.
4. Find the expansions of $\frac{9}{101}, \frac{50}{101}, \frac{90}{101}$.
5. Solve $10^j \equiv 50 \bmod 101$ and $10^j \equiv 90 \bmod 101$.
6. Solve for the least l such that $10^l \equiv 1 \bmod 9^2$. (Casting out 9s, Theorem 14, is helpful.)
7. Expand $1/9^2$ as a decimal.
8. Expand $1/2^2$ base 3.
9. Expand $1/3^2$ base 4.
10. Expand $1/4^2$ base 5. Make a conjecture about $1/(b-1)^2$ base b. (Glaisher 1873 see [10, 11]. See Sierpiński [20, p. 275].)

Section 11   Concluding Remarks

From the preceding sections one can see that there are a number of systems for handling fractions. All have their strengths and weaknesses, and all raise interesting mathematical questions.

The advantages of decimal notation for fractions are:

1. It is easy to compare two decimals to decide which is bigger.
2. Adding is not too difficult. (But try adding $0.\overline{789}$ to $0.\overline{67891}$.)
3. It is easy to approximate a number by a finite decimal with a given degree of accuracy if the decimal expansion is known.
4. The system is familiar and is compatible with our base-10 notation for whole numbers.

The disadvantages are:

1. Infinite process becomes involved even for simple fractions like $\frac{1}{3}$.
2. For great accuracy of approximation, many digits are required.
3. Multiplication is difficult. (Can you multiply $0.\overline{678}$ by $0.\overline{6789}$?)

Just as our analysis of the method of finding the decimal expansion of a rational number leads to a method of finding the decimal of any real number, the methods of finding the Egyptian-fraction expansion and the continued-fraction expansion can be defined so that they hold for any real number. Again, problems of how many ways a number can be written arise and can be handled at least as easily as with the decimal expansions. In either of these theories, it would again be necessary to make a completeness assumption.

The completeness assumption is essential to prove that there really exist numbers which have desired properties. For instance, how do you know if there is a real-number solution to $x^2 = 2$, $x^7 + x^5 + x^3 + x = 1$, or $x^2 = -2$, or how do we know that there is a number π which expresses the ratio of the circumference to the radius of a circle? If we can prove that there is some finite or infinite decimal, finite or infinite continued fraction, or finite or infinite Egyptian fraction which works (even if we can't write this expansion down), then completeness guarantees that the number exists. Thus it is not hard to prove, after developing enough theory to show that it makes sense to write it down, that

$$\sqrt{2} = 1 + \cfrac{1}{2 + \cfrac{1}{2 + \cfrac{1}{2 + \cfrac{1}{2 + \cfrac{1}{2 + \cfrac{1}{2 \cdots}}}}}}$$

One can further show that (by Lord William Brouncker, 1620–1687)

$$\pi = \cfrac{4}{1 + \cfrac{1^2}{2 + \cfrac{3^2}{2 + \cfrac{5^2}{2 + \cfrac{7^2}{2 + \cfrac{9^2}{2 + \cfrac{11^2}{2 + \cdots}}}}}}}$$

or (Euler)

$$e = 2 + \cfrac{1}{1 + \cfrac{1}{2 + \cfrac{1}{1 + \cfrac{1}{1 + \cfrac{1}{4 + \cfrac{1}{1 + \cfrac{1}{1 + \cfrac{1}{6 + \cfrac{1}{1 + \cdots}}}}}}}}}$$

where e is the base of the system of natural logarithms, even though no one knows how to write down its decimal expansion.

The first statement is equivalent to

$$\sqrt{2} + 1 = 2 + \cfrac{1}{2 + \cfrac{1}{2 + \cfrac{1}{\cdots}}}$$

Let us set $x = 2 + \cfrac{1}{2 + \cfrac{1}{2 + \cdots}}$, which we could do had we proved

that infinite continued fractions make sense. We thus obtain that x must satisfy $x = 2 + 1/x$, or $x^2 = 2x + 1$. Solving this equation, we obtain $x = 1 \pm \sqrt{2}$ as roots. But the number we want must be the positive root, so $x = \sqrt{2} + 1$.

For the theory of continued fractions, *every* fraction has a finite expansion, every square root of an integer has a periodic expansion, and every periodic expansion comes from the root of a quadratic equation with integral coefficients. (See Khintchine [13], Olds [16], or Moore [15] for details.)

136 EXERCISES

1. If $0.a_1^{(1)}a_2^{(1)}a_3^{(1)} \cdots \, ; \, 0.a_1^{(2)}a_2^{(2)}a_3^{(2)} \cdots \, ; \, 0.a_1^{(3)}a_2^{(3)} \cdots \, ; \cdots$ are a collection of decimals, find the smallest decimal $0.b_1b_2b_3$ which is larger than or equal to all these decimals. [Hint: Let $b_1 =$ the maximum of $a_1^{(1)}, a_1^{(2)}, a_1^{(3)}, \ldots$; let $b_2 =$ maximum of the numbers $a_2^{(j)}$, where j is taken such that $a_1^{(j)} = b_1$; etc.]

2. Prove that given any collection of decimals, there is a decimal $M.b_1b_2b_3 \cdots$ which is the smallest number greater than or equal to all of the given numbers provided there is some integer bigger than all the decimals in the collection.

*3. Prove there is a number x such that $x^2 = 2$. [Hint: Consider the decimals of all numbers y for which $y^2 < 2$, and apply Problem 2 above.]

4. Prove that there is no fraction p/q such that $(p/q)^2 = 2$; i.e., prove that $\sqrt{2}$ is irrational. [Hint: How many times (odd or even) does 2 occur in p^2, $2q^2$?] (See Rademacher and Toeplitz [19, Chap. 4] for two other simple proofs.)

5. Assuming we have proved that it makes sense, evaluate

$$x = 1 + \cfrac{1}{2 + \cfrac{1}{1 + \cfrac{1}{2 + \cfrac{1}{1 + \cdots}}}}$$

6. Evaluate $1 + \cfrac{1}{1 + \cfrac{1}{1 + \cdots}}$

7. Evaluate $2 + \cfrac{1}{3 + \cfrac{1}{2 + \cfrac{1}{3 + \cdots}}}$

References

1. Edward G. Begle, *Introductory Calculus with Analytic Geometry*, Holt, Rinehart and Winston, Inc., New York, 1954.
2. Truman Botts, "Problem solving in mathematics I and II," *Mathematics Teacher*, **58** (1965), pp. 496–501 and 596–600.
3. ———, "A chain reaction process in Number Theory," *Mathematics Magazine*, **40** (1967), pp. 55–65.
4. Florian Cajori, *A History of Mathematics*, The Macmillan Company, New York, 1893, rev. ed. 1961.
5. ———, *A History of Mathematical Notations*, Vols. I and II, The Open Court Publishing Company, La Salle, Ill., 1928.
6. ———, *A History of Elementary Mathematics*, The Macmillan Company, New York, 1917.
7. Leonard E. Dickson, *History of the Theory of Numbers*, Vols. I to III, Carnegie Institute of Washington Publication 256, 1918, 1920. (Reprinted Chelsea Publishing Company, New York, 1952.)
8. W. C. Eells, "Number systems of the North American Indians," *American Mathematical Monthly*, **20** (1913).
9. Karl F. Gauss, *Disquisitiones Arithmeticae 18*, trans. by Clarke, Yale University Press, New Haven, Conn., 1966.
10. J. W. L. Glaisher, *Nature*, **19** (1879), pp. 208–209.
11. ———, *Proceedings of the Cambridge Philosophical Society*, **3** (1878), pp. 471–473.
12. G. H. Hardy and E. M. Wright, *An Introduction to the Theory of Numbers*, Corrected 4th ed., Oxford University Press, London, 1962.
13. A. Ya. Khintchine, *Continued Fractions*, 1935, trans. of rev. ed. by P. Wynn, P. Noordhoff, Ltd., Groningen, Netherlands, 1963.
14. W. G. Leavitt, "A Theorem on Repeating Decimals," *American Mathematical Monthly*, **74** (1967).
15. Charles G. Moore, *An Introduction to Continued Fractions*, National Council of Teachers of Mathematics, Washington, 1964.
16. Carl D. Olds, *Continued Fractions*, Random House, Inc., New York, 1963.
17. Oystein Ore, *Number Theory and Its History*, McGraw-Hill Book Company, New York, 1948.
18. Hans Rademacher, *Lectures on Elementary Number Theory*, Blaisdell, New York, 1964.
19. ——— and O. Toeplitz, *The Enjoyment of Mathematics*, trans. by H. Zuckerman, Princeton University Press, Princeton, N.J., 1957.
20. Wacław Sierpiński, *Elementary Theory of Numbers*, trans. by A. Hulanicki, Panstwome Wydawnictwo Naukowe, Warsaw, 1964.

21 David E. Smith, *History of Mathematics, Vols. I and II*, Ginn and Company, Boston, 1923. (Reprinted Dover Publications, Inc., New York, 1958.)

22 ———, *A Source Book in Mathematics*, McGraw-Hill Book Company, New York, 1929.

23 B. M. Stewart, *Theory of Numbers*, 2d ed. The Macmillan Company, New York, 1964.

24 James J. Sylvester, *The Collected Mathematical Papers of James Joseph Sylvester*, Vol. III, Cambridge University Press, London, 1904–1912, pp. 440–445.

25 James V. Uspensky and M. A. Heaslet, *Elementary Number Theory*, McGraw-Hill Book Company, New York, 1939.

26 John Wallis, *Treatise of Algebra Both Historical and Practical*, London, 1685, Chap. 89.

Glossary of Symbols

$a \equiv b \mod m$, 99

AG (F_n) (affine plane over the field F_n), 262–272
   definition of, 271

AG $(2, F_n)$, see AG (F_n)

$\mathcal{C}$, 287
   special circle in AG (F_5), 285

e (number of edges of polyhedron or plane network), 8

f (number of faces of polyhedron or plane network), 8

F_n (field of n elements), 259, 260, 262

$\mathcal{G}$ (see grid figures), 153–160, 188

g.c.d. (m, n), 435, 436, 438

I_n (see integers modulo n), 255, 256

κ (see chromatic number), 72

l.c.m. (m, n), 435, 438

M_n, 111

M_p, 97, 114, 119, 121

M_{11}, 112, 122

M_{13}, 114, 122

M_{17}, 114, 122

M_{19}, 121

M_{23}, 119

M_{31}, 122

M_{61}, 123

M_{83}, 122

M_{89}, 123

M_{107}, 123

M_{127}, 123

M_{11213}, 123

$M_{16,188,302,111}$, 124

$n | b$, 88

$n \nmid b$, 88

$\phi(n)$, 120

Π, 91

$\Pi(p)$ (see Angle of parallelism)

$\pi(k)$, 106, 108

$q^\alpha \| n$, 91

Q.E.D., 182

$\mathcal{Q}$ area for every figure which belongs, 194
   definition of, 189
   figure not in, 201

$\mathcal{Q}^*$, 196

$\mathcal{Q}_t$, 188

$\mathcal{Q}_t^*$, 196
   definition of, 188

$\sigma(n)$, 124
   formula for, 126

$\sigma^*(n)$, 140

v, (number of vertices of polyhedron or plane network), 8

$[x]$, 405, 451

Index

Absolute geometry, 221–226
  fundamental axiom of, 222
Abundant number, 95, 136
  definition of, 86
Aeschylus, 213
Affine plane, 270, 272, 290, 291, 295, 297
  of order n, 291
Africa, the scramble for, 70–73
African number systems, 392
Alcuin, 84
Algorithm, 391, 435
Aliquot part, 81, 82
Al-Khowarizmi, 435
Amicable numbers, 140
Amicable pair, 139, 140
Analytical geometry, 248–254
Angle
  Hilbert's definition of, 245
  of parallelism, 234
  sum in a triangle, 223, 238, 240
Answering strategy, 325
Arabian mathematics, 445
Arabic notation, 94
Archbold, J. W., 314
Archimedean axiom, 225, 242, 247
Archimedean solid, 3, 7
Archimedes, 213, 415
Area, 147–209
  of a circle, 148, 186
  of congruent figures, 161
  of curvilinear figures, 148, 150
  fundamental properties of, 152
  of irregular figures, 148
  pathology of, 201–208
  of a rectangle, 152
  of rectilinear figures, 148, 150
  triangulated, 173
Aristotle, 139
Ass and haystack, 223
Asymptotic triangle, 235
Augmented number, 141, 142
Augmented unevenly even, 141
Aurelius Augustinus, 84
Axiom, 214
  for affine planes, 270, 289, 290
  of Archimedes, 225, 242, 247
  of completeness, 242, 247, 453, 471
  of congruence, 245
  of continuity, 247
  of Euclid, 214
  for finite hyperbolic planes, 305, 306
  of Hilbert (for plane geometry), 243–247

Axiom, of incidence, 244
  of order (or betweenness), 243, 244
  for projective planes, 293
Axiomatic method, 213
Azulai, A., 138

Babylonian fractions, 414, 415
  number system, 97, 392, 395
Bachmann, H., 78
Bagemihl, F., 76
Ball, W. W. R., 78
Barlow, P., 122, 143
Barrow, I., 212
Base-3, 400
Base-7, 403
Base-10, 411, 413, 414, 470
Base-12, 399, 406, 407, 414
Base-b, 402, 408, 413, 414
  place-notational representation, 397–405
Baston, V. J. D., 76
B. C. Cartoon, 85
Beck's Hex (*see* Hex)
Begle, E. G., 391, 474
Behnke, H., 78
Bernoulli, J., 465
Bertelsen, N. P., 107
Besicovitch, A. S., 76
Best worst, 372
Betweenness (*see* Axioms of order)
Bible, 81, 139
Binary digit, 54
  representation, 357
Binocular vision, 241
Blank, A. A., 241
Bleicher, M. N., 334
Blumenfeld, W., 241
Bolyai, J., 228–230, 303, 313
Bolyai, W., 228
Bonola, R., 313
Borsuk, K., 247
Botts, T., 423, 474
Bounded region, 177, 200
Bovillus, 113
Boyer, C. B., 147, 148, 149, 209
Bridge-it, 329
Bridges of Königsberg, 43–45
Brouncker, Lord William, 471
Buck, R. C., 209
Burckhardt, J. C., 465

Cajori, F., 84, 94, 143, 392, 474
Calculus, 149
Carathéodory, C., 149, 150, 209
Casting out 9s, 115, 403, 470
Catalan, E. C., 141
Cataldi, P. A., 114, 122, 124
Cayley, A., 55
Chinese-Japanese number system, 394
Chinese rings, 52
Chromatic number, 72
Chrome alum crystal (as regular octahedron), 16
Cigarettes, 76
Circle, 279–288
  in $AG(F_5)$, 280
  equation of, 252
  exterior point of, 284
  great, 217
  interior point of, 284
  perimeter of, 186, 187
Circuit
  Euler, 43, 44, 63
  Hamilton, 19, 43, 45–48, 50, 52, 54
Clock arithmetic, 255
Cohn-Vossen, S., 78
Colony graph, 71
Coloring problems
  four-color, 55–63
  map coloring, 55–75, 77
  pennies, 63
  ping-pong ball, 77
Common notions, 213
Common refinement, 154, 159, 167, 172, 184
Complete residue system, 115, 257
Completeness (*see* Axiom of completeness)
Composite number, 84
Computers
  Illiac, 123
  Lehmer, 109
  Swac, 123
Cone, 252
Congruence, 222
  axioms of, 245
  modulo a prime, 101
  modulus m, 99
  theorems for triangles (*see also* Axioms of congruence)
  theory of, 98, 101
Conic section, 252, 285

Continued fraction, 440–444, 470, 471
Convex polygon, 172–174
  polyhedron, 7, 8
  region, 175
    bounded, 200
Coordinate axes, 248
Coordinate geometry (*see* Analytic geometry)
Coordinate system, 248
Copernicus, N., 5
Country-colony map-coloring problem, 70–73
Courant, R., 78
Coxeter, H. S. M., 77, 78, 221, 313
Creation, 84
Crowe, D. W., 314
Crystals, occurrence of regular polyhedra in, 16
Császár, A., 36, 78
  polyhedron, 36, 37, 42, 66
Cube (*see also* n-cube)
  four-dimensional, 41
  perfect, 410
Cuboctahedron, 11
Cundy, H. M., 78, 272, 313
Cunningham, A. J. C., 123
Curve, Osgood's, 201–208

Dantzig, T., 84, 143
Davenport, H., 90, 143
Decimal, 391, 449, 450, 456
  algorithm, 451, 456, 457
  definition of, 448
  expansion, 471
  finite, 444, 445, 447
  fraction, 444–470
  nonrepeating, 444, 445
  notation, 470
  periodic, 456
  repeating, 444, 447, 455, 456
Deficient number, 84
Degenerate ellipse, 252
Degenerate hyperbola, 252
Degenerate parabola, 252
Degenerate polyhedron (*see* Tessellation)
Deltahedron, 21–30

Dembowski, P., 314
Democritus, 213
De Morgan, A., 55
Density, 406, 409, 410, 411
De Parville, A., 48
der Waerden (see van der Waerden, B. L.)
Desargues' theorem, 293
Descartes, R., 6, 137–139, 254
Diagonal, 177
   interior, 177
Dickson, L. E., 84, 139, 140, 143, 474
Digits, distribution of, 408–414
Diminished number, 141–142
Diophantine equation, 13, 22
Dirichlet, P. G. L., 105
Distribution of digits, 408–414
Divisor, 81, 83
   greatest common, 435, 436, 438
   number of, 93
   proper, 82
Dodecahedron, 3, 18, 63
Dominated (definition for matrix games), 368
Dorwart, H. L., 313
Doubly perfect number, 137
Dresher, M., 387
Duality in projective planes, 293
Dubins, L. E., 387
Duke of Württemberg, 5
Dynkin, E. B., 75, 78

Earth, 3, 5
Edge, W. L., 313
Eells, W. C., 392, 474
Egyptian fraction, 421–434, 471
Egyptian number system, 414, 415
Elements (five and regular polyhedra), 3
Elements of Euclid, 3, 213, 215, 308–313
Ellipse, 196–199, 252
Empire graph, 71
Equation
   of a circle, 252
   Diophantine, 13, 22
   of a line, 249, 251
   quadratic, 252, 279–288
Equilateral triangle, 216
   Euclid's construction, 216

Eratosthenes, 107–108
   sieve of, 109
Erdös, P., 434
Esau, 139
Escott, E. B., 140
Euclid, 93–95, 103, 213, 214, 216, 221, 435
   algorithm, 435–441, 446
   axiom, 215
   elements, 3, 213, 215, 308–313
   geometry, 215
   perfect number, 93–102, 111, 114, 123–124, 127
   photograph of Proposition 9, 212
Eudoxus, 213
Euler, L., 6, 12, 43, 45, 120, 122, 127, 140, 279, 463, 472
   characteristic, 32
   circuit, 43–45, 63
   conjecture on orthogonal Latin squares, 279
   formula, 6–12
      modified, 31, 32, 65
Evenly even numbers, 142
Evenly uneven numbers, 142
Eves, H., 272, 313
Exhaustion, method of, 324
Expectation (of a game), 360
Exterior-angle theorem, 220
Exterior line, 296, 300
Exterior point
   of a circle, 284
   of an oval, 298

Factor, 81
   tables of, 102
Factorization, 84–93, 109–124
   divisible by 7, 115
   divisible by 9, 115
   divisible by 11, 115
   divisible by 13, 115
Farey, J., 416
   fraction, 416–420
   sequences of order n, 416
   series, 418, 419, 424, 425, 442, 443
Farey-series method, 425, 426, 429, 433

Farey-series method, modified, 431
Felkel, A., 102
Fermat, Pierre de, 110, 113, 114, 117, 121, 137, 408, 463
    last theorem, 110, 440
Fibonacci, 94, 426, 430
Fibonacci method, 427–430, 433, 434
    modified, 431
Field (definition of), 258–261
Figure, 150
    building blocks, 153
    circle, 152, 185
    decomposed into triangles, 166
    ellipse, 152, 185
    grid, 153–160, 188
    interior of curve, 185
    intersection of two, 191
    polygon, 152
    right-angle, 159
    set difference of two, 193
    staircase, 161, 163, 189
    union of two, 190
Finite affine plane, 289–292
    analytic geometry (see also $AG(F_n)$), 262–272
    decimal, 444, 445
    field, 260
    geometry, 262–272, 289–308
    hyperbolic plane, 306, 308
    projective plane, 292–295
Fire, 3
Five-color theorem, 59
Fladt, K., 78
Football of pentagons and hexagons, 19
Forced win, 322
Forder, A. G., 247
Four-color problem, 55–63
Fraction, 414–416
    Babylonian, 414, 415
    Chinese-Japanese, 394
    continued, 440–444, 471
    decimal, 444–470
    Egyptian, 421–434, 471
    Farey, 416–420
    unit, 414, 441
Franklin, P., 78
Freudenthal, H., 24
Fundamental Theorem of Arithmetic, 87, 89

Games, 315–387
    balanced, 357
    cooperative, 385, 386
    inverse of, 323
    inverted-tree, 319, 322, 323
    losing, 340
    matching pennies, 365
    matrix, 365
    Nim, 340
    North Atlantic, 378, 379
    of chance (see Games of chance)
    positive-sum, 381, 382
    rank n of, 323
    reduced, 350
    Tic-Tac-Toe, 324–326
    tree, 318–326
        definition of, 319, 322
        node, 324
    tree of, 324
    trim of, 350
    value of, 372, 376
    war, 379
    winning, 340
    zero-sum, 366
Games of chance, 359–365
    definition of, 361
    expectation, 360, 362
    fair, 361
    favorable, 361
    payoff, 361
    roulette, 360
    throwing a die, 359
    tossing a coin, 359
    unfavorable, 361
    value, 362
Gardner, M., 78, 279, 313, 327, 387
Gauss, K. F., 98, 228, 240, 460, 465, 466, 474
Genesis, 139
Geometry (see Chapters 1 and 4)
    absolute, 221–226
    analytic, 248–254
    Euclidean, 215
    finite, 262–272, 289–308
    hyperbolic, 230–241
    non-Euclidean, 216, 219
    projective plane, 292–295
    spherical, 217–221
Gillies, D., 123
Giola, A. A., 143

Glaisher, J. W. L., 470, 474
God, 84
Goodwyn, H., 465, 467
Graph, 71–72
Great circle, 217
Greatest common divisor, 435, 436, 438
Greatest integer in x, 405, 451
Greek mathematics, 83, 127
Greek number system, 394
Grid figure, 153–160, 188
Guthrie, F., 55

Hadrian, 141
Hadwiger, H., 74
Haeckel, E. H., 16
Half-dollar, 76
Half-line, 232, 245
Half-plane, 245
Halsted, G. B., 16
Hamilton, W. R., 43
  circuit, 19, 43, 45–48, 50, 52, 54
Hanoi, tower of, 48–52
Hardy, G. H., 118, 121, 143, 213, 340, 387, 474
Hasse, H., 78
Heaslet, M. A., 439, 475
Heath, T. L., 213, 214, 216, 308, 310, 313
Heawood, P. J., 55, 56, 68, 69, 70–74, 78
Hebrew number system, 394
Hein, P., 327, 328
Hermes, O., 30
Hex, game of, 318, 327–339
  Beck's, 331–334
    Black to win is the natural outcome, 334
  White to win, 329
Hexahedron, 30
  exactly seven kinds of, 29
Hilbert, D., 78, 243, 313
  axioms for geometry, 150, 243
  definition of angle, 245
Hindu mathematics, 94, 415, 445
Hindu-Brahmi number system, 394
Holden, A., 16
Huckleberry Finn, 365
Hughes, D. R., 314
Hungary, mathematics in, 228
Hyperbola, 252

Hyperbolic, 217
Hyperbolic geometry, 217, 230–241
  fundamental axiom of, 217
Hyperparallel, 233

Iamblichus, 93, 139
Icosahedron, 3
Ideal line, 294
Ideal point, 294
ILLIAC computer, 123
Incidence (*see* Axioms of incidence)
Income tax game, 316
Induction, mathematical, 50, 85, 173, 174, 437
  course-of-values, 178
  hypothesis of, 176
  parameter of, 180, 323
Infinitesimal calculus, 149
Integer, 174
  greatest in x, 405, 451
Integers modulo n, 255, 256
Integration, 150
Intercept form of the equation of a line, 249
Interior point
  of a circle, 284
  of an oval, 298
Irrational number, 457

Järnefeldt, G., 273
Join (of two games), 340
Jones, B. W., 313
Jordan, C., 209
Jordan-curve theorem, 60, 338
Jumeau, Prior of St. Croix, 137
Jupiter, 5

Kanold, H. J., 136
Kant, I., 216
Kasner, E., 49
Kempe, A. B., 55

Kepler, J., 3, 5, 6, 229
Khaldun, Ibn, 140
Khintchine, A. Ya., 443, 472, 474
Koestler, A., 6
Königsberg, bridges of, 43, 44
Kreyszig, E. O. A., 209
Kronecker, L., 439
Kuhn, H. W., 387
Kustaanheimo, P., 273

Lambert, J. H., 102, 458, 460
Lamé, G., 439, 440
Latin square, 256, 274, 276
  orthogonal, 274, 275, 277
Least common multiple, 435, 438
Leavitt, W. G., 469, 474
Lebesque, H., 149, 209
Left parallel, 233, 243
Legendre, A. M., 106, 223
Lehmer, D. H., 107, 109, 123, 143
Lehmer, D. N., 103, 109
Leibniz, G. W. von, 457–460
Leonardo of Pisa, 94, 426, 430
Liber Abaci, 426
Limiting value, 409
Line
  at infinity, 294
  equation of, 249
  exterior, 296, 300
  ideal, 294
  secant, 284, 296, 300
  tangent, 284, 296, 300
Littlewood, J. E., 213
Lobachevsky, N. I., 149, 227–230, 232, 303, 313
Lucas, E., 48, 123
Luce, R. D., 387
Luneburg, R. K., 241

Map
  coloring problems, 55–75, 77
  country-colony, 70–73
  ordinary, 57, 58, 61

Map
  standard, 59, 61, 62
  on a torus, 65–67
Marienbad, 317, 318, 319, 340, 341, 358
Mars, 5
Maschler, M., 387
Mathematical induction, 50, 85, 173, 174, 437
Matrix, 366
  of payoffs, 366
Matrix game, 365–377
  application of, 378–380
  definition of, 368
Maximin, 372
Mayan number system, 395
McDonald, J. D., 387
Measure (*see also* Area), 147
  additive, 147
  measure theory, 149
  mismeasurement, 150
  quantitative, 147
  theory of, 149, 150
Mediant, 420
Mediation, 420
Meisell, E., 107
Mersenne, M., 96, 97, 110
Mersenne number, 97 (*see* M_n p. 477)
Mersenne primes, 102, 104, 111, 119, 127, 128
  divisor of, 121
  largest known, 128
  list of, 123
Meschkowski, H., 313
Method of exhaustion, 324
Midy, E., 467
  theorem, 467
Minimax, 324, 364, 372, 381
  theorem, 373
Mini-oval, 298
Mixed strategy, 370
Möbius strip, 76
Monotonic
  antitonic, 199
  isotonic, 199
  piecewise, 199
Moore, C. G., 443, 472, 474
Morgan (*see* De Morgan, A.)
Morgenstern, O., 387
Multiply perfect numbers, 137

Nachshon, Rau, 139
Nash, J., 327
Natural draw, 321
Natural outcome, 321, 322
    Black to win, 322
    definition of, 322
    draw, 322
    number, 391
    tree game, 323
    White to win, 322
n-cube, 3, 16, 40–48, 50, 54
    skeleton, 40
Negative, 258
Neptune, 6
Network, 8, 10
    derived from a map, 71
    Euler circuits in, 43–45
    of polyhedron, 8–12
    problems about, 142
Neumann (*see* von Neumann, J.)
Newman, J., 49
Nicomachus, 83, 93
Nim, 317, 318, 340–359
    balanced game of, 357
    game of, 340
    odd matchstick, 344
    reduced game of, 350
    winning strategy for, 358
Niven, I., 121, 143
Noah, 84
Noah's ark, 81
Non-Euclidean geometry, 216, 219
    perfect number, 124–136
Nonrepeating decimal, 444, 455
North American Indian number systems, 392
n-simplex
    6-simplex as Császár polyhedron, 42
    skeleton of, 42, 63
Number
    abundant, 95, 136
    amicable, 140
    augmented, 141, 142
    chromatic, 72
    composite, 84
    deficient, 84
    diminished, 141, 142
    doubly perfect, 137
    Euclidean perfect, 93–102, 111, 114, 123, 124, 127

Number
    evenly even, 142
    evenly uneven, 142
    irrational, 457
    Mersenne, 97 (*see* M_n, 477)
    multiply perfect, 137
    natural, 391
    non-Euclidean perfect, 124–136
    perfect (*see* Perfect number)
    positive whole, 2, 356
    prime, 84–93, 102–109
    rational, 457
    real (*see* Real numbers)
    relatively prime, 92
    squares of, 111, 114, 406
    systems (*see* Number systems)
    triply perfect, 137
    unevenly even, 141, 142
Number systems
    African, 392
    Babylonian, 97, 392, 395
    base b, 402, 408, 413, 414
    Chinese-Japanese, 394
    Greek, 127
    Hebrew, 394
    Hindu-Brahmi, 394
    Mayan, 395
    North American Indian, 392
    Roman, 94, 393
    three's, 87, 90

$\mathcal{O}$ (*see* Oval)
Octahedron, 3, 16
    exactly 257 kinds of, 30
    four-dimensional, 41
Olds, C. D., 443, 472, 474
Optimal strategies, 376
Order (*see* Axioms of order)
Ordinary map, 57, 58, 61
Ore, O., 55, 61, 78, 84, 143, 392, 435, 446, 474
Orthogonal Latin square, 274, 275, 277
Osgood, W. F., 206, 209
    curve, 201
Ostrom, T. G., 314
Oval, 296–303
    exterior point of, 298

Oval, in finite plane, 296, 297
   in planes of even order, 300–303
   in planes of odd order, 297–299
   interior point of, 298

Pair, amicable, 139, 140
   forward marker indicated by, 335
   next pair after, 335
   previous pair before, 336
   rearward marker indicated by, 335
Parabola, 252
Parallel class, 292
Parallel postulate, 214, 215, 221, 246
Pasch, M., 243
   axiom, 243
Peano, G., 243
Pedoe, D., 314
Pennies, coloring problem for, 63
Pentahedron, 29
Perfect cube, 410
Perfect number, 141
   definition of, 81, 82
   doubly, 137
   Euclidean, 93–102, 111, 114, 123, 124, 127
   even, 127
   listed, 83
   multiply, 137
   non-Euclidean, 124–136
   odd, 124, 128, 131–133, 136
   triply, 137
Perfect square, 405, 407, 410
Periodic decimal, 456
Pervusin, J., 123
Ping-pong ball, coloring problem for, 77
Plane (*see* affine, finite, projective)
Planets (and regular polyhedra), 5
Platonic solid (*see* Regular polyhedron)
Pluto, 6
Poe, Edgar Allen, 365
Poincaré, H., 303, 314, 394
   model of hyperbolic geometry, 304
Point, 264
   exterior, interior, 84, 298
   at infinity, 294
Polygon, 166–177
   circumscribed about a circle, 187
   convex, 172, 173

Polygon, divide into triangles, 172, 173
   inscribed, 187
   polygonal region, 172
   reflex, 172–174
   regular, 187
   triangulated, 173
Polygonal figure
   area of, 184
   triangulated, 184
Polygonal region, 177–185
   boundary, 177
   decomposed into triangles, 180
   hole, 177
   triangulated, 177
Polyhedron
   Archimedean solid, 3
   convex, 7, 8
   Császár, 36–39, 42, 66
   regular (*see* Regular polyhedron)
   seven-edged, 16
   with tunnels, 31, 35
   without diagonals, 31–39
Poselger, E. T., 465
Positive whole number, sum of distinct nonnegative powers of 2, 356
Postulates of Euclid, 213, 214, 310
Potato, cut into seven-edged solid, 16
Powers, R. E., 123
Prenowitz, W., 337
Prime factorization, 85, 87, 90
Prime numbers, 84–93, 102–109
   definition of, 84
   estimates of, 105
   Mersenne (*see* Mersenne primes)
   relatively, 92
Prime-number theorem, 106
Principle of induction (*see* Induction, mathematical)
Prize of $10, 77
Probability, 360
   chance event, 361
Projection, of polyhedron onto plane, 8, 9
Projective plane, 292, 293, 295
   of order n, 295
Puffball, 151, 201
Purloined Letter, The, 365
Pythagoras, 80, 81, 139, 213
Pythagorean society, 80, 139, 439

Quadratic equation, 279–288
Quadratic residue, 115, 121
Quadrilateral, Saccheri, 237, 238
Quantum mechanics, 273
Qvist, B., 314

Radar, 378
    application of Hamilton circuit to, 52–54
Rademacher, H., 78, 103, 143, 473, 474
Radiolaria, occurrence of regular polyhedra in, 15
Raiffa, H., 387
Rational number, 457
Rau Nachshon, 139
Real numbers
    least positive, 165
    lower number, 189, 194
    representations of, 162
    theory of, 162
    upper number, 189, 194
Recorde, R., 137
Reciprocal, 258, 259
Rectangle, 153–160
    sectioned, 157
Reduced game, definition of, 350
Reflex polygon, 172–174
    triangulation, 176
Region, convex, 175
Regius, Hudalrichus, 94, 113
Regular polyhedron, 3, 4, 6, 7, 12–20
    in crystals, 16
    in radiolaria, 15
    mystical significance of, 3
    of type $\{p,q\}$, 12
Relatively prime numbers, 92
Repeating decimal, 444, 447, 455, 456
Right-angle figure, 159
Right parallel, 233, 243
Ringel, G., 69, 70, 78
Robbe-Grillet, A., 387
Robbins, H. E., 78
Robinson, R. M., 123
Rohnius, J. H., 102
Rolf, H. L., 140
Rollet, A. P., 78
Roman numeral, 94, 393
Roswitha, 141

Rouse Ball, W. W., 78
Rudolff, C., 446
Ruffus, G., 94
Ryser, H. J., 314

Saccheri, G., 217, 226, 227
Saccheri quadrilateral, 237, 238
    angle sum in a triangle, 238
St. Augustine, 84
St. John, Christopher, 141
Sanford, V., 446
Sapientia, 141
Saturn, 5
    conquest of, 64–69
Savage, L. J., 387
Scientific American, 28
Secant line, 284, 296, 300
Seelhoff, P., 123
Segment (of a line), 245
Segre, B., 314
Self-evident truths, 213
Septahedron, exactly 34 kinds of, 30
Sequence has density d, 409
Seven-color theorem for the torus, 66–67
Shanks, D., 81, 83, 121, 124, 143
Sharing (in games), 383–385
Side payment, 381
Sierpiński, W., 121, 136, 139, 141, 143, 434, 469, 470, 474
Similar triangles, 239
Singer, P., 16
Slope-intercept form of the equation of a line, 251
Smith, D. E., 414, 435, 446, 475
Sodium chlorate crystal, 16
Solution of a game, 376
Sphere
    Kepler's explanation of planetary spheres, 5
    with handles, 31, 34
Spherical geometry, 217–221
    fundamental axiom of, 217
Spherical triangle, 221
    angle sum in, 219
Splitting method, 422, 425, 429
Square, Latin (*see* Latin square)
Squares of numbers, 111, 114, 406

Stäckel, P., 228
Standard form (of a positive integer), 91
Standard map, 59, 61, 62
Staudt (*see* von Staudt)
Stein, S., 90, 143
Steinhaus, H., 78
Stemple, J., 61
Stevin, S., 446, 447
Stewart, B. M., 422, 434, 475
Strategy (in game theory)
    answering, 325
    mixed, 370
    optimal, 376
    winning (for Nim), 358
Super-oval, 298
Süss, W., 78
SWAC computer, 123
Sylvester, J. J., 426, 427, 475
Szmielew, W., 247

Tangent line, 284, 296, 300
Tarry, G., 279
Tessellation, 14
Tesseract (= 4–cube), 40
Tetrahedron, 3, 16
    four-dimensional, 41
Thabit ben Karrah, 140
Thales, 213
Theaetetus, 213
Theorem
    exterior-angle, 220
    of arithmetic, 87, 89
Three's system, 87, 90
Tic-Tac-Toe, 324–326
Tietze, H., 67, 78
Toeplitz, O., 78, 103, 143, 473, 474
Tomato plants, 273
Tom Sawyer, 365
Topology, 32
Torus, 64, 65
    seven-color map on, 67
Tower of Bramah (*see* Tower of Hanoi), 49
Tower of Hanoi, 48–52
Tree games, 318–326 (*see also* Games)

Tree nodes, 324
Triangle
    area of, 163
    area of isosceles right, 161
    big, 168
    component, 167
    equilateral, 216
    inequality, 223
    little, 168
    similar, 239
    spherical (*see* Spherical triangle)
    tiny, 170
    triangulation of, 167, 172, 173, 180
Trim (of a game), 350
Triply perfect number, 137
Tucker, A. W., 387
Tutton, A. E. H., 16
Twain, M., 365
Tweedledum and Tweedledee, 385
Twelve-color theorem for countries with colonies, 70
Two games, join or union of, 340

Uhler, H. S., 123
Unevenly even number, 141, 142
Union, of two games, 340
Unit fraction, 414, 441
Universe, 3
    Poincaré's model of, 304
    symbolized by dodecahedron, 3–5
Uspenskii, V. A., 75, 78
Uspensky, J. V., 439, 475

Vaidya, A. M., 143
Valerio, Luca, 148
Value, of a game, 372
    definition of, 376
van der Waerden, B. L., 24, 313
Violets in the spring, 229
Vision (*see* Binocular vision)
von Leibniz, G. W. (*see* Leibniz, G. W. von)
von Neumann, J., 209, 368, 387
von Staudt, K. G. C., 11

Waerden (*see* van der Waerden)
Wallis, J., 459, 460, 475
War game, defense of two passes, 379
Whitney, H., 74
Williams, J. D., 387
Winken, Blinken, and Nod, 385, 386
Winn, C. E., 61
Winning strategy for Nim, 358
Wolfe, H. E., 313

Worst best (minimax), 324, 331, 364, 372
Wright, E. M., 118, 121, 143, 340, 387, 474

Youngs, J. W. T., 69

Zuckerman, H., 121, 143, 474

APPENDIX 2000

BY ANATOLE BECK

MICHAEL N. BLEICHER

DONALD W. CROWE

PAGES REFERENCED

49

At the time the first edition of this book was written, the United States was engaged in a war against the Vietnamese government located in Hanoi. There was always a danger that reaction to U.S. action there might result in war with the Soviet Union, an ally of the North Vietnamese. The danger that this might escalate into thermonuclear war was small, but not impossible. By contrast, no one expects that the Universe would last the 5.8 trillion years it would take for the Brahmins to complete their transfer.

55 *and* 74

As of 1977, the status of the *four-color problem* changed. In that year, K. Appel and W. Haken published a proof that four colors suffice. Although their proof aroused considerable discussion among mathematicians because of its essential reliance on large amounts of high-speed computer time, its correctness is now generally accepted. That is, the *four-color problem* is now the *four-color theorem*.

Recent information on the current state of improvements to the Appel-Haken proof, as well as statements of some new problems related to the four-color theorem, are contained in an article by Robin Thomas, "An update on the four-color theorem," in *Notices of the American Mathematical Society*, Vol. 45, No. 7, pp. 848—859 (1998). An idea of the complexity of the Appel-Haken proof is given by their own statement (quoted by Robin Thomas): "This leaves the reader to face 50 pages containing text and diagrams, 85 pages filled with almost 2500 additional diagrams, and 400 microfiche pages that contain further diagrams and thousands of individual verifications of claims made in the 24 lemmas in the main sections of text. In addition, the reader is told that certain facts have been verified with the use of about twelve hundred hours of computer time and would be extremely time-consuming to verify by hand." The improvements made by Thomas and his co-workers somewhat reduce the enormous numbers involved in the original proof, as indicated by his statement, "Our unavoidable set has size 633 as opposed to the 1,476 member set of Appel and Haken; our discharging method uses only 32 discharging rules instead of 487 of Appel and Haken." Nevertheless, part of their improved proof still uses a computer and cannot be verified by hand.

93

Section 3 correction: Theorem 26 is Proposition 36, Book IX of Euclid. Volume II, page 421 of the Dover reprint of the Heath Translation.

97

Theorem 34, line 14: Computers often use base 16 now.

103

Line 5: Of course, modern computers go far beyond these limits. For more information try www.utm.edu/research/primes/lists/.

107

For many more values of $\pi(k)$ see www.utm.edu/research/primes/howmany.shtml

108

Example 53: It should have ~~49~~ and ~~65~~

110

Caption: Fermat's last theorem was proved by Andrew Wiles in 1995. His proof is a *tour de force* of twentieth century number theory. It can now be said that everything Fermat claimed to be provable was. Many things which he merely conjectured were not true. The more romantic among us believe Fermat did have a proof with the methods of his day which is still awaiting our discovery. More information may be found at www.ams.org/new-in-math/fermat.html or at www.mbay.net/~cgd/flt/flt0.1.htm.

123

Much larger Mersenne primes are now known, and new ones continue to be discovered. See www.utm.edu/research/largest.html, www.utm.edu/research/primes/mersenne.shtml and www.scruznet.com/luke/mersenne.htm. If you have a PC, you can join the Great Internet Mersenne Prime Search (GIMPS), see www.mersenne.org.

136

For more on odd perfect numbers see www.its.caltech.edu/carnohan/perfect.html.

138

For more on multiplying perfect numbers see www.astsun.astro.virginia/~eww6n/math/multiperfectnumber.html.

140

For more on amicable numbers see www.vejlehs.dk/staff/jmp/knwnap.htm or www.astro.virginia.edu/~eww6n/math/amicabletriple.html.

327 *and* **386**

The years since 1969 have been momentous for John F. Nash, one of the inventors of Hex and a pioneer in the subject of game theory. Although his genius had been hampered during part of this time by some mental illness, he was also basically a very deep and inventive mathematician. One of his theories concerned the co-operative games like those shown on page 386. In 1994, he shared the Nobel Memorial Prize in Economics for that work, which has been seminal in the accomplishments of many who have pursued that subject. The Nash Equilibrium will be a permanent part of his legacy. You can read about him, and it, in *A Beautiful Mind*, by Sylvia Nasar, Simon & Schuster (New York) 1998.

332

Since the publication of this book, W. Charles Holland and Anatole Beck have added a new wrinkle to the proofs about who can force a win at Hex. Lemma 14 seems to prove that White can win at Hex with any first move. It does not say that! It says that White can win with any first move IF Black can win. That would be a contradiction, so Black cannot win. Then by Lemma 11, White must have a winning strategy.

Now, in the same way, Lemma 17 seems to prove that if the first move by White is into the corner shown, then the answering move by Black is a winning move. That is not so. What is shown is that IF White can still win, THEN the Black move is a winning move. Since the hypothesis and the

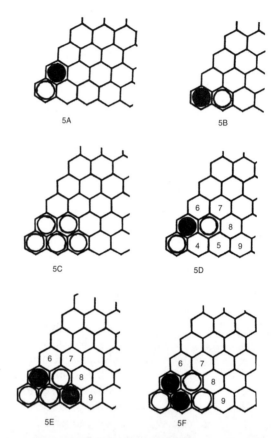

5A
5B
5C
5D
5E
5F

conclusion of that statement are contradictory, the hypothesis must be false. It is not shown that the move by Black is a winning move.

And in fact, it is not! The theorem with Holland shows that IF that move is a winning move, then White can still win after the move, which is a contradiction, and that contradiction kills the assumption. To see this, consider the position in Illustration 5A. On the hypothesis that Black's move was a winning move, Black should now have a winning strategy against any play by White. Thus, it follows that in the position shown in Illustration 5B with Black to move, White can force a win. *A fortiori*, if the position is as shown in Illustration 5C with Black to move, White can force a win.

White's strategy is then to place a stone as shown in Illustration 5D, pretending that the board is as it is in 5C, and play as though it is that way, except that as soon as Black places a stone in either space 4 or space 5, White places a stone in the other one, and continues to play as though both stones were white. They play until the board is full, at which point White, under the illusion that all five stones shown in Illustration 5C are white, imagines that he has won. But when he rids himself of that delusion, just as in the previous cases, it turns out that he has indeed won. Let us see why.

In the illusion, all five of the stones in Illustration 5C are white. But if none of the stones in the spaces numbered 6–9 are connected to the top of the board by a chain of white stones, then the imagined winning chain has nothing to do with the five stones in 5C. If space 9 is the only one containing a white stone linked to the top of the board, then the remaining five stones from 5C are superfluous. Thus, the only remaining cases to consider are those in which one of the stones in spaces 6–8 is connected to the top of the board by a chain of white stones.

In that case, we have either the situation in Illustration 5E or 5F, with one of the spaces 6–8 occupied by a white stone connected to the top of the board. But in this case, no matter which space that might be, the three stones from among the five in 5C which actually *are* white are sufficient to complete the chain from the top of the board to the bottom, and White has indeed won. Since the assumption that the position in 5A is one in which White is to move but cannot prevail against Black's supposed winning strategy, yields a conclusion which is a contradiction to that supposition, the supposition is false. Thus, Black has no winning strategy against best play by White if White is to move in 5A. It follows that the move by Black is not a winning move.

338

A variant of Hex constructed by Steve Alpern and Beck in 1991 is played on a cylinder or an annulus. In this version, the board (if it can be called that) is made up of rings of hexagons, side by side. Each ring consists of m hexagons and they are piled up n deep. We could think of an ordinary hex board with m rows n deep, in which we imagine that each space at the left side of the board is to the *right* of the rightmost space in the same row. The task is for White to build a chain from the top of the board to the bottom, as before, but Black must build a ring separating the top from the bottom.

We prove that, just as in the case of the ordinary Hex board, there is a winner (just one) when the board is full. We know that if m is even, however small, and no matter how large n is, there is a winning strategy for White, even if Black goes first. (The answer to every move by Black is a move which is $m/2$ spaces to the left or the right, which are the same space.) It is not known what is the case if m is odd. [*American Mathematical*

Monthly 98 (1991) pp. 365–372] This construction is valuable in proving a theorem in dynamics.

Another application of Hex is to be found in a paper by David Gale [*American Mathematical Monthly* 86 (1979) pp. 818–827]. He shows that it is possible to obtain one of the premier theorems of Mathematics, the Brouwer Fixed-point Theorem, from the ordinary theorem on the uniqueness of the winner in Hex.

396

Theorem 4, line 5 correction: Should read, such that $0 \leq a_0 \leq b-1$.

403

Line 19 correction: The line should read $\equiv \dot{1} + 0 + 5 + \dot{8} + \ddot{1} + \ddot{8} + 1 + 9 + 8 + 8 + 8 \equiv 3 \mod 9$ (The "1" is missing in the text.)

414

For more information consult *Mathematics in the Time of the Pharaohs* by Richard J. Gillings, 1972 MIT Press, Dover reprint, 1982.

424

It has been shown that the splitting method works. See Beeckmans, Laurent, The splitting algorithm for Egyptiona fraction. *J. Number Theory* 43 (1993), no. 2, 173-185.

429

M. Bleicher and P. Erdős introduce another algorithm in "Denominators of Egyptian Fractions," Ill J of Math. **8** 1977, which according to a paper of Yakota in the J. of No. Th. **24** in 1986 is close to "best possible," in one definition of that term.

434

A more recent paper on this topic is M. Bleicher and M. Ahmadi, "On the conjectures of Erdős and Strauss and Sierpiński on Egyption Fractions," IJMSS **7**, 1998. For more on the historical side see the paper by Keven Hansen in *The Proceedings of the Midwest Conference on the History of Mathematics*.

438

Line 4 correction: Should read g.c.d(-m,n) = g.c.d.(m,-n).

451

Theorem 88, line 2 correction: Should read, then $0 \leq a_i \leq 9$.

452

Theorem 90, line 10 correction: Should be $< N.a_1 a_2 \cdots a_n + \frac{1}{10^n}$

467

Theorem 128, line 2 correction: Should read, then $p_j + p_{j+t} = q$.